Macroevolutionary Systematics of Streptotrichaceae of the Bryophyta and Application to Ecosystem Thermodynamic Stability

—

Richard H. Zander

2017

Zetetic Publications, St. Louis

3p

Richard H. Zander
Missouri Botanical Garden
P.O. Box 299
St. Louis, MO 63166 U.S.A.
richard.zander@mobot.org

Zetetic Publications in St. Louis produces but does not sell this book. Amazon.com is a ready source, and any book dealer can obtain a copy for you through the usual channels. Resellers please contact CreateSpace Independent Publishing Platform of Amazon.

ISBN-13: 978-1974188680
ISBN-10: 197418868X

The image on the cover of this book is of the "clocks" or seed heads of the common dandelion. These represent the radiation of descendant species in dissilient genera. As hubs of scale-free and small-world networks, dissilient genera help sustain survival, through evolutionary and ecologic redundancy, of both the family Streptotrichaceae and the ecosystems of which that family is a significant floristic element.

The illustration on the title page shows the theoretical evolution of a dissilient genus through time, from a core progenitor on the left, through radiation of more specialized descendant species or lineages, to gradual extinction of most species, only leaving, in this case, a specialized descendant.

TABLE OF CONTENTS

PREFACE

The best new science is presented by someone who has worked on a tough problem for years, delving into inscrutable jargon-filled literature and wrestling with recondite technical methods, and has finally published a paper in which the hard-won solution is then suddenly "obvious" to readers upon publication. I have, in spite of much keyboard thumping over the past ten years, apparently not yet produced such an explanatory prodigy. I earnestly commend the present work, however, to the marketplace of ideas.

It is not enough to point out other's faults, but one must advance a method that overcomes them. A portion of the material in this book was originally submitted to a certain systematics journal. After some time I queried the editor, and he allowed as how six noted systematists had refused to review it and he was still trying to find someone to do so. Later I received a nice letter from him saying that the journal must decline the manuscript without review. The editor kindly quoted one potential reviewer as replying: "I seldom can even understand what Richard Zander does and so I would be a poor choice to review his work." Another refusnik wrote: "I have tried to review Zander's papers along these lines before and I have found them to be nearly unintelligible." Well ... hmmm. Scientific stances can be somewhat akin to religious beliefs. Apparently direct descent of a species from another species is an "out-of-context" phenomenon for phylogeneticists who are totally focused on shared descent revealed by nodes that cluster taxa sharing advanced transformative traits. Psychologically, this mental vice is called "epistemic closure," which differs from cognitive dissonance in that there is no sense of unease or moment of doubt.

Some reviewers of my papers have carped that they (cladists) have considered and rejected my ideas back in the 1980's, and I should go read the literature of that period for illumination of how I far have strayed. It is possible that the recent crop of budding systematists consider the long-running paraphyly controversy as simply an irredentist squabble between old men fighting old battles that are now irrelevant. This is hardly the case. I offer a way to do something new with powerful concepts involving process-based evolution.

Why call it "macroevolutionary systematics"? That is, a name more up-to-date and attractive to millennial students? Like, say, "quantum systematics"? (Not to be confused with V. Grant's 1963: 456, 1981: 155 quantum evolution.) How might that work? The quantum would then be the ultimate unit of systematics knowledge, i.e., the hint. I have in the past discussed this (Zander 1993: 85)—for instance:

> "A 'hint' of support is certainly not actionable alone, nor are even several hints impressive. Using 0.60 probability as representative of a hint, being just beyond totally equivocal, requires 0.60 to be used as a prior seven times in successive empirical analyses with Bayes' Formula, with no contrary information, to reach 0.96. The credible interval for very minor support is 0.60."

There is a statistical equivalent to a hint, the deciban. If we use one deciban (Zander 2014a: 4) as the equivalent of a basic unit of detectable knowledge, which has a Bayesian posterior probability of 0.55, then we have a quantum in Bayesian statistics, and willy nilly in systematics.

Technically, a deciban is the minimum clue that a hypothesis is true. It takes 13 decibans to reach 0.95 probability, which is the minimum probability recognized across scientific studies necessary to ensure actionable confidence in a simple hypothesis. But a hint is difficult to work with, as I found in my Bayesian study of *Didymodon* (Zander 2014a,b,c). The next highest level of information is the bit as used in information science, equivalent to 0.666 probability, or about 1 standard deviation. It is almost exactly equal to 3 decibans. One bit (3 dB) is the minimum information needed to make a decision between two alternatives, in this case whether the information is definitely good enough to work with, that is, something better than a hint. Using the bit as the next level of support above the deciban excludes about 1 standard deviation of variation or about 1/3 of potential hypotheses, while 4 bits is 0.94 probability, excluding nearly 2 standard deviations of variation. Thus, one bit is "working support." In the present work we will use the bit as a kind of superquantum. I am, however, too respectful of past study to promote the present research as quantum or even superquantum systematics, no matter how snazzy the phrase may be. So, "macroevolutionary systematics" must suffice to distinguish this method by emphasizing taxon to taxon evolution even though informational bits are much used.

Reviewers have occasionally complained that I use big words and unfamiliar concepts from other fields (e.g., Brower 2015; Schmidt-Lebuhn 2014; for reply see Zander 2014). Words conveying complex ideas and new concepts are necessary when introducing a new analytic method, and when criticizing older, cladistic methods.

ACKNOWLEDGMENTS

This book is dedicated to students of bryology worldwide who have been attracted to study of Pottiaceae, in part, by my 1993 Genera of the Pottiaceae. I particularly single out bryologists at the University of Murcia, Spain, who have recently greatly advanced the understanding of the family.

I thank Si He for translations of Chinese herbarium labels, and Pat Herendeen for helpful criticism. I am grateful to Patricia Eckel, who did the illustrations and Latin, and provided much encouragement. Peter Nyikos communicated much enjoyable discussion via e-mail. Jessica Beever helped determine the condition of the calyptra of the rarely fruiting *Leptodontium interruptum*. Kanchi Gandhi was generous with his aid in nomenclatural matters. John Atwood, Bryophyte Collections Manager at MO, was helpful in dealing with specimens and facilitating loans from other institutions. The curators of the herbaria mentioned in the taxonomic portion of this book are thanked for their help in obtaining collections for study. The Missouri Botanical Garden continues to provide much appreciated support for this research.

I extend pity and concern to those cladist reviewers of the now defunct original paper out of which this book developed, who could not figure out what I was saying. I can only hope the extended discussions, examples, metaphors, analogies, gnomes (aphorisms), diagrams, and illustrations in this volume will clarify and illuminate "stem-thinking" as a model for evolution as opposed to standard cladistic "tree-thinking," which remains tied to the rather old-fashioned hierarchical cluster analysis of phenetics. The test of whether the effort is valuable is "does this method work in practice?"

PRÉCIS

Biological systematics is the key to understanding and modeling the mechanisms of the present climate and biodiversity crisis. Genera are the basic dynamic units of ecosystems. Healthy ecosystems have many dissilient (radiative) genera. These genera each have a number of species or short lineages descendant from a core progenitor species.

A new moss family, Streptotrichaceae, is carved from the larger Pottiaceae, in part, on the basis of hitherto ignored traits, and demonstrates a complex dissilient structure. Sequential Bayesian analysis provides a means of determining optimal estimation of order and direction of evolution in a lineage. The Streptotrichaceae is supported by 130 informational bits, which include postulated missing links. The 28 species of the family are supported by 113 bits in the range (2–)3–5(–11) bits per species. There are 4 major scale-free networks totaling 51 bits. The contribution to ecosystem health by various taxa may now be compared by this metric of information theory.

Healthy ecosystems have multiple dissilient genera, which are modeled as hubs in scale-free networks. These networks in nature provide evolutionary redundancy through banking of taxa closely related to the critical and usually generalist core species and thus each other. Scale-free networks become small-world networks in the context of ecosystems wherein redundancy of habitat exploitation by similar species is common. Healthy ecosystems buffer and stabilize pathways of negentropic decrease, avoiding rapid energy flush. Ecosystems with more evolutionary and ecologic redundancy survive better in competition with other ecosystems and against biotic and physical catastrophes of thermodynamic disequilibria. Agents of the African savanna (humankind) have vigorously extended this rather specialized ecosystem inappropriately worldwide. Effecting a return to natural small-world networks is a restorative goal.

Richard H. Zander, St. Louis, 2017

PART 1: METHOD
Chapter 1
INTRODUCTION

Abstract —

Premise of research. Evolution-based classifications are more informative and predictive when based on direct, serial descent, with parsimony applied to both trait and taxon transformations.

Methodology. A protocol of new and revisited analytic methods in serial macroevolutionary analysis is detailed, and exemplified in a re-evaluation of the large genus *Leptodontium* (Pottiaceae, Bryophyta) and related genera.

Salient results. A new primitively epiphytic family, Streptotrichaceae, is described, based on peristome teeth grouped in fours, leaf base sometimes with hyaline fenestrations, laminal miniwindows often present juxtacostally, tomentum often ending in series of short-cylindric cells, widely spreading to squarrose leaves when moist, no central strand in stem, no epidermis on either side of the costa, hydroid costal strand absent, laminal papillae simple to branched or multiplex, basal cells strongly differentiated, and KOH reaction of lamina yellow. The new family is arranged as a series of dissilient (radiative) genera with isolated small or monotypic genera, which comprise scale-free networks. These, together with ecosystems, model a small-world network best suited for survival from thermodynamic catastrophes and competition from other ecosystems.

Conclusions. Macroevolutionary systematics is a robust and complete replacement for phylogenetics, particularly for molecular systematics as presently practiced. Synthetic techniques defining the new family include a focus on maximum parsimony of taxon transformations (minimizing redundancy of superfluous reversals and parallelisms), serial descent in context of Granger causality, a dissilient taxon concept (radiative group) equivalent to a scale-free network hub, a morphological clock, determination of probability of ancestry and descendancy in context of order and direction of evolution, sequential Bayesian analysis using bits and decibans, identification of traits that flag evolutionary processes as opposed to being evolutionarily neutral or nearly neutral tracking traits, and the value of small-world natural networks to ecosystem thermodynamic stability.

Substance of this book — This book is intended as a continuation of papers I have published over the past several years, with a detailed, worked out example of a new family composed of genera extracted from the Pottiaceae (Bryophyta), a family of mosses adapted to harsh or peculiar environments. I try here to not overly repeat arguments made in more detail or more vigorously elsewhere, see particularly Zander (2013; 2014a,b,c, 2016a). Some review is necessary, however, to avoid confusing and losing the newly interested, curious reader. The present application of the results of macroevolutionary classifications to Darwinian evaluation of ecological thermo-dynamic stability is an extension of the work of Brooks and Wiley (1988) and certain other authors dealing with entropic phenomena in nature.

The study involves the mosses, small plants with well-developed haploid gametophytes and diploid sporophytes. There is considerable variation in some species but thankfully the degree of variation is commensurate with the size of the organism. Just as phyletic constraint limits new traits to those tolerable in combination with the general anatomy and environmental interaction of the progenitor, variation in miniscule plants is also miniscule but apparently just as adaptationally effective. Thus, techniques in the mesocosm of vascular plant taxonomy are valid among the microcosm of the mosses, and moss species are usually quite distinctive.

In the study of direct descent, we are looking for transformations of traits between progenitors and descendants (Crawford 2010), not between nodes consisting of traits shared by sister groups. Given some outgroup or other polarization criterion, the method is simply minimization of parallelisms and reversals. Although this is done intuitively in standard classical systematics, the macroevolutionary systematic method offers a clearly stated cookbook series of steps, and provides Bayesian support measures with which to confront other researchers waving statistically certain cladograms that appear to show contrary results.

It has taken me 15 years to understand phylogenetics vis-à-vis classical systematics. The main area of difficulty, wherein we are "talking past one another," is the problem of evolutionary continuity. Continuity is modeled in phylogenetics by an unknown and unnameable shared ancestor giving rise to another unknown and unnameable shared ancestor, which gives rise to another … etc. Now that I think I do comprehend this empty axiom, the present book and my previous works (see Literature Cited) provide an alternative method of evolutionary analysis for orthosystematic classification purposes. Given that otherwise intelligent phylogenetic systematists have been mentally paralyzed by my discussions, I have here included a chapter explaining the present method in evolutionary systematics in the simplest possible terms, and have repeated the explanations throughout the book in different ways, showing efficacy in different contexts. This is intended to illuminate concepts invisible to some and difficult to grasp for others. I apologize to initiates for any tedium.

This method is here dubbed "macro-evolutionary systematics" to emphasize that this particular systematics is not the presently hegemonic field of phylogenetic systematics. *Macroevolutionary systematics* is using direct descent to model evolutionary change at the species level and above. The similar term, *macrosystematics,* on the other hand, is classification at the genus level and above, relevant to the present work because the dissilient genus concept is used instead of clades to recognize genera. Entropic macroevolutionary systematics simply reflects the use of informational and thermodynamic entropy concepts in dealing with biodiversity at the genus and ecosystem levels.

There is some mathematics in this book. Although taxonomists may have the reputation for being innumerate, this is changing and the math is simple, and if anything over-explained. Math has problems. Foremost are the concepts, which are often quite simple given a decent but usually hard-to-find explanation; then, there is math as a language, in which mathematicians vie for the most terse and unintuitive manner of making a statement. Statistics, to make matters more annoying, is not mathematics; it is basically physics. Statistics has its own problems as a language, with "words" invented by ancients that have become fixed as standard but are not intuitively understandable, and there are three (3) schools of statistics (Bayesian, Fisherian, Neyman-Pearson) that hegemonically battle for minds. Given the long-standing and well-reviewed (e.g., Gigerenzer et al. 1989) holy wars in statistics, this is really your problem, though I try to explain concepts three ways from Sunday.

Presentation of evolution as serial trans-formations of taxa, with evolutionary trees occasionally branching and positing unknown shared ancestors only when necessary, is not a new idea. A standard, though not complete, example of a macroevolutionary tree is that of the Equidae, the horses (or equids) (e.g., Hunt 1995). Well, why are evolutionary systematists now not as successful in receiving munificent grants, hordes of eager student followers, and other perquisites of establishment biology? I think it is because they have no computer applications. Surely, some code maven can work up software that will do what is presented in this work. When this finally happens, there will be interesting times!

Problems — There are three main problems that had to be solved in theory for analysis of direct descent of taxa.

(1) *Sloppy nesting.* By this is meant the fact that when, say, a species is designated as direct ancestor of another species, there are commonly minor traits that do not match well. This is allowable by conceiving of a species as changing anagenetically in minor traits (those apparently neutral) at the infraspecific level through millions of years. A progenitor, then does not have to match precisely an ideal base from which a descendant evolved through critical advanced trait transformations. Both progenitor and descendant may be spotted in spite of emergent neutral mutations affected by drift and bottlenecking, or single trait local adaptations.

(2) *Gaps in a series.* One expects a gradualistic step-wise transformation in taxa from progenitor to descendant. One can then insert, through inference, a postulated missing link in an evolutionary series to ensure a gradual series. This interpolation is a prediction (or retrodiction) of an actual taxon,

which may yet be discovered in nature or in fossil form.

(3) *Branching of series*. The dissilient species concept models the branching of short serries from a single core progenitor. Thus, a model of adaptive radiation allows multiple ends for short series, saving the theory of gradual transformation of one species into another, either anagenetically or in a population with minimum of two advanced adaptive traits linked by isolation of some sort.

These are discussed at greater length further on in the book.

Summary — The main features of macro-evolutionary systematics, as opposed to phylogenetics, are:

(1) *Higher resolution*. Phylogenetics cannot resolve direction of evolution between two sister groups, based only on information on shared descent.

(2) *Continuity*. Evolution is between taxa, not nodes on a cladogram, which are artifacts of the minimum spanning tree or Markov chain. Taxa evolve, traits do not evolve separately. Continuity of taxa also supports recognition of serial lineages, reflecting evolutionary change that is commonly fixed gradually at the species level.

(3) *Predictability*. Undiscovered or extinct end members or missing links of a lineage based on serial descent may be described to some extent.

(4) *Modeled evolutionary patterns*. The dissilient taxon is the primary model. The end result of analysis with direct descent is a colligated set of evolutionary radiation cores (dissilient taxa). Each lineage within a radiative group may be expected to be 1 to 3 or rarely 4 taxa in length, and end in an apparent evolutionary dead end, or sometimes in a taxon that proves to be generalist and central to another set of radiating lineages.

(5) *Support from information theory*. Support for order and direction of evolution is through bits, which may be summed because they are logarithmic, and easily translated to Bayesian posterior probabilities.

(6) *Taxa are well defined*. Each species is generally defined as having at least two otherwise independent traits linked by some evolutionary process, known or unknown. Each dissilient taxon

(a radiation core) constitutes an empirically derived taxon higher than species, such as genus, etc.

(7) *Macroevolutionary systematic classification has no artificial "principles."* No taxa are lumped or split to accommodate to cladistic aversion to paraphyly. Indeed, cladistics paraphyly is considered the essence of macroevolution as it helps reveal serial descent.

(8) *Better distinction between similar genera.* A genus definition using gross similarity of all species and gaps between genera is liable to problems when gaps between genera are small or weak. If one uses radiation from a core super-generative species as a genus definition, then similarities due to convergence of descendants may be better distinguished as belonging to one genus or another.

(8) Trait changes are identified as directly generated by an ancestral taxon, many of which are extant. This allows evaluation of traits that flag adaptational processes, those that are neutral or nearly so and useful as traits tracking evolution, and those that may be important as contributing to competition at the ecosystem level.

(9) The reality of dissilient genera allows a theoretical reframing of evolution in the context of an ecosystem. The dissilient genus is a collection of species evolutionarily very close (about 4 informational bits on average), and replacement to some extent is very possible. The species are similar or overlapping in habitat, and replacement would be nearly immediate on many cases. Thus a dissilient genus is self-healing, and this feature is extended toward the stability of the ecosystem by maintaining levels of biodiversity. One might replace the term "biodiversity" with "ecosystem health" as a better goal for action in the present extinction event. What is being extinguished through damage to dissilient genera are ecosystems. There is no easily explained tipping point in biodiversity study, but there is when describing ecosystem collapse.

The above elements of macroevolutionary systematics, and much else, are discussed and exemplified in this book in a reevaluation of moss species in and near the genus *Leptodontium* (Pottiaceae, Bryophyta).

Chapter 2
SIMPLE GUIDE TO MACROEVOLUTIONARY SYSTEMATICS

An alternative is presented here to phylogenetic analysis as a means of classification. The major problem with phylogenetics is fairly simple: (1) Macroevolution is correctly modeled by paraphyly. (2) Phylogeneticists reject paraphyly. (3) This leads to clumsy methodology (data restricted to shared traits), strained scientific philosophy (rejection of scientific induction and theory in general), and wrong classifications (unwarranted splitting and lumping of taxa). Finding fault is easy, but devising an adequate alternative has taken years. The aim of this book is to demonstrate a method of constructing a macroevolutionary tree based on expressed traits, and gauging the support for its structure. This is considered here a robust and complete replacement for phylogenetics as presently practiced, including molecular systematics. This method has applications in ecosystem modeling and thermodynamic stability.

Macroevolutionary systematics is a way to generate evolutionary models by (1) constructing a cladogram or equivalent based on *shared* descent, which clusters taxa evolutionarily by minimizing parallelisms, (2) arranging the taxa in optimal *direct* descent by minimizing reversals and superfluous postulated shared ancestors or nodes, and (3) interpreting the results with an analytic key focused on direct (serial) descent, which clarifies trait transformations and sums informational bit support for each taxon order. This maximizes information about evolution by generating support measures in terms of informational bits, which can be directly interpreted as Bayesian posterior probabilities. A caulogram (evolutionary diagram) emphasizing direct descent is an explicit theoretical model of evolution in the group.

Analogy with dominos — Consider an evolutionary tree of direct descent as made up of species that fit end to end like dominos. The rules of the game are simply that you can put a tile at the beginning or end of a line, or inserted inside the line between tiles. More than one tile can fit at the end of another tile such that the line of dominos branches. The idea is to match traits as best possible to model gradual accumulation of new traits and gradual elimination of old traits. This minimizes reversals and parallelisms at least when nesting is strong. Let's try it. Here are two sets of randomly organized species as dominos.

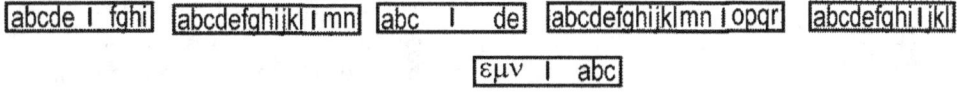

Figure 2-1. Above is a set of species, each species represented by a domino. Each domino-species has traits designated by letters. Each letter represents a unique advanced trait differing from a primitive character state not given. Within each species relatively primitive advanced traits are for demonstration purposes placed on the left side of the domino, just left of the vertical line; more advanced traits are on the right. (Nature inconveniently mixes these traits together.) The domino with the Greek letters is the outgroup. In an optimal evolutionary ordering, theory implies that as advanced traits accumulate they are transferred to descendants as conservative traits. This set will have no reversals if rearranged in optimal order.

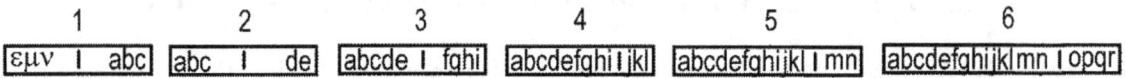

Figure 2-2. We need to place the dominos in evolutionary order. There, that was easy. We have lined up the species by matching a descendant's conservative traits with the advanced traits of its immediate ancestor. The dominos are numbered here in lieu of species names. The advanced traits of an ancestor may become the primitive or at least conservative traits of the descendant. Thus are traits conserved in the course of speciation. In this way, redundancy of traits that appear in a randomized order is minimized by assumed homology, and the lineage of dominos becomes an informational whole.

What is expected in a perfect lineage (Fig. 2-2) with conservative traits always surviving, is that there are no reversals back to primitive traits. If the optimum order above were any different this would add reversals. The evolutionary formula for Fig. 2-2 is **1** > **2** > **3** > **4** > **5** > 6, where progenitors are bold-faced and angle brackets indicate inferred direction of evolution. What would the data set be if the traits were represented as zeros for primitive or plesiomorphic and ones for advanced? In Data Set 2-1, below, we ignore

Data Set 2-1
```
[ 0          1        ]
[ 123456789012345678]
[ abcdefghijklmnopqr]
1 111000000000000000
2 111110000000000000
3 111111111000000000
4 111111111111000000
5 111111111111110000
6 111111111111111111
```

In Data Set 2-1, the first three lines, in square brackets, tell us about the data. The first 2 lines tell us that there are 18 characters (columns) and the third line lists the advanced traits of each character on each domino. The advanced trait character state is represented by a "1" in the data set and the primitive state (not given) by a "0". It also demonstrates that using a letter for an important advanced trait ignores the primitive form of that trait (the zero). If a reversal to a primitive form is

the unshared traits of the outgroup, species 1, "εμν". Also, this is not a clean cladistic data set because some columns are all of the same trait (same character state) and are therefore cladistically uninformative, but for our purposes this is evolutionarily informative.

adaptive, then the reversal is an evolutionarily important trait and may change the order of modeling direct descent. Let's see if it does in a few contrived examples. In Data Set 2-1, above, new, evolutionarily important traits are Italicized (again, note that the advanced traits of an ancestor become the primitive traits of the descendant).

Using PAUP* (Swofford 2003) (both neighbor-joining and maximum parsimony with bandb and 2000 reps bootstrap) the Newick formula of the cladogram for Data Set 2-1 (see Fig. 2-3) is (1,2),3,4,(5,6). Using shared traits alone, one cannot tell in Fig. 2-3 which of 1 or 2 is ancestor (other than telling the program that 1 is the outgroup), and one cannot model whether 5 or 6 is the progenitor or ancestor of the other. Additionally, neither the commas in the Newick formula nor the nodes in the cladogram it represents (Fig. 2-3) model direct descent. The set of dominos exemplifies a real evolutionary lineage, not a clade.

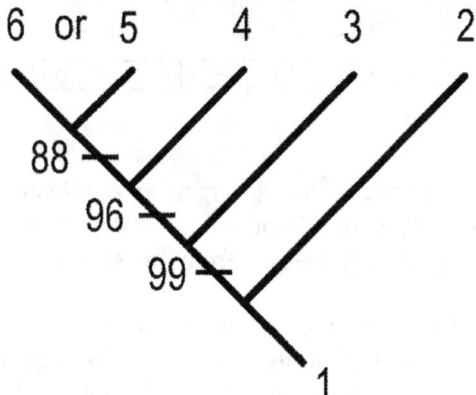

Figure 2-3. This is a cladogram of the data on which Fig. 2-2 above is based. Bootstrap percentages are given. Note that ancestry is not directly modeled, and the relationships of species 5 and 6 cannot be discerned on the basis of shared traits alone. Cladistic analysis cannot deal with 1 or 2 as progenitor because although they share traits "abc", those traits are shared by all taxa, and *not* sharing of traits "de" by 2 and the remainder of the cladogram is not cladistically decisive. The above cladogram, however, reflects *shared* traits of the taxa in Fig. 2-1. The evolutionary order of direct descent implied by the *cladogram* is similar to that of the domino model, but less informative, namely (1,2) > **3** > **6** > (4,5). This is less informative, as a hypothesis of evolution, than the domino arrangement.

Let's do it again but with a simpler data set, this time with five species and only two new traits occurring during speciation.

Data Set 2-1b, only five species and only two new traits per event

```
[ 0         1]
[ 1234567890]
[ abcdefghij]
1 1100000000
2 1111000000
3 1111110000
4 1111111100
5 1111111111
```

The evolutionary order by direct descent is **1** > **2** > **3** > **4** > 5. The Newick formula for the cladogram of shared descent is (1, 2), 3,(4, 5). The bootstrap support percentage (2000 replications) is high, 89 for both cladogram branches. An optimized order of direct descent should have the highest number of informational bits by adding all new traits and reversals after redundant traits are maximally

eliminated by recognizing that many traits are the same traits in other species with those in one species conservative and the other advanced. There are no reversals in Data Set 2-1b because it is ordered optimally.

Analogy with dominos, dealing with reversals — The above exercises were pretty straight-forward. But what if there were reversals in the data set? What if an advanced trait "h" turns back into its primitive form "**h**" (primitive or plesiomorphic character states are here singled out with **boldface type**). Consider the following species-as-dominos, lined up in what seems an optimal order of direct descent. As with Data Set 2-1b, we use only five domino-species and only two new traits per species to simplify. The difference is that all new advanced traits after the second speciation event revert to the primitive character state.

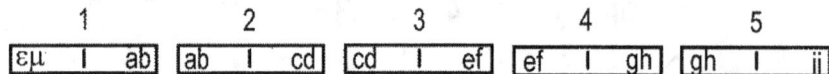

Figure 2-4. The five dominos-as-species are arranged to eliminate the redundancy of traits by overlapping as many as possible. As with all optimal domino orderings, this makes a new trait the same as a conservative trait modeling a carry-over from one species to another. See Data Set 2-2.

The arrangement of dominos in Fig. 2-4 ignores (as identical with those in the outgroup) all traits that are not new. The cladistic data set that represents the order of taxa above and shows primitive or plesiomorphic character states as zeros is:
Data Set 2-2

```
[ 0        1]
[ 1234567890]
[ abcdefghij]
1 1100000000
2 1111000000
3 0011110000
4 0000111100
5 0000001111
```

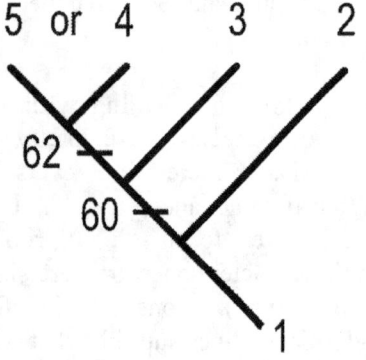

Figure 2-5. Cladogram of Data Set 2-2., for dominos in Fig. 2-4. Nonparametric bootstrap percentages are given.

In Fig. 2-5, above, the order of the dominos is to some extent preserved though the cladogram cannot order species 1 and 2 or 4 and 5 because only shared traits are analyzed by the PAUP* cladistic program. Bootstrap percentages are very low, however, compared to those for Data Set 2-1b, for which such percentages were 89 on the same branches. This indicates that reversion to primitive status of many traits when modeling direct descent with dominos introduces uncertainty in a cladogram, and by extension in the linear model using the domino metaphor.

Below is Data Set 2-2 again (as 2-2b), but with new advanced traits italicized and new reversals boldfaced. We can optimize by maximum parsimony of shared traits, or we can optimize direct descent by adding all trait transformations, including new reversals from new advanced traits, to see if this order is better than any other. The order in Data Set 2-2b gives 6 reversals (boldfaced) plus 8 new traits (Italics) = 8 total trait transformations.
Data Set 2-2b, annotated
```
1 1100000000
2 11110000000
3 0011110000
4 0000111100
5 0000001111
```

Now, in Data Set 2-2c, below, after changing order in data set 2-2b to reverse the order of the species, except with species 1 retained as the outgroup, we get 8 reversals (boldfaced) and 10 new traits (Italics). Adding all trait transformation yields 18 trait transformations. This is 10 more than with the order of data set 2-2b.
Data Set 2-2c, order reversed
```
1 1100000000
5 0000001111
4 0000111100
3 0011110000
2 1111000000
```

A new data set 2-3d, below, with a new order of taxa with domino 5 in middle, gives 14 new traits and 8 reversals, or 22 trait transformations, 14 more than in Data Set 2-2b.

Data Set 2-2d, terminal taxon moved to middle

```
1  1100000000
2  1111000000
5  0000001111
3  0011110000
4  0000111100
```

Data Set 2-2b is optimal ordering by minimizing trait transformations. The optimal order of direct descent is thus a form of maximum parsimony. With more reversals tolerated in a model, the number of apparent new traits increases, thus *minimizing reversals is critical*. This is because orderings with many reversals generate apparently new traits, which because of homology are not really but are redundant. The optimal (however uncertain) order for direct descent is indicated by the order with minimum number of reversals and minimum number of trait transformations. The minimum number of reversals is ensured by the order with the maximum immediate overlapping of new traits, i.e., elimination of redundancy. What this comes down to is that you want the lowest number of from total trait transformations) when initially ordering species so you can eliminate false, redundant information solely due to wrong evolutionary order. This is much like maximum

parsimony. When in doubt, the lowest number of reversals is decisive. (The redundancy of the sentences above is on purpose to emphasize this new way of using trait homology.)

Branching — In Fig. 2-3, branching evolution is modeled. Here only new traits are used, so as to make numbers of important traits small for demonstration purposes. There are two sets of clusters of short branches. These branches come off progenitors that are generalist in expressed traits and produce in nature often highly specialized descendants. Each cluster of branches that radiates (dissilient, exploding) from one progenitor is a natural grouping higher in rank than species, i.e., a genus. So species 3 and 9 are progenitors of two genera, respectively. The coherent ancestral lineage in line of direct descent (less branches) extends thus: **1 > 2 > 3 > 8 > 9**. (Remember, boldface means implication of progenitor status.) The formula for the lineage including branches is **1 > 2 > 3** (> 4, 5, (**6 > 7**)) > **8 > 9** (> 10, 11, 12). In the dissilient genus, the species all share a central set of traits, which is expected, but that fact is now modeled here as process-based evolution. Fig. 2-4 models what happens when a species is missing (a "missing link").

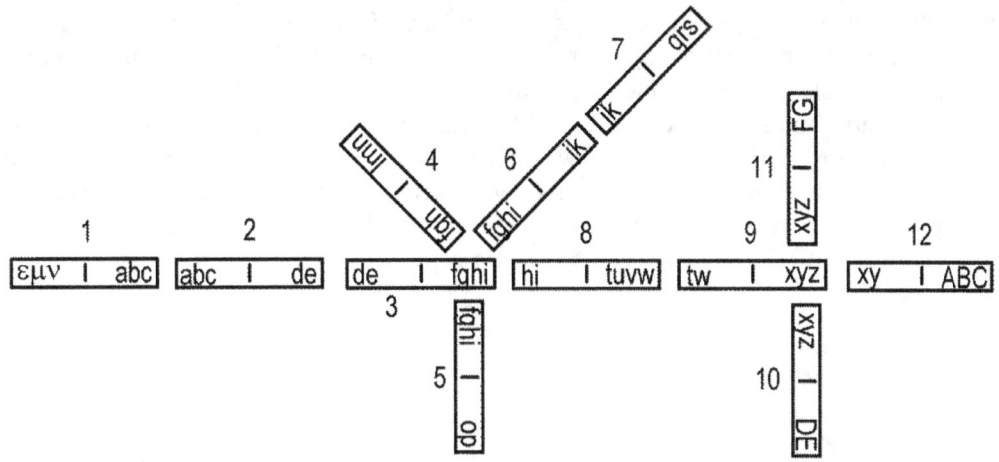

Figure 2-3. In this diagram, branching evolution with only new traits given is modeled. Most lineages have numbers of short branches. These branches come off progenitors that are generalist in expressed traits and produce often highly specialized descendants. Each cluster of branches radiating (dissilient, exploding) from one progenitor is a natural grouping higher in rank than species, i.e., a genus.

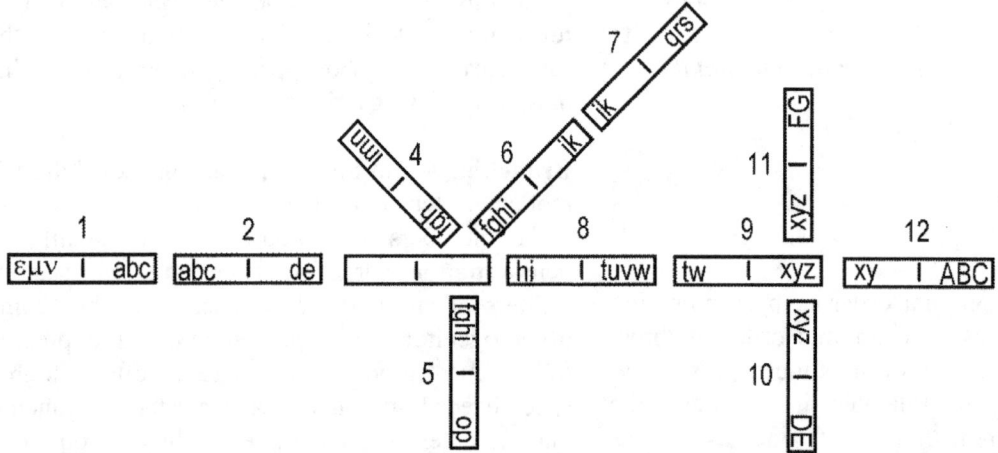

Figure 2-4. This is the same arrangement as Fig. 2-3, but species 3 is missing (the blank domino), either extinct or not yet discovered. It can be predicted and inserted in the lineage, however, as a generalist direct ancestor of species 4, 5, 6 and 8. The missing link would be inferred to have the traits defgh, which connect and complete the domino lineage. Such interpolation is a prediction or retrodiction that can allow intelligently informed searches in nature or fossils/subfossils.

Gaps — Where do we fit a species that has a lot of advanced traits? Morphologically isolated species crop up in any large-scale taxonomic project.

In Fig. 2-5(A) we have species 4 with 15 traits but candidate ancestral species has only seven matching (homologous) traits, and the whole lineage is clearly an accumulation or development of the traits of species 4. So even though there are eight new advanced traits, the position seems well chosen. Maybe one might name species 4 to a new monotypic genus which explains the several new traits as derived from an complex of an extinct ancestral species and its mostly extinct descendants. In Fig. 2-5(B), however, species 6, with eight new traits is only connected with the remainder of the lineage by two conservative traits. Clearly it is to the advantage of the analyst to search for another lineage that will allow species 6 to match more conservative traits.

In Fig 2-5(C), we have such a lineage, and species 6 (at the end) matches six of its traits with those of species 9. If lineage (C) is available, then species 6 can be rightly placed. If (C) is not available, then (B) is a closed causal group for species 6, and we are stuck with the inference that species 6 may possibly, though less than probably, be descendant of species 5. If (C) is available, however, then the closed causal group is larger, and the extra choice is helpful.

Although all groups are closed at some level, we strive to make the causal groups studied (i.e., taxonomic groupings) large enough to maximize choices yet avoid the chaos of too much, often uncertain data.

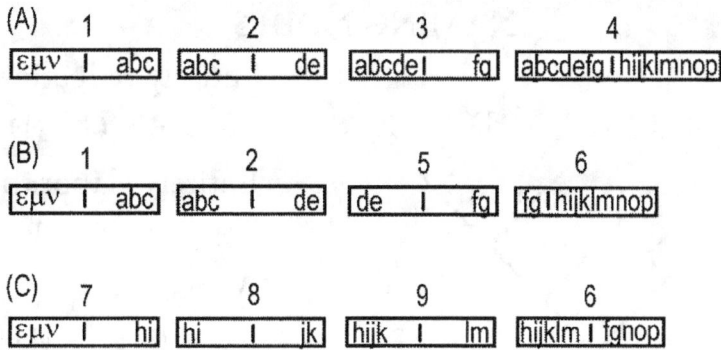

Figure 2-5. Where do we fit a species that has a lot of advanced traits? In (A) we have species 4 with 15 traits but candidate ancestral species has seven identical traits. So even though there are eight new advanced traits, the position seems well chosen. In (B), however, species 6, with eight new traits is only connected with the remainder of the lineage by two conservative traits. In (C), species 6 (at the end) matches six of its traits with species 9. See discussion for explanation.

Wrestling with cladograms — The *method of macroevolutionary systematics,* most simplistically, is to provide the nodes of an evolutionary tree with the names of extant or inductively inferred taxa using an analytic key, which is simply a multichotomous key emphasizing nested distinctions.

The *object of macroevolutionary systematics* is to nest extant species (or higher taxa) in sometimes branching series, from a primitive (generalized) taxon radiating in one or more series to advanced (more recent elaborations or reductions). Postulating unknown shared ancestral taxa is sometimes necessary to ensure gradual steps in evolution. Such nesting is acceptable in a *closed causal group,* such that *post hoc ergo propter hoc* (the first instance must cause the second) is at least probabilistically true given no other close alternatives (relatives) exist.

The *first way* to generate a diagram of serial (direct) descent is to take a cladogram, evaluate each pair of sister groups to determine which is most probably the ancestral taxon, and do this until the cladogram is collapsed into a caulogram, with some ancestral nodes perhaps left as shared unknown ancestral taxa. Naming nodes on a cladogram may seem simple, and indeed parsimony or likelihood is a good guide to general groupings. But details are problematic because the generation of a cladogram is mechanical, but discursive reasoning is necessary to evaluate the import of non-phylogenetic data (those not informative of possible shared descent, but of serial descent). "Mechanical" because a cladogram is generated in a black box with a dichotomous representation of shared descent of traits, which is only by charity a model of evolution, because evolutionary continuity in a cladogram is from node to node. Nodes are artifacts of the optimization process and only notionally represent shared ancestors.

NAMING NODES

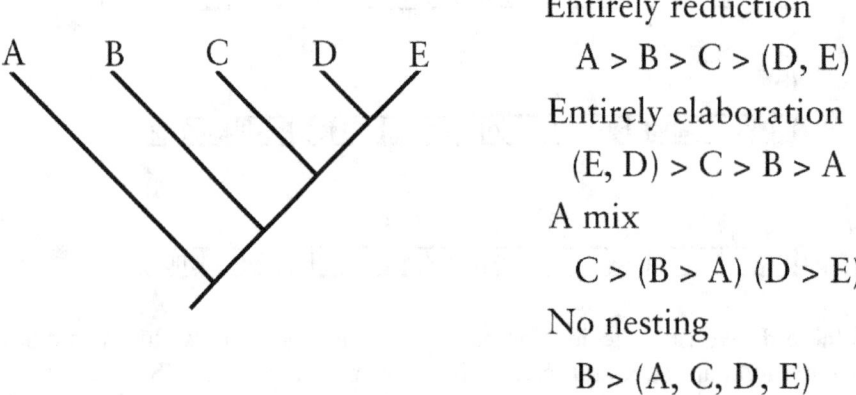

Entirely reduction

A > B > C > (D, E)

Entirely elaboration

(E, D) > C > B > A

A mix

C > (B > A) (D > E)

No nesting

B > (A, C, D, E)

Figure 2-11. Naming nodes is not simply identifying the node with the nearest terminal taxon. This is because much important data on serial nesting is not in the cladistic data set. If reduction is the main evolutionary theme, the nearest terminal taxon may be indeed the same as the node. If elaboration is the theme, the order of serial evolution could be reversed. More commonly, one may expect a mix of elaboration and reduction. Or there may be no descendants from any one of the putative descendants.

The *second way* to generate a caulogram is to envision the study group as including one or more generalized progenitors, each surrounded by a set of descendants. Such descendants may also (1) have descendants of their own, or (2) prove to be generalized in a different manner and have a set of descendants of their own. Use an outgroup to help polarize sets of traits associated with taxon transformations. Minimize reversals of traits and assumptions of new traits; this maximizes information by minimizing redundancy of traits, i.e., optimizes nesting. Forcing all taxa into cladogram nodes is possible but requires us to invent not unknown shared ancestors but unknown terminal OTUs. See Figs. 2-12 and 2-13 for graphic explanations.

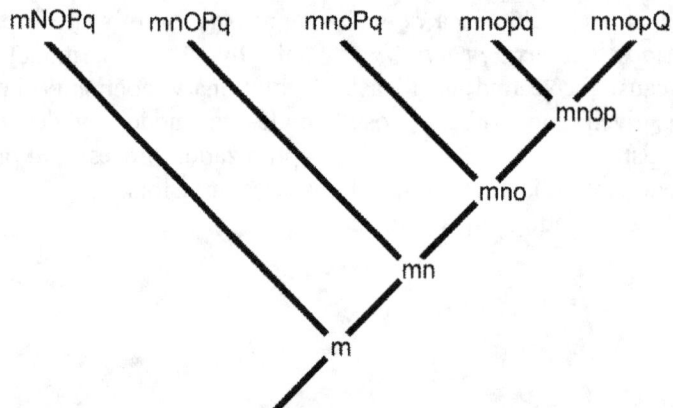

Figure 2-12. A cladogram showing traits of nodes. Each node includes all the traits shared by terminal taxa subtended distally in the cladogram. The trait Q is added as an autapomorphy for one of the terminal taxa to distinguish the two. In this simple cladogram, the nodes are not nameable to any taxon, but phylogenetically refer to traits that a shared ancestor should have in the maximally

parsimonious cladogram. Thus, "naming nodes" means forcing an extant taxon into a node, not discovering what unknown species the node actually represents.

FORCING TAXA INTO CLADOGRAM NODES

Advanced traits are capitalized, plesiomorphic traits are lowercase. "Unknown" descendants are terminal.

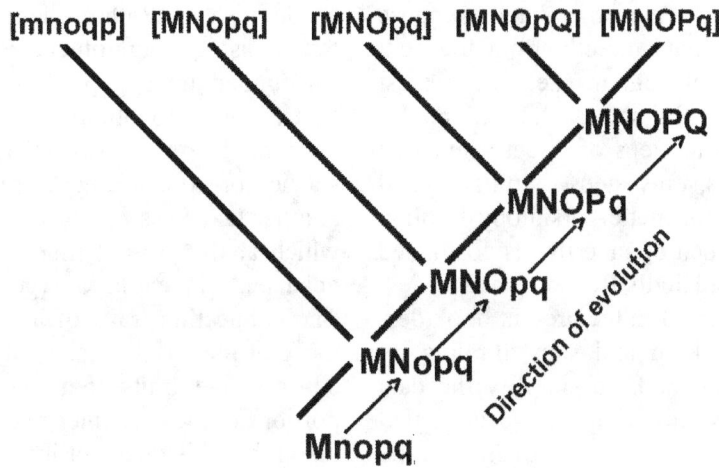

Figure 2-13. It is possible to treat extant taxa as cladogram nodes and see how their evolution affects the terminal taxa that cladistic methodology requires. This fully resolved cladogram is constrained to show the gradual evolution of the extant taxa placed in the nodes. Primitive traits are plesiomorphic and are here given as lowercase letters. These evolve into an advanced state in a different species with a different combination of traits, see arrows. Note that all deduced descendants require reversals to ensure the cladistic logic of the nodal traits, and the fourth and fifth (terminal) descendants are nonsense taxa required to form the cladogram node that is a combination of P and Q evolving into two descendants through reversals.

How did we get here? — Cladistic science was quite pleased to create classifications based on maximum parsimony of synapomorphic trait transformations. *Cladistics* created a dichotomous tree of species sharing the optimally largest number of homologous traits, which, of course, contains a lot of evolutionary information. A cladogram does not directly show either branching or serial descent because the axis of a cladistic "lineage" is a series of gradually less inclusive trait combinations. Continuity in a cladogram is implied by the shared traits but this is not directly translated into shared descent of taxa. *Phylogenetics* tried to improve on the cladistic structure by injecting an evolutionary dimension called recentness of shared descent. What

descended? The branches. From what? The nodes. Each node was identified solely as an unknown shared ancestor. Thus, cladistics focused on "branch-thinking" or clade relationships, while phylogenetics introduced the additional element of "stem-thinking" or ancestral relationships on a phylogram. The length of a branch on a phylogram was proportional to the distance from the hypothetical shared ancestor.

But the logic of the postulated unknown shared ancestor of phylogenetics was never followed through. Let's use parsimony of trait trans-formations in the context of a cladogram that has ancestors at the nodes. Throughout the present book, a species is considered to have at least two traits linked by some evolutionary process, known or unknown, otherwise one trait may be the result of an allele floating in one species' complex gene pool. If the unknown shared ancestral species was different from both sister groups and also from other unknown shared ancestors, then it took *two trait transformations* to generate it from the last unknown ancestral species, and *two trait reversals* to generate each sister group (the separately generated descendant species), totaling an additional six trait transformations not included in parsimony analyses. On the other hand, suppose both sister groups derived directly from the next lower species in the cladogram (i.e., the nearest neighbor)? We save six steps. But … phylogeneticists do not count steps of direct speciation in parsimony analysis, only steps of traits shared by sister groups. Phylogenetic parsimony is blind to the intellectual burden of superfluous postulated entities, but I hope the Kindly Reader is not.

The methods presented in the present book deal with both branch-thinking and stem-thinking by relying on cladistic methods to simplify the data through hierarchical clustering of species that minimizes redundancy of traits, and eliminating hypothetical shared ancestors except when the situation calls for prediction of a missing link in branching linear chains. Cladistics models evolution with shared apomorphies, phylogenetics with unnamed shared ancestors, and macro-evolutionary systematics with named ancestors through direct descent. Serial macroevolutionary generation of one taxon from the other is paraphyly, pure and simple. Although paraphyly is avoided in thought, word, and deed by phylogeneticists, it is the fundamental informative desideratum of macroevolutionary systematics.

Direct descent of supraspecific taxa — It is quite possible that serial descent of genera, families, and other taxon groups larger than a species can be accomplished. Rather than treat a supraspecific taxon as an agglomeration of many species surrounded by a gap of some sort in expressed traits, in macroevolutionary systematics the critical element for a supraspecific taxon is the core species or core lineage, which is connected with another taxon, as opposed to specialized branches which at the present time seem to be specialized dead ends in evolution (geologic time will tell). The connecting traits of a supraspecific taxon are those of the basal species. Such species best retain conservative traits that are clues to connective portions of the core lineage of a putative progenitor taxon, but this does not include specialized taxa of branches of extinct basal portions of the core lineage. This is applicable to any set of sets (genera, families, orders) that shows the dissilience of radiate evolution and thus is worthy of an empirically based scientific name.

Chapter 3
JUSTIFICATION FOR A NEW SYSTEMATICS

Short Statement

Cladists have long based their methods on a simple justification (Ashlock 1974): "A cladogram is a hypothesis, the best explanation of the distribution of characters, be they morphological, behavioral, or other, in the organisms under study, using all of the facts available." It is no longer the best explanation, and a better explanation is conveyed by a better model of evolution, namely direct descent. I have been informed by reviewers of my papers that I should read the older literature of cladistics, which would show me where I have lost the true path. One example I found (from Engelmann & Wiley 1977 on "The place of ancestor-descendant relationships in phylogeny reconstruction") reads thusly:

"Autapomorphies are identified because they supposedly differ from the hypothetical ancestor. But, the ancestral morphotype is not a scientific statement, rather, it is a simple summary of characters derived from a cladistic hypothesis. An autapomorphy can only be identified by rejecting the alternate character as synapomorphous. We conclude that autapomorphies cannot refute ancestor-descendant relationships. Therefore, given our initial assumptions, ancestor-descendant relationships based on morphology are not objective statements when applied to fossil populations or species."

Is this compelling? In the present work I present methods of identifying the order and direction of evolution of serial descent of morphospecies, as a basis for devising a well-supported classification. I commend this to the reader's attention as a replacement for cladistics. Given that most researchers now use phylogenetic techniques as "orthopraxy" (following Imre Lakato's research programme ideal of not having to justify first principles) rather than "orthodoxy" (following and vigorously justifying methods when queried), I hope a simple change in assumptions may be amenable.

Cladistics uses a modification of clustering technique to analyze evolution of traits of a group of taxa, this being a microevolutionary study of a macroevolutionary phenomenon. Cluster analysis is grouping by similarity, with various modifications one can use to deal with odd features of data such as outliers. Cladistics uses trait transformations rather than overall similarity, which is an advance as it clusters more in line with the theory that traits of species are more rare and taxa more specialized or at least different as species evolve away from some central taxon that is close to an outgroup taxon. But, because the nodes in both cluster analysis trees and cladistic trees merely signal levels of similarity, overall or trait transformation, evolution is not modeled by one node transforming into another higher in the tree. A node in a cluster or cladistic tree represents or models nothing in nature (see Tal & Wansink 2016 for why this is impressive anyway). If the point of evolutionary analysis is to analyze evolution, phylogenetics misframes the procedure and produces instead a dichotomous cladogram of maximum parsimony or likelihood.

How does that work? A cladogram is like a dichotomous natural key. Here is a simple cladogram of four taxa, including an outgroup X similar to Species A, and the formula is X, (A (B,C)). Phylogeneticsts often state that B and C have a shared ancestor, represented by the node between B and C. The nodes do not show in the Newick formula for the cladogram, but do in the dogbone cladogram:

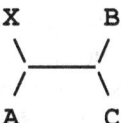

Figure 3-1. Dogbone cladogram.

Species A does not share the ancestor of B and C, but instead possibly shares an ancestor with X. Exactly what are these ancestors at the nodes? Let's examine a dichotomous natural key that is the equivalent of the above.

The natural key simply places the more odd or unusual (interpretable as "advanced") taxa farther along in the key, and the most generalist taxa or those most like nearby groups at the beginning of the natural key. An example of a simple natural key is the following:

1. Similar species in genus M as outgroup .... Species X.
1. Species in genus N 2
    2. Leaves green, reproducing with flowers .... Species A
    2. Leaves bluish or reddish, reproducing asexually 3
        3. Leaves bluish, reproducing only with tubers .... Species B
        3. Leaves reddish, reproducing only by fragmentation .... Species C

The above dichotomous natural key clusters the taxa in the same way as the cladogram. Rotate the natural key 90° counterclockwise to simulate a tree-like cladogram. It places advanced, highly specialized species B and C in a couplet after Species A, to some readers at least implying that A is less advanced, more primitive, and perhaps ancestral to them both. Thus, "natural" means evolutionary, in this case like a genealogy of one taxon generating another but limited to descendants generated either one at ta time (laterally dichotomous like Species A) or in pairs (terminally dichotomous like Species B and C).

So … is there a shared ancestor for Species X and A, and another for Species B and C implied in the natural key? The phrases "Species in genus N" and "Leaves blue or red, reproducing asexually, applies to both Species X and A, and to Species B and C, respectively, and does not distinguish any other species known or unknown. The nodes as vaguely described in the natural key are entirely unhelpful in modeling an implied shared ancestor for Species X and A, or for Species B and C. The same is true for the two nodes of cladogram X, (A (B, C)). There is no descriptive model for the hypothetical shared ancestor of X and A, or of B and C. The concept of unknown and unknowable shared ancestors is a fantasy designed to give the dignity of an evolutionary tree to a cladogram. If there is no shared ancestor that gives rise to another taxon, exactly modeling monophyly along a series of species is impossible. A cladogram generally clumps taxa by evolutionary relationship, but does not model monophyly, no matter how resolved or well supported the branches may be.

The present volume proposes an evolutionary model with actual taxa at the nodes of an evolutionary diagram (the caulogram), not just at the ends of branches. This is possible because, for all but the more ancient taxa, the progenitors of descendant species commonly are extant.

Monophyly can be exactly modeled. There is no taxon definition native to cladistics (other than the "phylogenetic species concept" which is unsatisfactory because it depends on trait monophyly or undemonstrable taxon monophyly), another reason there is no model of serial speciation events.

A method that cannot falsify a hypothesis is just as unscientific as a hypothesis that cannot be falsified. Microevolutionary analysis of trait changes from node to node alone cannot falsify hypotheses of taxon transformations. Cladistic analysis of shared traits is not a test or model of serial descent nor can it distinguish monophyletic groups in any but the most broad aspects. Here, "analyze" means to organize information into easily conceived and measurable components, and put these back together into a process-based theory that is congruent with other theory and predictive of the results of evolution.

Mathematics is a valuable and perfect (or nearly so) art. It is rooted in reality by dimensions, e.g., units like meters and volts. In practice, sometimes 1 plus 1 is not equal to two. One apple plus one anvil is at most two "things." In phylogenetics, a progenitor taxon plus its descendant taxon does not equal two sister groups, even relative to the "ancestral node," which is simply a mathematical construct of a cladistic optimization procedure. A "sister group" implies there is a shared mom somewhere, and this may not be so, as one sister may be the other's mom. Phylogeneticists may argue that a progenitor must have zero branch length, but this is true only in the most parsimonious solution. This is because there may be evidence in sister groups of a progenitor-descendant relationship that requires a less parsimonious cladistic tree.

Those who cling to hypothetico-deductive Popperian methods are exposed to the prospect of imposing the structure of their analytic method on their apprehension of nature. A simple (silly) case is: Nature may be organized into named species; species may be alphabetized; ergo evolution is alphabetic. More relevant (far less silly) case: Species are generated from other species gradually; gradual evolution may be organized along the lines of a bifurcating nested dendrogram; ergo, using cladistics or Markov chain Monte Carlo, evolution may be viewed as a bifurcating nested tree of life; not-so-ergo, evolution *is* a bifurcating tree of life.

A deduction from this reification is that evolution proceeds from one node to another (from one set of nested traits to the next); then it follows that similar trees generated from molecular data model evolution of these nested nodal traits, and, because there is more data in DNA, molecular analyses of molecular "kinds" (strains or species are equivalent) rule. As one can see, building error on error leads to rejection of evolutionary theory that tries to explain all relevant data.

Two Cladistic Problems

1. Phylogenetics uses only evidence of shared traits to make a tree of relationships. Evidence of serial or branching linear descent of taxa is ignored. This blindness requires strict phylogenetic monophyly that no taxon can generate another taxon at the same rank or higher. The result of using only shared descent is low evolutionary resolution of even well-resolved cladograms, and multiplication of over-lumped or poorly conceived cryptic taxa. Consider **A** > **B** > C, where species A speciates B and B speciates C. (The angle bracket indicates direction of speciation; boldface is a progenitor.) The cladogram for this is A(B, C) given that B and C share at least one trait advanced over those of A (unless there is a reversal). In this case, information is lost when data on serial descent is ignored.

2. Parsimony analysis often uses minimum spanning trees and Markov chains to analyze phylogenetic relationships of both morphological and molecular data. The minimum spanning tree is much used in electronics to answer the question: "what is the least amount of wire to connect a number of points." The minimum tree between two species is a direct line, that for three species is three spokes, that for four species is either four spokes if all are equally divergent, or a dog-bone arrangement (see Fig. 3-1, 3-2 below) if they are more similar as pairs. The nodes are required for the minimum spanning tree and mean little to a *cladist*, but are invested with evolutionary significance by *phylogeneticists* as indicative of unknown shared ancestry and recency of descent. This directly requires an ad hoc assumption of "reciprocal monophyly" in which two sister lineages quickly or at least gradually divest themselves of traits of the unknown putative (mathematically reified) shared ancestral taxon.

The nodes of minimum spanning trees are not evidence of anything in nature, thus such a tree does not model evolutionary processes. A cladogram based on a minimum spanning tree lacks evolutionary resolution because the cladogram does not distinguish between progenitor and descendent in a sister group, even when relevant information is available.

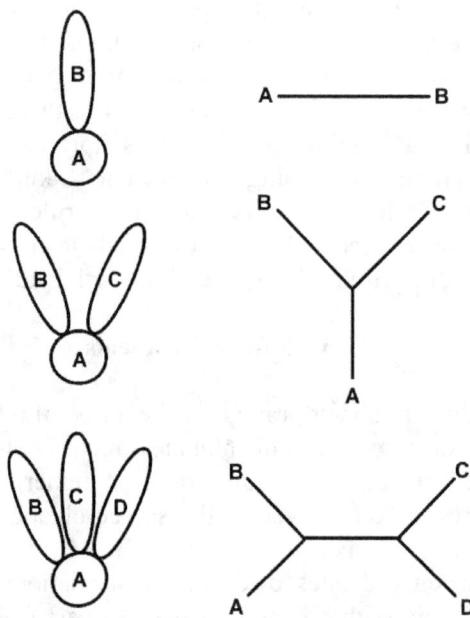

Figure. 3-2. Caulogram on left, minimum spanning tree cladogram on right. The spanning tree nodes are artifacts of the analysis. Caulograms are always rooted because they model the order and direction of evolutionary transformation of taxa, but cladograms may or may not be rooted.

Phylogenetic analysis using Markov chain Monte Carlo techniques commonly requires a first order Markov chain to do the heavy mathematics and statistics. This is a dichotomous branching tree where splits are determined only from information on the result of the last split. In the case of *molecular data*, given that both ancestral and descendant species continue to diverge through mutation of the tracking sequences, a dichotomous tree should reveal the direction and order of cladogenesis of extant molecular races of species. But not of morphospecies undergoing selection. When molecular races go extinct, a portion of critical data is lost on the serial generation of descendant species from a progenitor species. If only a recent molecular race of the progenitor is extant, that progenitor winds up at the top of the cladogram. If only an ancient molecular race of the progenitor is extant, the progenitor finds itself at the bottom of the cladogram. If two or more molecular races of the progenitor are extant, we have molecular paraphyly, which is an invitation for phylogeneticists to either lump species or name at the species level the paraphyletic molecular races. A detailed discussion on this contretemps is

given by Zander (2013: 51–65). Another problem is when an unrelated taxon is included in the molecular data set of several other taxa, and by sheer chance is placed within or terminal with the unrelated group, rather than clearly isolated or at the base of a rooted tree.

In both Markov chain and parsimony analysis, the disappearance of the ancestral taxon or its necessary unimportance (e.g., calling it an ancestral population rather that a species) is necessary for the analytic method to work, thus a "hard science" method with statistically well-supported results determines evolutionary theory (now called phylogenetic theory), not the other way around as it should be. Dichotomous trees with unnamed and unnamable nodes must only rarely model evolution of species. Repeatability of results is not the only criterion of good science, particularly when the results are patently wrong. These two simplistic means of generating apparently well-supported "hard science" trees, (1) grouping by shared descent alone (ignoring serial descent), and (2) node anonymity through Markov chains and minimum spanning trees (ignoring adaptive radiation), is why classical taxonomy is not now being well appreciated or properly funded. The more mechanical, simpler, more exact-looking method using computers and DNA data is unfortunately far more attractive.

Another way to look at this is that the shortest distance between two species is a straight line. If a serial lineage of three or more species was clearly nested by increasing specialization (accumulation of new traits in a single lineage of descendants from one generalized progenitor), then a straight line connecting each species is the minimum spanning tree. Inference of branching spanning

trees (or Markov chains) is unnecessary. The cladistic insistence on the use of only shared traits allows the imposition of an unnecessary method (cladistics was birthed from phenetics) on the true simple linear order. This creates false entities (nodes and connecting internal branches) that model nothing and are not evidence of anything. This also involves a burden of unnecessary special mathematics and statistical techniques that represent only smoke, mirrors, and magical arm-waving.

Simple Explanation

Analogies are always imperfect but let us try one close to home. Consider a maternal germline lineage, or series of moms. Each mom is generated by a previous mom and generates one or more

moms until we come to a final daughter, or sis, not yet a mom. The caulogram (a model of direct descent) for two moms with progeny can be compared to equivalent cladograms in Fig. 3-3.

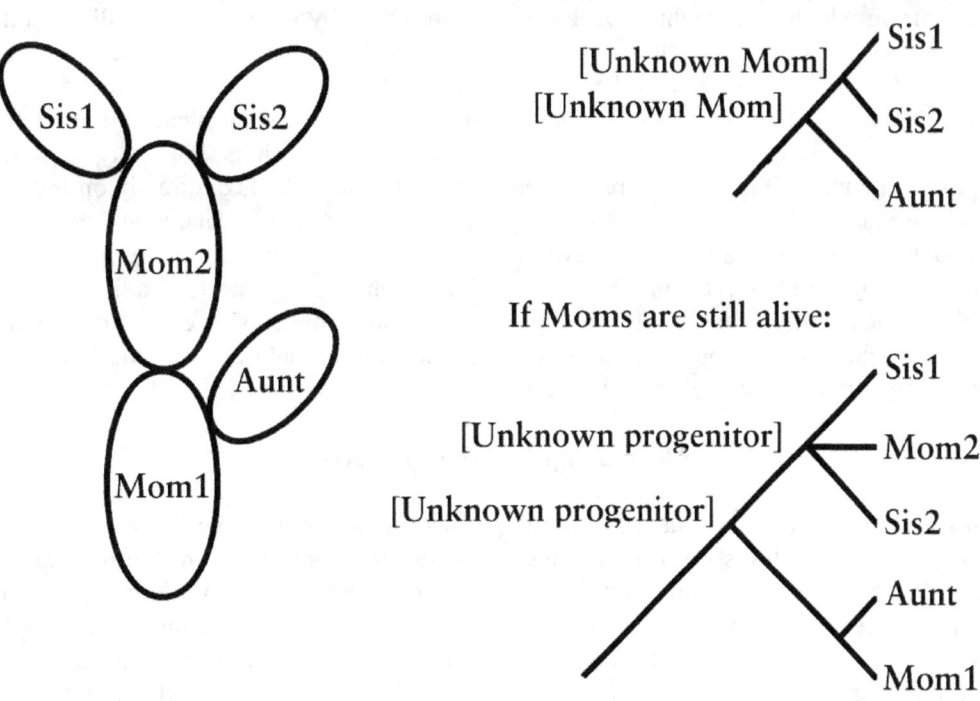

Figure 3-3. On left is a caulogram of a maternal lineage, with evolutionary formula **Mom1** > (Aunt, (**Mom2** > (Sis1, Sis2))). On right are two phylogenetic cladograms, the top if the moms are absent (extinct), and the lower if moms are extant. The top cladogram is evolutionarily correct, and the nodes mean something, but the bottom is false and lacks resolution of serial evolutionary processes.

In Figure 3-3, on the left is a caulogram of a maternal lineage **Mom1** > (Aunt, (**Mom2** > (Sis1, Sis2)). Think Grandma generating Aunt and Mom, then Mom generating two Sisters. The nodes of an equivalent cladogram correctly model the evolution only if the moms are not extant. Is that ever the case? That is, are potentially differently named ancestral taxa of extant taxa ever entirely absent such that cladograms are mostly the same as evolutionary trees? This is doubtful, because the efficacy of taxonomy itself is dependent on survival of closely related species.

Also, macroevolutionary analysis of, say, the moss genus *Didymodon* (Zander 2013, 2014a,b,c) indicated that the most recent species and at least one or commonly two species in the same serial lineage are extant, that is, may be lined up as a series of two or three closely related species with an evolutionary relationship of gradually increasing advanced traits. Thus, there is a sliding window through time of, say an average of five million years, during which any one species survives and during which one might expect two speciation events without extinction of the ancestral taxon. This may be true for other taxa, but it is yet early days for macroevolutionary analysis.

If the two "moms" (Fig. 3-3) are alive, however, and are part of the data set, then the cladograms have nodes that mean nothing beyond an indication of shared traits (revealing cladistics' origin in hierarchical cluster analysis). That is, evolutionary continuity in a cladogram is between sets of shared traits (nodes). The splits then do not distinguish the direction of evolution between Grandma and Mom or Aunt, or Mom and the Sisters. If there are living moms then the phylogenetic distance between nodes is extremely biased if the nodes and internal branches between them are taken to be evolutionarily significant. The analogy is somewhat strained but I think effective. And it is important because, in my taxonomic experience, "extended families" are the rule.

In addition, the cladograms in Fig. 3-3 have the cladistic formulae (Sis1, Sis2) Aunt, or (Sis1, Mom2, Sis2)(Aunt, Mom1). The formulae can entirely replace the cladograms. Where in the formulae are the nodes? For that matter, where are the branches? The tree dangling below the line of taxa on the top of a cladogram (or on the right) is not an evolutionary tree but is instead the remains of the raw minimum spanning tree or Markov chain used to optimize the arrangement of the parenthetical sets of taxa. It is wrong to think of a phylogenetic tree as a detailed evolutionary tree.

How does one tell if the ancestral taxon represented by the nodes are still around and have wrongly been made part of the phylogenetic data set? If all the ancestral taxa are dead and gone, there would be no problem as the cladistic nodes would represent them. The way to find out is to attempt to nest the taxa directly on each other such that postulation of a shared unknown ancestor is less parsimonious than simple serial trans-formations as a kind of radiation of gradually increasing change. This can be done with the help of an outgroup and an Analytic Key, explained in the Methods section.

More Complicated Explanation

Phenetics uses a number of different methods of grouping taxa by shared traits to find clusters (often a dendrogram) of maximum similarity. Cladistics introduced a way to group by transformations of character states to create a dichotomously branching non-ultrametric (i.e., has variable branch lengths) tree. The clue that cladistics is, like phenetics, hierarchical cluster analysis, is that the nodes model nothing, they are simply places in a dendrogram that signal that more distant taxa share the same advanced, nested traits.

In cladistics, the simplest way to organize branching to minimize trait changes is called the most parsimonious tree. Each node in the tree, in *cladistics* is simply part of the mechanical method of grouping taxa with a minimum of trait changes into pairs of "sister groups." In *phylogenetics*, evolutionary explanations are added to the cladistic method such that tree nodes are termed "shared ancestors" probably going extinct after generating two new species, and branch lengths become significant as measures of amount of evolutionary change (from the node). Thus, phylogenetics attempts to add a causal connection to trees of shared descent.

In both cladistics and phylogenetics, species from one genus that branch out of the middle of a group of species of another genus must be lumped into that enclosing genus. This is a principle of cladistic classification called "strict phylogenetic monophyly." The assertion of the present study is that focusing entirely on changes in traits is a microevolutionary technique (e.g., for population genetics) improperly used to explain macroevolutionary changes (species to species transformations). That is, individual allele changes in the context of one species or population, compared with changes and fixation of sets of character states between two different species (or higher taxa).

The lumping and splitting of taxa to remove paraphyletic groups is probably due to there being no empirical method of distinguishing taxa in cladistic technique. In addition, the direction and order of descent is poorly resolved because the continuity is compromised by nodes not being taxonomically named when such inference is possible. If the traits represented by a node simply summarize those traits that are fit all or most taxa distant to the node on the cladogram, then the node represents no species at all, and at most may be named as a higher taxon solely in a mechanical

fashion, with no reference to evolutionary theory. The word "paraphyly" is simply an evasive designation of evidence for serial descent ignored by the cladistic method which focuses entirely on shared descent. "Paraphyletic" is somewhat disparaging phylogenetic cant for what is generally known as ancestors involved in macroevolution, that is, a label for a group from which one or more other groups at the same (or higher) taxonomic rank have apparently evolved. "Para" implies faulty, wrong, amiss, or merely similar to the true form. But suppose we stand the disparagement on its head. Classical systematists, observing the same phenomenon of one taxon embedded cladistically in another, see this as perfectly natural. The resulting phylogenetic lumping and splitting of taxa may be termed "cacophyly," meaning a bad evolutionary model where good taxa are unwarrantedly split or reduced to synonymy. A way to deal with these contrasting viewpoints is a kind of dialectic (Aubert 2015) that conciliates the two in a pluralistic framework based on modeling direct, serial descent, in which case paraphyly does not need to become cacophyly.

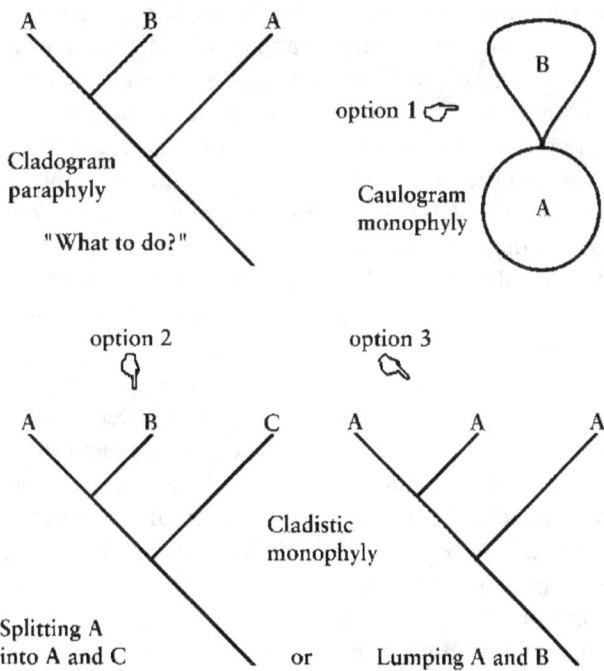

Figure 3-4. Upper left: We begin with two taxa, A and B, split in a cladogram with A paraphyletic. In morphology this may be due to two species of one genus (A) separated by a species in a different genus (B), or in molecular study, by two molecular races of one taxon (A) separated by a molecular race of another taxon (B). Option 1: paraphyly in a cladogram becomes monophyly in a

caulogram. Option 2: Cladists split A into two species, A and C. Option 3: Cladists lump A and B into one species, A (or whichever has the earlier taxonomic name). In Option 1, strict phylogenetic monophyly is saved with options 2 and 3, but then perfectly acceptable paraphyly becomes cacophyly.

Technical Discussion

Cladistic analysis is here considered incomplete because traits do not evolve, taxa do. A *cladogram* is the result of a cluster analysis using trait transformations rather than mere similarity; a cladogram tracks traits not taxa. Such transformations encourage cladists to aver that standard cluster analysis lacks a direct evolutionary dimension while cladistics recognizes such by optimizing on shared advanced traits. The continuity in a cladogram, however, is though nodes characterized by traits shared distally on the cladogram (except for reversals), but that in a *caulogram* consists of a series of taxa.

Cladistics polarizes characters but evolutionary systematics polarizes taxa. Sometimes a cladogram approximates a caulogram, but often not. The inner nodes of a cladogram are solely of trait transformations on a minimum spanning tree not of taxon transformations. A dichotomous tree needs to be re-interpreted as an evolutionary tree (evolution is not necessarily dichotomous but may be simply serial or multichotomous or any such combination). Thus, a cladogram is empty, being a field of universals without particulars, or context without content. For example, the evolutionary formula may be A > B > C and actual taxa provide continuity. A cladogram with formula A(B, C) has no continuity in terms of taxa. Because a cladogram can be entirely expressed as a nested set of parentheses and taxon names, the nodes on a cladogram are actually the same as parenthetic sets and do not produce a causal model.

Given no details about the putative shared ancestors, a cladogram is "too vague to be wrong" (Ferris 1997: 67), i.e., falsifiable. A few major corrective principles in macroevolutionary systematics are (1) If two cladistically sister taxa are easily modeled as one generating the other by elaboration or simplification of rare or unique traits or trait combinations, then the cladistic "shared ancestor" is taxonomically the same as the one sister taxon generating the other. (2) Transformations at the taxon level follow the Dollo rule that complex traits are not re-evolved. (3) Examples of one generalist taxon radiating a set of specialized descendants signals an empirical example of a taxon higher than a species, i.e., a "dissilient" genus. (4) Analyses based on expressed morphological traits are more effective than molecular analyses because extinct molecular paraphyly renders DNA study inexact, truly informative heterophyly (paraphyly or short-range phylogenetic polyphyly) is uncommonly sampled (Zander 2008), and any Bayesian support must take into account negative Bayesian support for contrary morphological results.

The major difference in analysis between morphological and molecular data is that extinct or cryptic paraphyly is identical in extant morphospecies, but variation is sometimes confoundingly different between extant molecular races. Evolution itself has removed informative molecular races needed to correct molecular cladograms that apparently detect non-monophyly. Such is not a major problem with morphological taxa though of course some morphological races informative of direction and order of evolution may be unknown. Because no attempt is made by cladists to taxonomically name the nodes of a cladogram, much less reinterpret the cladogram as a caulogram, the actual resolution of molecular cladograms doubtfully provides any better resolution than the average distance of paraphyly in the group at that taxonomic level. Basically, though, the fact that there are often many molecular races and these races may each generate descendants of different morphospecies, means that order of generation of branches in molecular cladograms is limited to order of generation of molecular races, not morphospecies.

There is much ado about total evidence in the literature, and this is discussed further elsewhere in this book. The biological species concept is an example of how some facts are well dealt with through analysis informed by data on sexually isolated groups. The biological species concept is known, however, to fail in certain groups, but at least it well explains some of the data and some of the processes of evolution for some of the taxa. Cladistics, on the other hand, uses only shared descent (eschewing serial descent) as a paradigm for analyzing evolution, and thus creates biased results all of the time.

Scientists are inured to the small samples involved in molecular analyses, probably due to the large cost of sequence analysis some decades ago. Massive intraspecies sampling is, however, no longer expensive, and all but very new and very old species may be expected to exhibit some degree of heteroplasy (multiple molecular strains or races). In some cases (e.g., Alonso et al. 2016) molecular races from a single species may generate their own descendants of distinct species, a process that generates paraphyly or even phylogenetic polyphyly. Such paraphyly may be at species, genus or family level, and is often masked in published cladograms by name juggling following the phylogenetic principle of strict monophyly (one taxon cannot generate evolutionarily another taxon of the same or larger taxonomic rank, and classification must reflect this).

In phenetics, the terminals of dendrogram branches are called OTUs (operational taxonomic units). But with molecular taxonomy, the exemplars actually represent molecular races or strains, because there is much evidence for the ubiquity of such in modern analyses with many exemplars of each species. The terminals of molecular cladograms might be better named OSUs (operational subtaxonomic units), with classification based thereon viewed askance.

Molecular analysis is valuable for testing for large degrees of evolutionary distance and for bracketing of paraphyly that implies serial evolution of the embedded taxon from the paraphyletic taxon. It has several problems including (1) small samples, (2) heteroplasy (radiation of molecular strains or races that separately generate their own species-level lineages), (3) all problems associated with cladistics analysis using morphological data, (4) conflicting studies, (5) reuse of data contributes to small samples and absence of experimental replication, (6) no native taxon concept, and (7) the ancestral molecular strain continues to mutate thus pseudoextinction is valid (at the infraspecies level) in molecular analysis leading to difficulty in serial modeling for sister groups of such strains. Heteroplasy makes molecular analyses analogically presbyopic (far-sighted) in that evolutionary relationships of closely related taxa are poorly discernible.

Modeling speciation — Evolution in the groups dealt with here (Streptotrichaceae and Pottiaceae) is twofold.

(1) Anagenetically, species develop (in theory) series of populations that reflect a reduction of trait expression through parsimony of photosynthate in less mesic (cool, wet, well lit) environments, e.g. hyperoceanic areas. Taxa in hyperoceanic and high-montane habitats seem to retain well-expressed sets of characters (larger and broader leaves, more and bigger laminal marginal teeth, bigger laminal cells, poorly developed stem sclerodermis) while the same taxon in more difficult environs are more streamlined, with some traits much reduced or lacking, while other traits associated with drier environments (thicker laminal cell walls, leaves narrowed above, stronger papillae) are accentuated. When a species' spectrum of variation is broken into two or more groups by differential environmental change, the broken or gapped range of morphological variation is like a description of the gapped environment, using evolutionary theory as a language.

(2) Speciation also is through addition of specialized traits for asexual reproduction (e.g., distribution of gemmae or fragile leaves) or survival in harsh environments (e.g., water absorbing tomentum or trigonous laminal walls). The two types of evolution occur through direct descent, and cannot be modeled through analysis of shared descent. In cladistics, evolutionary continuity is modeled through a series of transformations of traits shared by two or more species, but in evolutionary systematics it is modeled through evolutionary processes of reduction and elaboration of ancestor to descendant taxon involving all trait transformations, shared or unique. The latter is here considered far more instructive of evolutionary process than is cladistics.

Caulograms and cladograms — A caulogram (commagram, Besseyan cactus) diagrams both direct, serial descent, and shared descent. A caulogram may include inferred, interpolated ancestors as "missing links" at various taxonomic levels when data is available. This models evolution as a dynamic causal process. A cladogram, however, is restricted to shared traits and methodologically necessary postulated unknown and unnameable shared ancestors at every node, and is a static representation of evolutionary relationships, with little power of prediction.

A *cladogram* represents all relationships as branches. Distance measures may be provided but surety of relationship in terms of Bayesian posterior probabilities is not the same as an emphasis that some clade is really different from another. A clade may include a quite different taxon which is synonymized merely to avoid paraphyly. A cladogram may have many lineages converging at the base as simply a set of nodes. A cladogram is analyzed with an optimal set of nested transformation of traits that apply to all parts of the cladogram. A *caulogram*, on the other hand, does emphasize when taxa are quite different by illustrating dissilient genera, and identifying missing links (unknown, extinct, or otherwise lost taxa of intermediate evolutionary order) when these may be inferred. The inference of trait transformation during optimization is limited to just two taxa; there is no need to postulate evolutionary action-at-a-distance.

A clear and concise explanation of the difference between cladistics and evolutionary taxonomy was given by Mayr (1981).

Justification for a New Synopsis

It has been simplistically explained that taxonomy generally results in studies that are purely descriptive, not comparable to experimental science with its statements of past work, methods, results, and discussion. I have in the past pointed out that taxonomy, even the most minor contributions ("Hey! I found such and so species in a county in which it was heretofore and hitherto unknown!"), are part of the great Linnaean endeavor of discovering, describing and classifying the Earth's biological diversity. Recent developments have also deeply involved systematics in evolution, ecology, and other more experimental or quasi-experimental (Cook & Campbell 1970) sciences, while classification and demonstration of evolutionary uniqueness is critical to biodiversity decisions in conservation. We can now extend that justification for taxonomic (faunistic and floristic) study to an even broader context:

Systematists are contributing to the conservation of humanity by monitoring the present extinction event, predicting what will replace the damaged biosphere, and determining methods of preserving or enhancing life-supporting, buffering biodiversity during the transition to a new, more precarious habitat for all life.

Each of many small taxonomic projects, including monographs, floristic and faunistic treatments, studies for species and habitat protection, is a benchmark for this through-time characterization of the human-caused extinction event. The purpose is to find the several or many best means of mitigating the worst effects of over-population, over-development, and global warming. How many deaths associated with the e-tinction event, from war, starvation, disease, and global weather upheaval, it will take to give respect to the many-faceted efforts of conservation-oriented systematists to delineate the problems and find solutions. And there is the obvious moral dimension, in which concern for universal survival is pitted against disconcertingly popular self-centered worship of mammon. Given the breakdown of fragile ecosystems and massive swings in monocultures, we should preserve the biotic multicultures we need for survival.

Why preserve multicultures? Because, as discussed at length in the chapter on ecosystem survival near the end of this book, dissilient genera form hubs in scale-free networks, where evolutionary and habitat redundancy allow immediate (in geologic time) repair of thermodynamic ecosystem damage due to loss of species, while a train of rare genera within the family provide long-term replacement of genera degraded by cyclic climate change. Cladistics and its application of the principle of strict phylogenetic monophyly destroys proper appreciation of critically important scale-free and small-world networks in nature. Systematics is central to planning to mitigate or at least prepare for the ongoing sixth extinction event.

Chapter 4
PREVIOUS STUDY

Streptotrichaceae — The family Streptotrichaceae has been presaged for quite a while by various students of the Pottiaceae—as Leptodontioideae by Hilpert (1933), and Leptodontieae by Chen (1941) and Saito (1975). Zander's (1993) study of the genera of the Pottiaceae recognized on cladistics grounds the Leptodontieae (1993: 45) as including two clades, first, *Hymenostylium, Reimersia,* and *Triquetrella.* and, in a second, *Leptodontiella, Leptodontium, Streptotrichum,* and *Trachyodontium.*

The Leptodontieae of Zander (1993) was cladistically distinguished by a node characterized by the following traits: central strand absent, ventral costal epidermis absent, and laminal cell walls flattened. The first clade, with *Hymenostylium, Reimersia,* and *Triquetrella,* was distinguished by stem section triangular. The second clade, incorporating most of the present Streptotrichaceae, was distinguished by many trait transformations: stem hyalodermis present, leaves dentate or denticulate, 4–6 cells across width of costa, costa flattened, abaxial costal epidermis absent, stomates absent, and spores more than 15 μm in diameter. Of course these are basal traits, which may be viewed as central tendencies as there are often reversals and anomalies overwhelmed by other evidence, but this central tendency is clear and, with additional traits recognized here, notably the tomentum ending in series of short-cylindric cells, points to the present concept of the Streptotrichaceae.

In the present work, the first clade is also eliminated because *Hymenostylium* (now *Ardeuma*) has hygric habitat, semicircular costal section, abaxial costal epidermis present, perichaetial leaves similar to the cauline, and peristome absent; *Reimersia* has narrowly acute to acuminate leaf apices, costal epidermis present abaxially, laminal papillae absent, perichaetial leaves similar to the cauline, and peristome absent (the two concavities in the distal portion of the leaf base are probably not conservative); *Triquetrella* occasionally has a stem central strand, has an abaxial costal epidermis, and spiculose laminal papillae (rare in *Leptodontium*). These are largely traits common to the Pottiaceae.

Leptodontium — A study of the moss genus *Leptodontium* (Müll. Hal.) Hampe ex Lindb. has reviewed the New World species (Zander 1972), and included many Old World species in synonymy and discussion. The present work was originally an effort to finish that study by incorporating the Old World species, but was extended when *Leptodontium* was found to be part of a well characterized segregate of Pottiaceae, in fact, a new family. The genus was in the 1972 monograph divided into four sections, in part recognized here as separate genera.

Since 1972, the have been a genus segregated, *Leptodontiella* (Zander & Hegewald 1976), with *L. tricolor* resurrected from 1972 synonymy in the same paper. Frahm and Schumacker (1986) contributed a valuable "type revision" of European *Leptodontium* species. Many species of the genus were covered in a checklist of the mosses of Central Africa (Born et al. 1993), also in another checklist for sub-Saharan Africa (O'Shea 1995), and in one for tropical Andean countries by Churchill et al. (2000). A catalogue of the mosses of central Latin America was issued by Delgadillo et al. (1995). Major works that included considerable information on *Leptodontium,* and which were of particular value in the present study, were publications of Sharp et al. (1994) on Mexico, of Churchill and Linares (1995) on Colombia, of Catcheside (1980) on South Australia, of Magill (1981) on South Africa, of Li et al. (2001) on China, of Aziz & Vohra (2008) on India, of Norris and Koponen (1989) of Papua New Guinea, and of Eddy (1990) on Malesia. Many smaller works were published since 1972, including a particularly fine summary of the thatch-roof ecology of *Leptodontium gemmascens* by Porley (2008). Two important students of the Pottiaceae who have published a number of relevant small papers are T. Arts and P. Sollman. Avid and productive researchers in Pottiaceae at the University of Murcia, Spain, have not yet turned their complete attention on *Leptodontium,* but doubtless it is a matter of time. Other papers on

Lepdotontium are cited in the taxonomic section of this book.

Leptodontiella —This genus was based by Zander and Hegewald (1976) on *Leptodontium apiculatum,* a small moss of tree limbs and rock at high elevations in Peru. The gametophyte is similar to that of *Leptodontium flexifolium.* Previously known only in sterile form, fruiting specimens were immediately distinctive in the very short seta (rare in *Leptodontium*) and the short urn. The distinctive sporophytes of *Leptodontiella* exaggerated the tendency for *Leptodontium* species to group their teeth in fours (two pair of two rami each), in some specimens further split into up to 50 rami.

Streptotrichum — Herzog (1916) described *Streptotrichum ramicola* from branches of shrubs and trees in high elevations in Bolivia. Like *Leptodontiella,* the peristome teeth are usually 32 and arranged in eight groups of four rami, but up to 42 rami may be differentiated in some capsules. The teeth are elongate and have differentiated anticlinal walls. The gametophyte is quite like that of *Leptodontium araucarieti.*

Trachyodontium — Yet another species collected on shrubs at high elevations (Steere 1986), this time from Ecuador, fit this relationship. The gametophyte of *Trachyodontium* is distinctive, with a unique cartilaginous leaf border. The peristome teeth are of 64 elongate rami arranged in 16 groups of four rami.

Modern Classification

Classification of the Pottiaceae at the subfamily and tribal level was reviewed by Zander (1993: 50) and reframed in a phylogenetic context. For non-botanists, remember that the ending -aceae means a family, "-oideae" means a subfamily, and "-eae" means a tribe, these given in order of hierarchical classification rank.

The classification of Saito (1975) was:

```
┌Trichostomoideae
┤ ┌Barbuleae
 ├Eucladieae
 ├Leptodontieae
 └Pottieae
```

That of Chen (1941) was:

```
     ┌Pleuroweisieae
     ├Eucladieae
     ├Trichostomeae
     └Tortelleae
  ┌Cinclidotoideae
  ├Barbuleae
  └Hyophileae
 ┌Leptodontioideae
 ┌Pottieae
 └Merceyeae
```

That of Hilpert (1933) was:

```
 ┌Leptodontioideae
 ├Barbuloideae
┤Pottiaceae
 └Cinclidotaceae
```

That of Walther (1983) was:

Leptodontioideae

    Leptodontieae: *Leptodontium, Tuerckheimia, Triquetrella, Rhexophyllum, Luisierella, Leptodontiella, Streptotrichum.*

Thus, *Leptodontium* and its relatives have been recognized by these experts on the Pottiaceae as a subfamily of some significance. Since then, Walther (1983) in Syllabus der Pflanzenfamilien, Laubmoose, recognized the subfamily Leptodontioideae with the single tribe Leptodontieae, consisting of *Leptodontiella, Leptodontium, Luisierella, Rhexophyllum, Triquetrella, Tuerckheimia,* and *Streptotrichum.*

My own classification (Zander 2006) was based in part on the Zander (1993) morphological cladistic study.

Pottiaceae
    Timmielloideae
    Trichostomoideae
    Barbuloideae
        Barbuleae
        Bryoerythrophylleae
        Hyophileae
        Leptodontieae
    Pottioideae
        Pottieae
        Syntrichieae
    Merceyoideae

At the genus level, in 2006 I arranged *Leptodontium* and its relatives as follows:

Barbuloideae subfamily
    Leptodontieae tribe
        Triquetrella
        Reimersia
        Hymenostylium (now *Ardeuma* because the type has been excluded)
        Rhexophyllum
        Trachyodontium
        Streptotrichum
        Leptodontiella
        Leptodontium

One can see that various authors, including myself, have associated *Leptodontiella, Leptodontium, Trachyodontium,* and *Streptotrichum* with various and sundry similar taxa, including *Ardeuma* (as *Hymenostylium), Luisierella, Reimersia, Rhexophyllum, Triquetrella,* and *Tuerckheimia.* In the present chapter on outgroup analysis, none of these was found to be particularly of value as a nearest-relative outgroup.

    The most up-to-date family-level classification of the Pottiaceae is that of Goffinet et al. (2008), as modified online at *http://bryology.uconn.edu/classification.* Taxa now in Streptotrichaceae as established in the present volume are rendered in boldface. That system is duplicated below, but with the addition of *Ardeuma,* recently described.

ORDER POTTIACEAE M. Fleisch.
Pottiaceae Schimp.

Acaulon Müll. Hal., Acaulonopsis R. H. Zander & Hedd., *Algaria* Hedd. & R. H. Zander, *Aloina* (Müll.Hal) Kindb., *Aloinella* Cardot, *Andina*J. A. Jiménez & M. J. Cano, *Anoectangium* Schwägr., *Ardeuma* R. H. Zander & Hedd., *Aschisma* Lindb., *Barbula* Hedw., *Bellibarbula* P. C. Chen,

Bryoceuthospora H. A. Crum & L. E. Anderson, *Bryoerythrophyllum* P. C. Chen, *Calymperastrum* I. G. Stone, *Calyptopogon* (Mitt.) Broth., *Chenia* R. H. Zander, *Chionoloma* Dixon, *Cinclidotus* P. Beauv., *Crossidium* Jur., *Crumia* W. B. Schofield, *Dialytrichia* (Schimp.) Limpr., *Didymodon* Hedw., *Dolotortula* R. H. Zander, *Ephemerum* Schimp., *Erythrophyllopsis* Broth., *Eucladium* Bruch & Schimp., *Exobryum* R. H. Zander, *Fuscobryum* R. H. Zander, *Ganguleea* R. H. Zander, *Geheebia* Schimp., *Gertrudiella* Broth., *Globulinella* Steere, *Guerramontesia* M.J.Cano, J.A.Jiménez, M.T.Gallego & J.F.Jiménez, *Gymnostomiella* M. Fleisch., *Gymnobarbula* J.Kučera, *Gymnostomum* Nees & Hornsch., *Gyroweisia* Schimp., *Hennediella* Paris, *Hilpertia* R. H. Zander, *Hydrogonium* (Müll. Hal.) A.Jaeger, *Hymenostyliella* E. B. Bartram, *Hymenostylium* Brid., *Hyophila* Brid., *Hyophiladelphus* (Müll. Hal.) R. H. Zander, *Indopottia* A.E.D.Daniels, R.D.A.Raja & P.Daniels, *Lazarenkia* M.F.Boiko, *Leptobarbula* Schimp., **_Leptodontiella_ R. H. Zander & E. H. Hegew.**, **_Leptodontium_ (Müll. Hal.) Lindb.**, *Ludorugbya* Hedd. & R.H. Zander, *Luisierella* Thér. & P. de la Varde, *Microbryum* Schimp., *Mironia* R. H. Zander, *Molendoa* Lindb., *Nanomitriopsis* Cardot, *Neophoenix* R. H. Zander & During, *Pachyneuropsis* H.A.Mill., *Phascopsis* I. G. Stone, *Picobryum* R. H. Zander & Hedd., *Plaubelia* Brid., *Pleurochaete* Lindb., *Pottiopsis* Blockeel & A. J. E. Sm., *Pseudocrossidium* R. S. Williams, *Pseudosymblepharis* Broth., *Pterygoneurum* Jur., *Quaesticula* R. H. Zander, *Reimersia* P. C. Chen, *Rhexophyllum* Herzog, *Sagenotortula* R. H. Zander, *Saitobryum* R. H. Zander, *Sarconeurum* Bryhn, *Scopelophila* (Mitt.) Lindb., *Splachnobryum* Müll. Hal., *Stegonia* Venturi, *Stonea* R. H. Zander, *Streblotrichum* P.Beauv., *Streptocalypta* Müll. Hal., *Streptopogon* Mitt., **_Streptotrichum_ Herzog**, *Syntrichia* Brid., *Teniolophora* W. D. Reese, *Tetracoscinodon* R. Br. ter, *Tetrapterum* A. Jaeger, *Timmiella* (De Not.) Schimp., *Tortella* (Lindb.) Limpr., *Tortula* Hedw., *Trachycarpidium* Broth., **_Trachyodontium_ Steere**, *Trichostomopsis* Card., *Trichostomum* Bruch, *Triquetrella* Müll. Hal., *Tuerckheimia* Broth., *Uleobryum* Broth., *Vinealobryum* R. H. Zander, *Vrolijkheidia* Hedd. & R. H. Zander, *Weisiopsis* Broth., *Weissia* Hedw., *Weissiodicranum* W. D. Reese, *Willia* Müll. Hal.

Pleurophascaceae Broth. — Type: *Pleurophascum* Lindb.
Pleurophascum Lindb.

Serpotortellaceae W. D. Reese & R. H. Zander — Type: *Serpotortella* Dixon
Serpotortella Dixon

Mitteniaceae Broth. — Type: *Mittenia* Lindb.
Mittenia Lindb.

Taxa that have been merged with Pottiaceae without any more reason than that they are included in a Pottiaceae molecular clade are *Cinclidotus* (Cinclidotaceae), *Ephemerum* (Ephemeraceae), and *Splachnobryum* (Splachnobryaceae) (as discussed by Zander 2006). The dispositions may be correct or they may represent dissilient families generated from Pottiaceae, but for now their inclusion in the Pottiaceae may be considered an artifact of mechanical taxonomy and its associated principle of strict phylogenetic monophyly. Further study is needed.

Chapter 5
METHODS

Macroevolutionary systematics is the use of inferred transitions or transformations between evolutionarily linked sets of traits (i.e., to develop a predictive classification of organisms that are hierarchically similar in all traits). The present work is focused on parsimony of serial evolutionary *taxon* transformations, with that of serial *character state* transformations being secondary though part of the analytic process. Given that a fairly new protocol is involved, details and explanations are called for. The method is a modification of that of Zander (2014a,b,c,d), which introduced Bayesian sequential analysis in developing a true evolutionary tree tagged with probabilities of taxon transformation for *serial descent* (direct ancestor-descendant relationships) (Zander 2014e). This is here refined in combination with exemplified formal criteria for distinguishing taxa, and for support for order and direction of serial evolution by adding the logarithmic bits from information theory.

The method is explained here in necessarily somewhat tedious detail but boils down to a few steps:

In short:
(1) Use a cladogram and outgroup analysis (or equivalent) to polarize and group species by evolutionary relationship.
(2) Rearrange the dichotomously branched species diagram by nesting them serially as ancestor-descendant, including branching of two or more descendants from one ancestor.

Longer explanation:
(1) Nesting species in one another such that redundancy in the guise of parallelism and reversals is minimized by identifying homology. Generalist progenitors grade into more specialized descendants.
    (1a) A morphological cladogram together with outgroup identification is a good way to initially order species by distance from some outgroup or central tendency.
    (1b) Or, start anywhere and attach similar species together into lineages of increasing specialization (away from some central tendency or outgroup), attaching species onto or into the lineage, and making branches when the species do not attach readily to ends of lineages.
    (1c) If you have molecular cladograms of your group, the branch order may be cautiously helpful. Heterophyly (bracketing of one species by two branches of another species) implies a progenitor-descendant relationship. Species that are far distant on a molecular cladogram are doubtfully closely related but species close on a molecular cladogram may be closely related or the same because of past, often extinct heteroplasy (multiple molecular races or strains).
(2) An analytic key is used to optimize nesting by minimizing the number of cladistically methodologically required shared ancestors. Cladograms minimize parallelism and reversals in shared traits, but the aim is now to minimize these in direct descent, that is, autapomorphic traits (those found separated and unshared in a cladogram) or unique traits not in the phylogenetic data set of shared traits are also found useful. In macroevolutionary systematics, sharing of traits is not as sister groups, but as progenitor and descendant. Unknown shared ancestral species are postulated only when necessary. Geographic disjunction, when not artefactual because of poor collecting, can be a good trait, given that it is an indicator of genetic isolation as good as morphological disjunctions.
(3) A caulogram (Besseyan cactus) of stem taxa is devised as an evolutionary tree.
(4) Probabilistic support for order and direction of evolution is assessed using the bits of information theory. This is much as was done for decibans by Zander (2014a,b,c). *Order* is measured as the support for one taxon before or after another in a lineage. *Direction* is measured as support for the

entire lineage by summing bits assigned to all trait transformations in a lineage. The direction of evolution, if well-supported, will help firm the integrity of a supra-specific taxon, that is, its evolutionary coherence. The noise of traits that apparently occur randomly (not matching order of taxa based on critical traits associated with possible recent selection, or which are apparently random in occurrence in related groups) in both the group studied and in related taxa must be minimized.

(5) Classification of genera and above is based on recognition of occurrences of taxic radiation (dissilience) from a core generative species, whether adaptive or not. When two or more species emerge from one, this knot of evolutionary activity is a candidate for an empirically based higher taxon. If a species is somewhat morphologically isolated it may be a recently evolved new direction of evolution or an isolated remnant of a larger lineage. If similar to an extant dissilient genus progenitor, it is probably the former; if morphologically quite distant and specialized, it is probably the latter.

(6) Gradual, stepwise evolution as a rule-of-thumb is matched by an expectation of gradual, stepwise extinctions. Thus, taxa with no or very few near relatives are more ancient than groups with many near relatives, given somewhat equal rates of extinction. Therefore, *Trachyodontium* is ancient, but *Streptotrichum* and *Leptodontiella* are less ancient and grade into extant species.

The aim is to establish the following: (1) when a taxon fits nicely in a series (sweet spot of 2 to 4 or occasionally up to 6 trait changes) or when a taxon does not belong at all (too many differences or major tracking traits absent), (2) the order of a series (apprehension of gradualism in both generation and extinction of species), (3) when branching occurs (no differential overlap of traits), and (4) when to postulate unknown shared ancestors as missing links (as when two apparent descendants have shared traits not in the progenitor). The traits chosen are either tracking traits or new unusual because advanced traits. Of very limited use are traits that pop up randomly in the studied group and in related groups (these are treated as noise).

Analysis — *Leptodontium* species (see Zander 1972) and species of associated genera were evaluated as to distinctiveness of trait combinations. Copious material was available from MO and elsewhere of these taxa for study. All specimens cited in this work have been examined unless clearly noted as "(not seen)."

Ranges of measurements for comparison were aimed to be about two standard deviations, or 95 percent of total variation, within which maximum expression approximates the geometric mean (Zander 2013: 130).

A practical species criterion was used, namely a minimum of two otherwise independent traits were considered necessary to infer a shared evolutionary process of some sort linking these traits in multiple individuals and populations. This is because one trait alone may be controlled by a single allele floating in one species' gene pool. This minimum of two-trait per species criterion allows empirical testing of the distinction of two closely related species by evaluation of the degree of independence of the two (or more) trait pairs. If a grid is constructed with traits of one species on the left and traits of the other on top, the observed frequencies of combinations will approach random if the species are not distinct, and will be stable if distinct. In most cases, species are obviously rather distinct, but in difficult cases a chi-square analysis will help.

With the availability of many collections made since the New World revision of *Leptodontium* (Zander 1972), variation and intergradation were assessed by standard omnispection and description. Species distinction from similar species was determined by sorting into groups and evaluating coherence of variation in morphology. This is done with the aid of conservative tracking traits, being those that are rare among nearby lineages, and new or potentially adaptive traits. The latter are those that occur in association with specialized or recent environments. Of course, new traits may spread among descendant species, and the adaptive become conservative if the physiological burden is tolerable in other, new environments. Determination of nearby lineages, of course, depends on a preliminary evaluation of serial ancestry. In difficult cases, a cladogram may be of aid, although such requires identification of serial lineages difficult to distinguish when rendered as dichotomous trees.

More particularly, Crawford (2010) gives some guidelines, while many indicators of progenitor-descendant relationships are limited to particular taxa and specialized habitats. Firstly, one segregates out the apparently advanced species as potential descendants. Typical "tells" for *descendants* are those that indicate a dead end in evolution or a very new direction of evolution. Examples are: polyploidy (with no evidence of diploidization), asexual reproduction, special morphology, unique features, narrowly adaptive traits, great complexity, and rare or recent environments. Features that may indicate *progenitor* status are: comparatively generalist morphology, ancient habitats, multiple morpho-types, and central position in knots of descendants undergoing radiation. Molecular data may be of aid in that heterophyly (molecular paraphyly and phylogenetic polyphyly) implies direction of evolution, the paraphyletic taxon being the progenitor of the embedded taxon. Groups of radiation are identified as descendants in a cloud of usually short lineages around a shared generalist morphotype. The generalist morphotype may be hypothetical (i.e., designated an unknown shared ancestral taxon) if an extant species is not at hand. A caulogram is then constructed connecting the radiative groups in a branching series. Classi-fication at supraspecific level is done by naming the clots of radiation in the caulogram.

In molecular systematics, sequences are chosen so that their mutation rates match or sample the rate of evolution. Too much change in DNA overwrites the information, too little is phylo-genetically uninformative. Likewise, morpho-logical tracking (conservative) traits should be chosen as those exhibiting some degree of persistence in series of speciation events but not so much that there is no change. In addition, horizontal exchange of genes may be a problem. Tardigrades how been found (Boothby et al. 2015) to have up to 17.5 percent of foreign sequences in their genome, ca. 6000 genes from mainly bacteria but also plants, fungi and single-celled organisms. Boothby et al. (2015) suggest that in tardigrades, famous for toleration of extreme stress such as long desiccation, DNA breaks into small pieces, and during rehydration, cell membranes and nucleus become temporarily permeable, facilitating DNA transfer. Given that the family of mosses, Pottiaceae, studied here is also resistant to dehydration in harsh environments, molecular data must be carefully examined for foreign bias.

Synthesis — The use of Bayes' Formula is essential in evaluating credibility of one-time events, such as speciation. I have discussed Bayesian analysis at length in other publications (Zander 2013, 2014a,b,c). If you cannot deal easily with Bayes' Formula and algebraic calculation of conditional probabilities, try the Silk Purse Spread-sheet, available online at:

*www.mobot.org/plantscience/ResBot/phyl/
silkpursespreadsheet.htm*

or use a simple graphical manner of doing the same analysis is using the "four-fold table" (see Levitin 2016: 109ff).

Sequential Bayesian analysis (or "empirical Bayes") uses the probabilistic result from a calculation with Bayes' formula as a prior for another calculation with Bayes' formula with additional data. This can be done simply by adding decibans (dBs) or informational bits. These are both logarithmic units of probability, the former to base 10 and the latter to base 2, and each kind may be added to a probabilistic total. Applications of sequential Bayes analysis in systematics are discussed at length by Zander (2014b). At least one trait more advanced or derived (e.g., specialized, rare or unique) is needed to infer a direction of evolution in terms of direct descent. Brooks (1981) and Brooks et al. (1986) make similar use of informational bits for phylogenetic analysis, where the minimization of redundancy in traits with maximum parsimony was tested by calculating maximum Shannon entropy (also known as informational entropy). It apparently did not catch on among systematists, but was instead firmly replaced by likelihood and Markov chain Bayesian analyses with molecular data. Sequential Bayes analysis (or "empirical Bayes") is much like sequential sampling algorithms (Bonawitz et al. 2014) and requires a discrete hypothesis space, i.e., the additional information must be directly relevant to the same hypothesis or set of hypotheses.

One bit equals almost exactly three decibans (dBs). Traits (as trait transformations) are here assigned bit values (rather than dBs as per Zander

2014a,b,c), with the goal being the sum of 5 bits for ca. 0.95 (actually gives 0.97) Bayesian posterior probability (BPP) total support for direction of evolution of one taxon and another. A level of 0.95 BPP is that level of certitude needed for confidence in the evolutionary order of two taxa. Short of 0.95 BPP, by analogy with Bayes factor assignments by Jeffreys (1961, see also Zander 2014b), 2 bits gives "substantial" support, 4 bits "strong" support, while more than 7 bits gives "decisive" support (for particularly critical studies). Note that a molecular conclusion about species relationships at 1.00 BPP (statistical certainty) may be falsified (refuted) by a contrary morphological conclusion at merely more than 0.30 BPP (i.e., minus one bit) because this results in a Bayes factor of less than 3 for the molecular study (Jeffreys 1961).

Bits are awarded in determining the *order* of evolution of two taxa relative to an outgroup or the previous member of a lineage. Bits may be summed across all taxa in a lineage to determine support for the *direction* of evolution of a lineage. In phylogenetic cladograms, support measures in BPPs are given at nodes for the *rest of the clade*. The equivalent in bit analysis is to sum the bits for the order of evolution for the *rest of the lineage*, then translate to BPP. Care is taken to eliminate or move elsewhere taxa that are so different as to add a large bit count. The morphological macro-evolutionary analysis incorporating theoretic gradualism of speciation events can show evolutionary nesting as strongly supported as are molecular clades in phylogenetic studies.

Deciban values as used in the previous study of *Didymodon* (Zander 2014a,b.c) can by translated to Bayesian posterior probabilities (BPPs) using Table 5-1 (see also Zander 2014b).

Table 5-1. Equivalency of *positive* bits and decibans (dB) with Bayesian posterior probabilities (BPP). The decimal fractions of bits are equivalent to adding one or two dBs to each bit, where dB are equivalent to poor data or "hints" that cannot be ignored.

Bits	dB	BPP
0	0	0.500
0.33	1	0.557
0.67	2	0.613
1	**3**	**0.666** or nearly 1 S.D. (0.683)
1.33	4	0.715
1.67	5	0.759
2	**6**	**0.799**
2.33	7	0.833
2.67	8	0.863
3	**9**	**0.888**
3.33	10	0.909
3.67	11	0.926
4	**12**	**0.940** or nearly 2 S.D. (0.955)
4.33	13	0.952
4.67	14	0.961
5	**15**	**0.969**
5.33	16	0.975
5.67	17	0.980
6	**18**	**0.984**

6.33	19	0.987
6.67	20	0.990
7	**21**	**0.992**
8	**24**	**0.996** or 0.99+
9	**27**	**0.998** 3 S.D. (0.997)
10	**30**	**0.999**
20	**60**	**0.999999** (odds of 1 million to one)

Table 5-2. Equivalency of *negative* bits and decibans (dB) with Bayesian posterior probabilities (BPP). If there are only two alternatives, subtract the BPP from 1.00 to get the support for that alternative.

Bits	dB	BPP
0	0	0.500
-0.33	-1	0.442
-0.67	-2	0.386
-1	-3	0.333
-1.33	-4	0.284
-1.67	-5	0.240
-2	-6	0.200
-2.33	-7	0.166
-2.67	-8	0.136
-3	-9	0.111
-3.33	-10	0.090
-3.67	-11	0.073
-4	-12	0.059
-4.33	-13	0.047
-4.67	-14	0.038
-5	-15	0.030
-5.33	-16	0.024
-5.67	-17	0.019
-6	-18	0.015
-6.33	-19	0.012
-6.67	-20	0.009
-7	-21	0.007
-8	-24	0.004
-9	-27	0.002
-10	-30	0.001
-20	-60.000	0.000001

Further explanation may be valuable here. A bit is equivalent (almost exactly equal) to three decibans. One bit plus one deciban equals either 1.33 bits or 4 decibans (Table 5-1). (Although a decimal fraction of a bit at base two may seem an abomination, these may be added and so are acceptable.) A deciban is used to designate the minimum detectable change in confidence in evolutionary direction and order, and is used for poor information that is nevertheless information (i.e., a "hint"), perhaps extreme values of a quantitative character or a qualitative variable trait that is diagnostic when present. A deciban is quite like the notion of a "just noticeable difference" in epistemological psychology (van Deemter 2012: 177).

A bit is used to designate the minimum confidence necessary to decide that the information is indeed an indicator of order and direction of evolution, and is used for good information (i.e., a "datum"). Although any one bit of information may be good *as information*, it may not be enough to make a decision that allow one to construct a well-supported caulogram. Thus, one needs 13 dB to reach 0.95 BPP or 4 bits to reach 0.94 BPP and 5 bits to exceed that to 0.97 BPP. If calculating mainly by bits and using dBs for additional hints, then 4.3 bits reaches 0.95 BPP. See Table 5-1.

Implied Reliable Credible Interval — An alternative manner of BPP calculation may be used when the advanced traits of a species are all or in part combined into one complex trait or presumably adaptive feature. Because they all depend on one another and cannot be viewed as accumulating randomly as linked only through isolation, any one of the adaptationally linked traits has more reliability as an indicator of evolutionary order and direction than presumably neutral traits (i.e., those that are tracking traits capable of existing in several environments). Generally speaking, morphologically isolated taxa are probably old, and most of their traits are probably neutral or nearly so, able to survive many kinds of environments, and are a motley tagalong from a now extinct evolutionary lineage. Such ancient traits if they match across generic boundaries may be used to explore the "taxonomic abyss" of mostly extinct species. A kind of evolutionary "uniformitarionism" may be invoked reflecting an expectation of gradualism in traits of new species, i.e., no or little evolutionary saltation.

For recent taxa in unusual habitats, particularly taxa with many near relatives, new traits may be considered all or almost all signals of adaptation, except those traits that appear scattered in relatives without correlation with particular environments, which may be considered neutral, that is, randomly fixed upon isolation and tolerated in a new habitat.

For new, unique or complex traits that may be decomposed into what are apparently adaptationally linked features, the Implied Reliable Credible Interval (IRCI) formula (Zander, 2013: 59) may be applied. This is the chance that at least *one* of the two traits is correct, that is, correctly indicates order and direction of evolution. It is simply one minus the product of the chances of each of all concatenated linear evolutionary arrangements being wrong (where the chance of being wrong is one minus their Bayesian posterior probability). One may use the Silk Purse spreadsheet for ease of calculation (Zander, 2013: 86, 163). The IRCI values are higher than those from sequential Bayes (i.e., adding bits together), and seem more reasonable given the importance and rarity of the complex trait, see Table 5-3.

Table 5-3. Bayesian equivalencies of informational bits and IRCI probabilities. If two traits or more are apparently not neutral (part of a complex trait, or new and rare) then the IRCI bit equivalent may be used. Summing informational bits is used for probable neutral traits (those scattered in many taxa and many habitats).

Bits	BPP	ICRI Probability	IRCI Bit Equivalent
1	0.666	0.666	1
2	0.799	0.888	3
3	0.888	0.963	5
4	0.940	0.988	6
5	0.969	0.996	8

Bayes factors — Although adding bits of all trait transformations as evidence of descendancy can provide high credible intervals for a progenitor, it remains possible that the descendant with lowest probability of descendancy (PD) may be the progenitor of the putative progenitor (reversing serial position). A *Bayes factor* (BF) calculation provides a measure of this possibility. It is useful when two different probabilities are derived using somewhat different data. The usual calculation of probabilities involves generating different probabilities of different *hypotheses* all of which explain that one data set, with the probabilities for all alternatives adding to 1.0, or unity. In computational systematics, both maximum likelihood and Bayesian Markov chain Monte Carlo methods calculate the probabilities the *data set* for different hypotheses. The idea is that the probability of a data set is exactly that of the probability of the hypothesis. When the different hypotheses are based on somewhat different data, we may again get probabilities that do not sum to 1.0. In this case, there are methods that use maximum likelihood or Bayesian probability. Well, how much more probable should the maximum be than that of the next most probable hypothesis to be to be credible?

To deal most simply with this, the Bayesian prior should be 0.50. As noted by Zander (2014a), for deciban calculation the prior is always 0.50, and thus the BF for two probabilities is simply the highest BPP divided by the lowest. The same rationale is true for bits—divide the BPPs of the two bit counts. Following Jeffries (1961), support from a BF from 3 to 10 is substantial, from 10 to 100 is strong, and more than 100 is decisive. A more stringent criterion was proposed by Kass and Raftery (1995), where support from 3 to 20 is positive, from 20 to 150 is strong, and more than 150 is very strong. (The key words are those of the authors.) Substantial support would be, for example, a BF with a 3:1 minimum ratio, such as 90:30 BPP or 99:33. Strong support in the first scale, for example, is a BF with a 10:1 ratio of BPPs, such as 99:9, or 99.9:9.99.

Bayes factors involved in sequential Bayesian analysis can show false positive evidence when differences of support between models are very small and sampling is sparse (Schönbrodt 2016; Schönbrodt et al. 2015). In the present case, however, the order of evolution of two taxa is supported massively by the bits contributed by a third taxon that is clearly derived from one of them. A core generative species with two separately derived descendant species is also better supported as core, because all bit counts can be added without insult to theory.

Direction of dissilient evolution is presumed to be away from a generalized ancestral taxon of ancient, broad or relictual habitat, towards specialized descendant taxa of recent habitats, by reason of taxic Dollo irreversibility of multiplex traits. Traits of specialized taxa are in cladistic terms autapomorphies, being unique or scattered on a cladogram and of little use in sister-group analysis but important in the present study. Adding the bits from all newly occurring traits provides a measure of confidence that some one particular species is an immediate descendant.

Fixed taxon transformations are not simply chance conglomerations of traits, but trait combinations are strongly affected by natural selection and are ultimately deterministic after chance variation is also fixed. Consider a roulette wheel with results totally controlled to reflect a pre-determined order of numbers obtained from a table of random numbers. Distinguishing this from truly random results generated by the wheel would be impossible (although published random number sets are usually pre-cleansed of all chance number sequences that appear to be non-random, and these can be distinguished), yet analyses of any introduced bias can be done using standard statistical methods. Thus, bit assignments are defensible in that random events when fixed are seldom without biases.

Estimations of bits for support of direction of evolution follow approximately the following schema based on the relative rarity (estimation of degree of autapomorphy) of the trait locally (within the immediate larger group, particularly of probable progenitor siblings). Positive bits are assigned to traits of descendants to reflect the level of support for the direction of evolution away from a putative progenitor. That level of support is based on rarity of the trait, measured by 1 bit assignments to each new trait in a serial lineage. These are *coarse priors* (Zander 2013: 85). Adding bits of traits of one species indicates probability of descendancy (PD) that that species is a descendant. Adding the PDs of all descendants is the probability of ancestry (PA) of the putative progenitor of them all. This is acceptable because each additional putative descendant with evidence of being a descendant adds to the credibility of the generalized putative ancestor actually being ancestor, given gradualist or step-wise macroevolution. That there is no credible justification for placing species in that group elsewhere renders it a *closed causal group* (Zander 2014b) of progenitor and its immediate descendants. This justifies the addition of bits of all descendants to create a PA for the progenitor. With only one descendant, the PA of the progenitor is, of course, the same as the PD of the descendant. For two taxa, these BPPs are 1.00 PA for the putative progenitor, and 1 minus the PD of the descendant (which provides its possible PA allowing for necessary reversals).

A closed causal group is quite like the concept of supervaluation in epistemological psychology (van Deemter 2012: 162). According to van Deemter, "… if you are uncertain about something, you may still be certain about anything that does not depend on the way in which your uncertainties are resolved. In other words, you call something true if and only if it is true under all the different ways in which your uncertainties might be resolved …," i.e., all ways of making it completely precise. This is also called an "existence result." The order of a series of ancestors and descendants may be uncertain at a particular level, but the direction of evolution of the group may increase with additional taxa because there is nowhere else to place them outside the group. This is something like the simplex method in which a local optimum may be applied to the whole group, such as results of analysis of one side of a polygon may be applied to the whole (Odifreddi 2000: 121).

Unknown shared ancestors may be postulated when two descendants of one progenitor share an advanced trait, but are otherwise equivocal in order of evolution. Rather than assuming that the advanced trait is evolved twice, an unknown shared ancestor between the progenitor and the two descendants reduces this scenario to generation and fixation of only one advanced trait. This is more compelling if two descendants share two or more advanced traits not in the presumed progenitor.

There is a limit to what should be tolerated in measures of statistical certainty (Gunn et al. 2016 and see Cohen 1994). The common phylogenetic assignment of unity or 1.0 BPP as support

measures is problematic. When such measures are involved in Bayes factors, the simple division gives an understandable fraction. But when two measures are involved in Bayes formula, the seemingly minor difference between 1.0 and 0.99 BPP is overwhelming in favor of the first.

More about Bayes factors — The Bayesian approach to hypothesis testing was developed by Jeffreys (1961) for scientific inference. Jeffreys dealt with comparing predictions made by two competing scientific theories. With Bayes factors, analysis consists of representing the probability of the data according to each of the two theories, and Bayes's theorem is used to compute the posterior probability that one of the theories is correct. Bayes factors evaluate the evidence in favor of a scientific theory, including the null hypothesis, and allow a way to incorporate additional information in hypothesis evaluation (Kass & Raftery 1995).

Bayesian statisticians have determined that the probability of one event must be at least three times that of its alternative to have support. Thus, 0.80 divided by 0.60 is 1.33, therefore no real support for the hypothesis with 0.80 BPP. But 0.90 is real support for two hypotheses in which one has 0.90 BPP and the other 0.30 BPP (because 0.90 is thrice 0.30 and the BF is then 3). In the Bayesian sense, as long as priors are 0.50 for all probabilities, simple division of the posterior probabilities is sufficient. Again, in macroevolutionary systematics, the support for direction of evolution in a lineage of direct descent—an indication of its coherence—is the sum of the bits supporting the order of each taxon in the lineage. That sum is translated to BPPs. This is allowable because bits are logarithmic and can be added. Adding bits is equivalent to sequential Bayes analysis, in which additional information is added by taking the posterior probability of one analysis, and using that as the prior for the next.

Consider the serial lineage $A > B > C > D$. Let's say descendant B has only one advanced trait, and B has then 1 bit or 0.666 BPP of being advanced over the base of the caulogram. Progenitor A has no advanced traits. What if A were the descendant and B the progenitor? What is the support for this reversal of order in the lineage? We use Bayes factors.

According to García-Donato and Chen (2005), the Bayes factor depends on the choice of the prior distribution and, in particular, it is extremely sensitive to imprecise or vague priors. Imprecise priors are due to the randomness of the data, and the sampling distribution is important. What about sequential Bayes, then? Are the series of changing priors (using the previous posterior as the new prior) introducing error? Each new prior is the complete replacement for all past analysis. Each sequential element adds accuracy and credibility to the result. Analysis of the order of two taxa or the direction of evolution of three or more taxa involves the same number of taxa, as bits are simply added in a different sequence for each model; thus we are not comparing models with different levels of imprecision due to different priors. An expectation of adding to 1.0, or unity, is thus acceptable.

The probability of B being descendant of A is 0.666 BPP, while that of A being descendant of B is the reverse, or minus one bit, or 0.333 BPP (BPPs taken from tables 5-1 and 5-2). To get the BF, divide 0.666 by 0.333, and we get a BF of 2, not impressive. What if B was supported by 2 bits? Then, from Tables 5-1 and 5-2, we divide the 2-bit BPP of 0.799 by the -2-bit BPP of 0.200, and get a BF of 4, *which is good support*.

A two-bit difference is, therefore, a desideratum for determining order of evolution. *Given that species are here required (usually) to have at least two traits not linked in other species, two bits are commonly available to determine evolutionary order of linearly nested species.* As I have pointed out in the past (Zander 2014b), "the perceived value of traits carrying information about direction of serial evolutionary transformation are the unacknowledged criteria for perceived level of species delimitation." Even if some taxa may have low BFs (say, 3–4 BF) as to order, the entire lineage generally has a high BPP of being correct in that the bits assigned to each species are added, then the BPP is read off Table 5-1.

Given that in the present case the probabilities of the order of any two species always sum to 1.00, the BPP may be substituted for the BF. Thus, a BF of 4 is 0.80 BPP.

Analytic model — Macroevolutionary systematics uses direct descent to model evolutionary change at the species level and above. Macrosystematics is classification at the genus level and above, in the present work the dissilient genus concept is used instead of clades to recognize genera. The basic unit of taxonomy is the species, yet the fundamental unit for macrosystematic analysis is the dissilient genus, which radiates descendant species from a core progenitor. A series of dissilient genera comprises a lineage of (usually) core species of genera generating core species of other genera, with most descendants in each genus dead-ending in highly specialized habitats. Problematic relationships that require inductive logical inference include: (1) When a single species is worth recognizing as a dissilient genus in potentio. (2) When a species is quite different from the core species and requires interpolation of a missing link. (3) When two species are equally distant from a joint progenitor species and a shared ancestor may be postulated as an intermediate. (4) When a species is apparently such a derived member of an extinct group that it is difficult or impossible to place it in a lineage of genera. See Fig. 5-1 for a clarification.

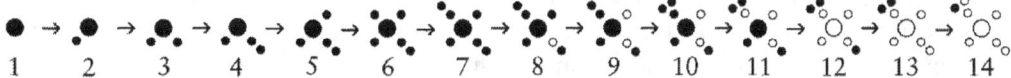

Figure 5-1. Evolution of an example dissilient genus through time. Solid dots are extant species of that stage of serial evolution, circles are extinct or unsampled species. 1: Core generative species derived from a previous lineage (not shown). 2–7: Gradual accumulation of descendants, of which two descendants (6 and 7) have descendants of their own. 8–14: Gradual extinction of species in the dissilient genus. 8: Inference is possible of the descendant that generated another descendant. 10: A descendant generates a second descendant, and an extinct descendant may be inferred from a highly specialized descendant. 11: Shared ancestor of two descendants of a descendant can be inferred. 12: Core species goes extinct and may be inferred from traits of extant descendants, the more the better. 14: Isolated specialized species is all that remains of the dissilient genus, and inference of the traits of even the core species of its genus is difficult.

Analytic key — Although "natural key" has been given various meanings, what is essentially the same is presented here as an *analytic key*, which is an indented prose diagram showing and explaining inferred evolutionary transformations among similar taxa as expressed in direct ancestry. A simply serial ("monochotomous") or multi-furcating analytic key was generated from expressed traits of the species studied here. This tree minimizes both trait changes and numbers of non-extant shared ancestors that must be postulated. No or minimal reversals is the criterion for optimizing parsimony of postulated unknown shared ancestral taxa. The analytic key models, in prose, speciation events of serial descent through gradual accumulation of advanced traits, whether by anagenetic change or a series of abrupt but minor cladogenic events (with survival of progenitor).

Bit support for descendancy was calculated for each taxic transformation by adding bits assigned to new individual traits. This synthetic method is simplified from that of Zander (2014a,b,c) to include only sequential Bayesian analysis using bits instead of decibans (also, see explanation of Bayes factors above). The Implied Reliable Credible Interval (IRCI) formula used in that previous study was found to be rarely necessary because of the overall high credible levels.

Do not worry that your analytic key seems to have very vague deep ancestors, often nameable only at high taxonomic ranks. Cladograms seem so exact in having one basal stem and only one taxon level (usually species) as exemplars. This is misleading. The taxonomic abyss caused by general extinction of whole lineages and survival of only a few poorly representative, specialized remnants is shared by both cladogram and caulogram. Morphologically isolated species may prove to be valid as monotypic genera if they are generalist and somewhat advanced over a similar generalist central progenitor in a dissilient genus; they are then potentially dissilient. Or, a morphologically isolated species may be weakly similar to an extant member of its larger group, and somewhat isolated, in which case it is probably a remnant of a once dissilient genus now mostly extinct and does represent a distinct genus; molecular analytic may be of aid in determining evolutionary distances.

Continuity and arranging order of series — The aim of macroevolutionary systematics is to create evolutionary trees of serial descent by optimizing on continuity of taxon transformations. Optimal continuity maximizes the evolutionary information by minimizing redundancy (assigning homology and identifying homologous traits in the taxa as identical), and maximizing the number of bits supporting the order and direction of evolution. Order is simply whether one species comes before or after another in a lineage. Direction is the general arrangement of the lineage, for which continuity is paramount given an assumption of gradual transformation. Order and direction both assume that advanced end members of lineages have the most rare or unusual traits compared to the entire lineage or to an outgroup, while evolutionary continuity requires small steps in transformations, particularly with minimum reversals of traits. Confirmation that species evolve gradually in a group is a caulogram demonstrating gradually stepped serial change.

The order of serial transformations is based on the concept of gradualism (Darwin's version of Linnaeus' rule: "Natura non facit saltum", Paterson 2005) in evolution. Gradualism does not necessarily imply anagenetic change (gradual change of one species into another) but refers to series of speciation events each consisting of few morphological changes. That few changes are assumed per speciation event is theoretically in line with the concept of phyletic constraint, in that major change is usually unviable or rapidly deselected in the same or nearby habitats, given the general morphology. Gradualism is preferred as an analytic assumption because species are often observed in clouds of similar morphology, indicating speciation is usually not through giant steps or if it sometimes is, it is not identifiable by present methods. Also, modeling gradualism can eliminate false information from redundant traits. This does not mean that punctuational speciation does not occur, but that the species resulting from punctuational speciation are not particularly different from each other, simply that they occur in spurts.

For instance, placing an advanced taxon before (in linear evolutionary order) a less advanced taxon makes the information of the less advanced taxon useless in inferring direction of evolution. Given that a reversal is involved, the hypothesis of increasing patristic distance is falsified and a *negative bit* must be assigned to each newly reversed trait. Reversals confuse estimation of the order of evolution. This only applies within a genus—it is as if evolutionary order is reset with a new core generative species. It also applies only when calculating bit support for the order of two taxa, not for initial placing of order of the taxa in a lineage by evaluating numbers of trait transformations for which a reversal is a new trait and you want to minimize reversals.

If a taxon has two advanced traits and one reversal, one bit is subtracted from two to gauge support. All advanced traits are explained with the simplest manner when the traits are arranged to eliminate or minimize giving informational status to redundant traits rendered identical by descent from an ancestral taxon with those traits.

For a complete, causally based, explanatory model, all evolutionary information has to be *used*, thus serially nested arrangements of taxa with increasingly advanced traits are the most powerful explanations. Using all the information also best clarifies the message, which is the direction and order of macroevolution. Inference of serial descent founded on gradualism in species transformations lends theoretical coherence to the analytic key.

Method of arrangement — The above material is not digression but important methodological background. This is the Methods chapter, however, so let us plunge into how to arrange taxa by direct descent. There are two phases, ordering by penalizing for superfluous parallelisms and redundant reversals, which produces an optimal caulogram of direct descent. Then, support for the order and direction of evolution is gathered, and in this phase parallelisms and reversals in the optimal caulogram are not penalized.

First, then, we try to develop an optimal caulogram. See Figs. 5-2 and 5-3 below, and the examples further on in this chapter. For phase two, see Fig. 5-4, and the chapter discussing the analytic key.

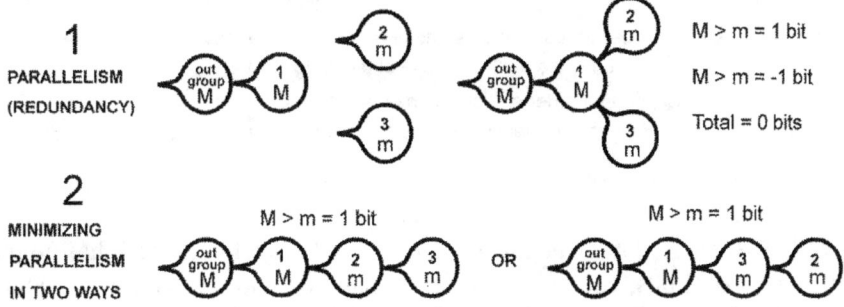

AFTER MINIMIZING PARALLELISMS AND REVERSALS,
NUMBERS OF TRAIT TRANSFORMATIONS LEFT OVER ARE INFORMATIVE

Species are numbered 1, 2 and 3.
Characters are M and N, with lowercase m and n for advanced states.
There is a penalty of minus one for a superfluous parallelism,
and minus one for a superfluous reversal.

Figure 5-2. The fundamental method in analysis by direct descent is given in Figs. 5-2(1) and 5-2(2). All information contributed by unnecessary, superfluous parallelisms and reversals is redundant, and must be eliminated. Assigning minus 1 bit to parallelisms and reversals helps determine order of evolution. Fig. 5-2(1) asks where should species 2 and 3 be attached. If they attach to species 1, then the superfluous parallelism of "m" cancels all information. In Fig. 5-2(2), if they are put in linear order, there is no penalty for parallelism and both ways of ordering are equally optimal, adding to 1 bit. The symbol > means "transforms to."

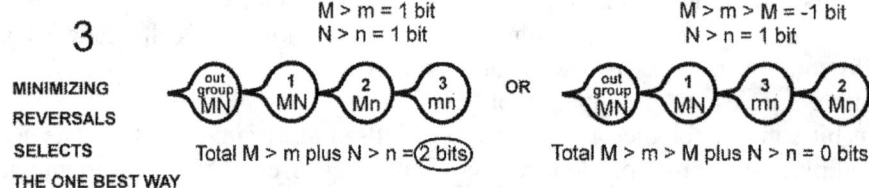

Now suppose species 2 had states Mn and species 3 had states mn.

3

MINIMIZING
REVERSALS
SELECTS
THE ONE BEST WAY

$M > m = 1$ bit
$N > n = 1$ bit

$M > m > M = -1$ bit
$N > n = 1$ bit

OR

Total $M > m$ plus $N > n = $ (2 bits)

Total $M > m > M$ plus $N > n = 0$ bits

4   Alternatively, suppose species 2 had states Mn and species 3 had states mN.

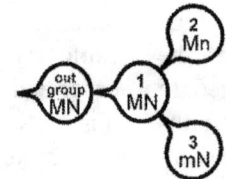

This is optimum with total 2 bits from $M > m$ and $N > n$.
Any other order sums to zero bits.

5   Suppose we had a third character O with advanced state o.

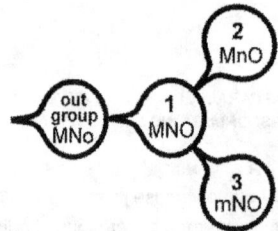

The parallelism of O is not superfluous because if the order is not branched
then the reversals of $m > M$ and $n > N$ would decrease the bit total.

Penalties for parallelilsms and reversals are only made
during the analytic phase of direct descent.
Support for order and direction of evolution is the sum
of all transformations on an optimal caulogram.

Figure 5-3. In Fig 5-3(3), the left caulogram is the optimal model because it has no parallelism, no reversals, and support of 2 bits, equivalent to a Bayesian posterior probability of 0.80. Figs. 5-3(4) and 5-3(5) discuss acceptable parallelisms. Support for order and direction of evolution is determined by number of bits interpreted as Bayesian posterior probabilities. Any leftover parallelisms and reversals in the optimal configuration will then add to other trait transformations as support for order and direction of evolution during creation of an analytic key. This support is restricted to the dissilient genus (explained later). See Methods section for how to sum support and translate to Bayesian posterior probabilities.

SUPPORT FOR ORDER AND DIRECTION OF EVOLUTION

Award 1 bit for each trait transformation
on optimized cladogram.

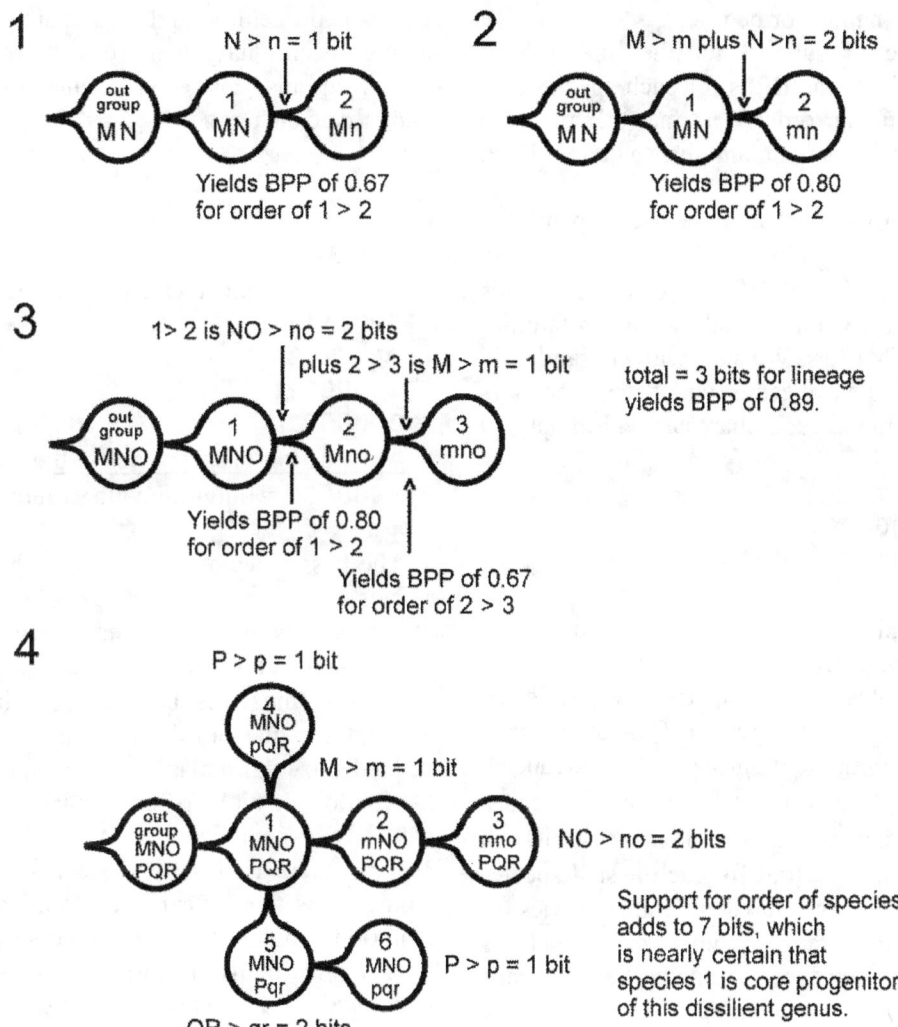

Figure 5-4. How to calculate support for order and direction of evolution. Order means any two species arranged linearly, modeling direct descent. Direction of evolution means away from the putative core progenitor. Part 1 models a core progenitor (species 1) giving rise to species 2 through transformation by 1 character state. Part 2 is the same but two traits transform. Part 3 shows three species in order. Part 4 evaluates a dissilient genus with species 1 as core progenitor, for which Bayesian support of 0.992 BPP is extremely large.

Figures 5-2 and 5-3 show phase one which models direct descent on an optimized caulogram. In Figure 5-4, Bayesian posterior probability (BPP) support is calculated for order (one species giving rise to another) and direction of evolution (being general agreement that the lineage is coherent towards increasing or advancing transformation away from the core progenitor, even if the order may be reversed in some elements).

The IRCI formula may be applied to increase BPP if one trait is part of a recognized trail complex such that if one trait is informative of the evolutionary order then so are the others. The formula, which (again) is simply the product of the chances each order is *incorrect* subtracted from one, was devised to determine the change that *at least one* of a number of things happening or being true will in fact happen or be true.

In the case of support for the direction of evolution, we have the BPPs for each species pair being in a particular order. The fact that the core species generates a descendant gives a certain BPP.

Detailed examples — Suppose we expand the explanation by using either a 0 or a 1 for two character states of each of five characters. This may be more easily understood by readers familiar with cladistic data sets. We start with six randomly arranged species in a genus with five characters shared among them, each character with binary (0 or 1) states.

```
00111, 00000, 00011,
11111, 01111, 00001
```

We also find a clearly related outgroup for that genus with, say, character states the same as one of these species (plus many other traits not the same because it is a different genus). Call those traits that are shared with the outgroup 0, less advanced. Thus, the formula 00000 for a basal species, 00001 for a species with one advanced trait, and add one additional 1 trait for each nested species in that lineage so we have a gradual series of evolution of taxa towards minimum primitive traits and maximum advanced traits ("trait" means character state). In the example of Data Set 1, we have no reversals, and full nesting of both phylogenetically informative shared advanced traits (most of them) and evolutionarily informative unshared advanced traits (1 in first column).

Both morphological and DNA traits are susceptible to analysis of serial nesting when postulating au unknown shared ancestor is unnecessary.

Example Data Set 1: Note that each of the five columns (1–5) represents a character with two character states 0 and 1, each termed here a "trait." This is full nesting but without the nodes required by cladograms.

But if two descendants agree that the core progenitor is indeed that, how can we gauge the additional support beyond just adding the individual trait transforms? The additional species represent *separate tests* agreeing that the core species is progenitor, and are not simply equivalent to one species having more trait transformations (which in any case contravenes expectation of gradualistic change).

```
  12345
F 11111 = most advanced species is F
E 01111
D 00111
C 00011
B 00001
A 00000 = basal species of the genus is A
X 00000 = outgroup with different traits
```

The basal species will probably be a generalist species, with little adaptive baggage, possibly widely distributed or existing in widely separated sites in ancient habitats and lands. The species with rare traits (the ones having 1's with more 0's below them in the lineage) are usually highly adapted to particular environments, local in distribution in more recent habitats or lands, and may have specialized or asexual reproduction. Because there is usually noise from minor fairly neutral traits that reveal order, one can doubt that evolution would commonly reverse at the taxon level from highly adaptive species into more generalized ones entirely cryptically. Multiple tracking traits that are selectively neutral are not selectively eliminated. Thus, arranging the species in this manner models a gradualist, highly nested serial evolution of extant taxa.

Suppose one of the middle species were apparently missing. The series can tolerate a certain amount of extinction of intermediates, or one can assume two traits evolved at once. Either explanation is allowable with little insult to the model. If the two traits were very rare, however, an intermediate species with one or the other of the traits might be *predicted* to be found in the future, or *retrodicted* to have occurred in the past. That is:

Example Data Set 2:

```
     12345
  F  11111 = most advanced species is F
  E  01111
       either D 00111 or D 01011 may be
       inserted here as an estimation if evolution
       of two traits (1 in each column) during
       the same speciation event is improbable. D
       is then a shared unknown ancestor, but
       differs from a cladogram node in its
       predictive value.
  C  00011
  B  00001
  A  00000 = basal species of the genus is A
  X  00000 = outgroup
```

Assignment of support. — What support is there for the arrangement of gradual accumulation of advanced traits preferably with no reversals?

An entropic Shannon bit technically implies there are two choices at 0.50 Bayesian posterior probability, but once the choice is made the bit represents 0.67 BPP, and is here termed an informational bit. One informational bit is assigned to a new serial instance. See Table 5-1 for bit, deciban and Bayesian posterior probability equivalents.

Example Data Set 3: No outgroup is used. Primitive species in the lineage are identified by omnispection of nearby groups, which have mainly 0 as traits, not 1 (advanced traits).

```
     12345
  F  11111   1 bit, the last 4 1's are redundant
  E  01111   1 bit, the last 3 1's are redundant
  D  00111   1 bit, the last 2 1's are redundant
  C  00011   1 bit, the last 1 1 is redundant
  B  00001   1 bit, no redundancy
  A  00000   0 bits for the basal species
```

A–F, respectively, are species. 1 and 0 stand for the two character states of each character (in each column). The boldfaced 1's indicate the only informative traits for each species that particular series given the redundancy (derived homologous nature) of the other 1's in the species. In this case

the traits are ordered against a hypothetical outgroup that has the traits of species A in addition to its own distinctive traits.

The total number of bits is 5, which can be interpreted as 0.97 Bayesian posterior probability (BPP) support for the evolutionary direction of this series of species (see discussion in the chapter on theory, and Table 5-1). Each species differs from the next lower in order by 1 bit, leading to a BF of only 2, so details of the order in this example are poorly supported. Thus the resolution of the example is 2 species distance. In practice, a species requires two distinctive traits, and these are commonly advanced compared to those of congeners. Thus, a BPP of 0.8 (2 bits) divided by a BPP of 0.20 (-2 bits) gives a BF of 4, or good support for the order of evolution, and a resolution of 2 species distance.

An *informational bit* when applied to a known element of knowledge is considered equivalent to a minimum confidence needed to make a yes or no decision. A *deciban* (dB) is considered equivalent to the minimum support for a decision possibly detectable by humans (a "hint"), and is one third of a bit. Using dBs instead of bits for all traits is selling the information short, however, given theory that includes Levinton's (1988: 217) evolutionary ratchet, and taxon-level Dollo constraint (Gould 1971) on reverse-evolution of entire species.

Another way to evaluate the series in Example Data Set 3 is to optimize on numbers of bits. The correct order maximizes the number of bits available with the given information because all information is already counted and redundancy is already minimized (see Data Set 5).

Example Data Set 4: Outgroup available, even if hypothetical.

```
     12345
  F  11111   1 bit, the last 4 1's are redundant
  E  01111   1 bit, the last 3 1's are redundant
  D  00111   1 bit, the last 2 1's are redundant
  C  00011   1 bit, the last 1 1 is redundant
  B  00001   1 bit, no redundancy
  A  00000   0 bits for the basal species
  X  00000   0 bits for the outgroup
```

Total bits remain 5 for the evolutionary direction of the group ABCDEF, but this use of an outgroup makes the polarization more credible as the species X is now included.

Now, take the above data and reverse it:

Example Data Set 5: Reversed order of species A–F relative to the outgroup X leading to greater redundancy.

```
    12345
A  00000   -1 bits
B  00001   -1 bits
C  00011   -1 bits
D  00111   -1 bits
E  01111   -1 bits
F  11111   5 bits
X  00000
```

For the series F–A in Data Set 5, adding reversals and new traits yield 0 bits. This is, then, compared to 5 bits when the species series is not reversed. The reversals seem to confer additional information but are redundant. Redundant data is in fact to be treated as negative bits or negentropy. The total entropy and negentropy for the order of Data Set 5 is zero bits. The order of taxa is then not supported by any positive bit difference between the species, except for F, which has 5 bits in support of its order, that is, the order **X** > F. The correct order (Example Data Set 4) will recognize the minimized redundancy of cladistic clustering of homologous advanced traits when maximizing information expressed in bits. Various permutations of order provide different bit totals between zero and 5 bits.

The nesting provided by cladistic software does minimize redundancy in the context of evolving traits (not species) dichotomously (not multifariously), but adds its own redundancy by introducing shared unknown ancestors when these are often unnecessary. It is entirely possible to adapt cladistic software to modify cladograms into quasi-caulograms by recognizing when a node is taxonomically the same as one of the sister groups. This may be done by including autapomorphies in the data set, and seeing which of two taxa that are sister to each other has the most autapomorphies.

That taxon is then most advanced in the lineage as a serial hypothesis.

Minimizing the redundancy of hypothetical nodal taxa in is quite different from the maximization of redundancy associated with minimizing the number of steps in parsimony analysis (as described by Brooks 1981 and Brooks & Wiley 1988: 278 in the context of entropy), because in the present case primitive traits are not expected to re-evolve from an advanced trait and thus in the Brooks and Wiley sense become themselves redundant and uneconomical because increasing the number of steps.

Adding a species from a different genus to this series would increase the bit count by its odd or rare traits but could be identified as foreign because it would not also equally minimize redundancy. It cannot be nested to reduce redundancy. The true series "climbs a ladder of 1,s" to provide such nesting of advanced traits (see Example Data Set 2).

What about when another group branches off of the outgroup? The outgroup X is a basal end member of species group XHI of three taxa (see below). That new species group branching next to ABCDEF is now a separate lineage.

Example Data Set 6: Three groups attached at the point of five same traits, all 0's, these primitive in ABCDEF and XHIJKL, but advanced in MNOPQR. Both A–F and X–L are nested sets of taxa having taxa with advanced traits increasingly distal in the lineage. So is M–R, but in that data set 9 is primitive and 0 is advanced. (Remember "1" simply means an advanced character state, not identity of all 1's, ditto 2's and 9's.)

```
    12345        12345
F  11111
E  01111
D  00111
C  00011    I  01101
B  00001    H  00011
A  00000    X  00000


R  00000    R is ancestral taxon for both
               ABCDEF and XHIJKL.
Q  90000
P  99000
O  99900
N  99990
M  99999
```

In Example Data Set 6, above, for taxa M through R 9 is a primitive trait, and 0 is advanced for each character column.

The advanced traits of the progenitor group M–R are the primitive traits of derivative groups A–F and X–L. The set of primitive nulls changes from 9 to 0 and thence to 1 among the three data sets.

There are now 5 two-state evolutionarily significant characters (5 columns of traits for each of the 18 species). The two groups, ABCDEF and XHI are divergent branches attached at the shared set of 0 nulls, that is, at A and X where the important primitive traits of both A–F and X–I are the same as the advanced traits of M–R.

In this scenario the advanced species R becomes the outgroup for both A–F and X–I. In reality, R will probably be too specialized to be a generalist widely distributed progenitor. In the *Didymodon* studies (Zander 2013; 2014a,b,c), the generalist full-featured progenitors themselves generate other generalist full-featured progenitors, but this may not always be the case.

Is everything so easy to calculate in macroevolutionary systematics? Sometimes two taxa vie for the same place in a linear sequence and the line must be broken by positing an unknown shared ancestor. At times some traits seem to skip over taxa, appearing in, say, every other taxon in the lineage; these we might treat as poor data and assign them only 1 deciban (0.33 bit) per null taxon, or discount them entirely as noise based on observation of variation within taxa.

Limitation of certainty — We can truncate the total summed bits in any analysis at 8 as no one set of traits should endow statistical certainty. Why are eight bits enough? At this level of analysis, 3 standard deviations (3 sigmas) are near enough to statistical certainty given the data that additional bits are mathematical arm-waving. Just as calculating pi to the billionth decimal place implies measurements at the quantum level, there are rational and realistic limits to representations of statistical certainty. For instance, representing statistical certainty in molecular systematics publications as unity seems deliberately innocent. In addition, the more traits in a descendant that are different from its putative progenitor, the more likely there is some missing link or maybe a different relationship should be investigated. The

key is gradualism, and two to 4 or 5 traits difference is the sweet spot for expected gradualist change from a progenitor to its immediate descendant.

Only if minor traits that are noise or fairly neutral but remain valuable as tracking traits show that there is a major reversal from advanced to primitive central traits should gradualism from primitive to advanced species be doubted. And quasi-neutral traits may become important upon long-distance dispersal to another, similar habitat with somewhat different selection regimes. There is no mechanical taxonomy to evaluate this.

Data sets in practice seldom have more than a series of 4 taxa in length in a group, and the maximum bits contributed by any one species is usually 3. Most caulograms, however, have many branches. This adds import to any particular species in supporting estimated direction of evolutionary transformation.

Unknown shared ancestral taxon — Occasionally an ancestral taxon apparently generates two descendants, minimally each sharing one new trait and each also with one unique new trait. There are two immediate scenarios, in Model 1, the ancestral taxon generated both descendants with one parallel new trait shared by both: **A** > (B01, C09), or in Model 2, the ancestral taxon generated one of the descendants which then generated the other descendant **A** > **B01** > C09, or **A** > **C09** > B01.

Model 1 requires parallelism in generation of trait 0 in both descendants B and C, and such parallelism may be doubtful, perhaps because the habitats of B and D are quite different and 0 is apparently a new, adaptive trait. Particularly if there are more than one new advanced traits in both B and C, then the most parsimonious solution is to advance the existence of a presently unknown intermediate taxon generated by A and immediately ancestral to B and C. This model would have the macroevolutionary formula **A** > **X0** > (B01, C09). This assumes that parallelism and quick change of one new trait to another new trait is a less common scenario than simple serial addition of new traits at each speciation event. The postulated unknown shared ancestral taxon X is thus generated by theory (gradualism), is predictive (we can search for X in nature or fossils), simplifies the model (is parsimonious in the case of multiple traits), and testable (we might actually find X).

Value of serial analysis with bits — There is variation in how the traits of new species match up with the environment. Some traits are clearly adaptive, either by some known process or simply by abundant instances of correlation. Others are apparently adaptively neutral (or nearly neutral). The neutral traits lag behind the adaptive traits upon speciation, because they are for various reasons less exposed to selection and so appear in the new species, slowly changing only with mutation and drift. Both adaptive traits and neutral traits are important for systematics. The adaptive traits (at whatever taxon level) are particularly useful in discerning and describing species. The tagalong neutral traits are valuable for tracking evolutionary paths and describing genera and families. One can recognize neutral traits as those that may reverse with no correlation with other traits; they seem uncorrelated with habitat changes among species; and they appear scattered in some parts of a caulogram. Yet neutral traits often are shared among species such that they identify serial evolution by their accumulation. Note here that an adaptive trait may become an evolutionarily informative quasi-neutral trait when speciation occurs between much the same habitats and the traits persist.

The differential selective laggardness of neutral (and nearly neutral) traits contributes to a *decoherence of information* about evolution. That is, adaptive and neutral or nearly neutral traits separate, with the latter lagging like the long train of a bridal gown, and often persist through multiple speciation events. This is in systematics equivalent to an increase in informational entropy and contributes to an argument against reversal of a taxon, upon speciation, to the complete traits exhibited by a previous taxon. The dissipative effect on information contributes to the theory that speciation is not reversible (Brooks & Wiley 1988: 81, 204), and traits as a set are not commutative (set A → set B but not set A ← set B), i.e., order matters. Reversals when they happen contribute to chaos, whether the new combination of traits has selective advantage or not.

Biological aspects of catastrophe theory (Saunders 1980: 12, 105) imply that whatever process is being tracked through time, when it is affected by a catastrophe, it may reverse itself for a while. Thus we may expect reversals of some traits, but not all. Of some possible analytic value is the fact that tracking traits that are less responsive to selection are less likely to reverse during a catastrophe. Thus, matching of dissilient events and reversals may well be able to identify a catastrophe, such as a glaciation or even just abrupt restriction of range expansion by effects on serially generated descendants. The Pleistocene glaciations may well have written their message in the serial lineage of certain groups. Caution is advised because catastrophe theory, as is well known, is more generative of hypotheses than solvent of problems (see discussions, Gilmore 1981: 447, 644; Saunders 1980: x, 129).

There has been much written about information theory and entropy and their applications to evolution (see Brooks & Wiley 1988, and Weber et al. 1990 for examples) but have either focused on the second law of thermodynamics or population genetics depending on the authors' personal training in science. There has been, heretofore, no direct entropic application for evaluating order and substance of serial transformation of taxa, and how serial patterns in taxa affect thermodynamic stability. For instance, in reference to speciation, Harrison (1990: 68) wrote that his "analogy for evolution is the dynamic behavior of a simple chemical system that forms a number of patterns simultaneously in the same space. These would all be superimposed in a polychromatic waveform that one might call 'white pattern.' To resolve it into its separate sinusoidal components would need a Fourier analysis, that is, a kind of spectroscopy." And (1990: 71)

"Thus, my threshold criterion is simply the condition needed for the entropy of the universe to increase. If the products removed are externally reprocessed into reactant, the whole thing may be formulated as an irreversible cycle. ... Satisfaction of the second law ... is the norm throughout biological development. I do not see anything in evolution that is thermodynamically any more mysterious or anomalous."

I will present a simple covering theory involving systematics and ecosystems, and both informational and thermodynamic entropy later in this book.

A more biologically oriented discussion of evolution and entropy was offered by Wiley (1990: 175): "When we examine available phylogenies we observe that evolution at the level of species is quite irreversible (Dollo's Law) in the following sense: We do not observe the same species evolving, going extinct, and then reappearing again at some later date when environmental conditions are right for it." Wiley discusses the Brooks-Wiley model of nonequilibrium evolution (Brooks & Wiley 1986), pointing out that the

> "irreversible behavior of certain biological systems has a quite different basis from that exhibited by purely physical dissipative structures. … biological systems exhibit order and organization based on properties that are inherent and heritable. Physical dissipative structures lack these characteristics."

Wiley is referring to Pirogogine's theory of disspative structures and thermodynamics of irreversible phenomena (Odifreddi 2000: 71). According to Wiley (1988: 177), Shannon's information theory differs in application to biology in that in biology there is a physical basis to the information. The main feature of evolution is "cohesion," the capacity to transmit information to subsequent generations, which is lacking in purely physical systems. "Dissipation in biological systems is not limited to energy and mass. It also takes the form of information dissipation." Wiley gives examples for cellular differentiation and evolution within populations focusing on allele frequencies. For speciation, the examples are couched in terms of phylogenetic relationships, namely cladogenesis, and information has to do with the numbers of connections on directed graphs. Given that phylogenetic relationships are not serial relationships, applications of this viewpoint have not emphasized scale-free networks as emergent phenomena with a selective function (Brooks & Wiley 1988: 125).

In sum, high entropy in information theory means no further information can be extracted from a message (Gleick 2011: 230). A genuinely disordered message is high entropy, too, but an encrypted message that merely looks disordered has low entropy. (A bit confusing.) The ideal cryptogram in nature has been decoded into "plain text," and the jumbled specimens are sorted into species and genera on a caulogram. Apparent chaos implies an encrypted message and has low entropy. High entropy means that a set of species has been decoded into a branching evolutionary series. The measurement of entropy is in terms of bits, which may be translated into Bayesian posterior probabilities. The best evolutionary diagram is one with the highest number of informational bits, that is, the highest informational entropy.

Microscopy methods — Slide preparation is critical. The extra time spent preparing a decent microscope slide of the specimen saves much annoyance and duplication of effort. In a couple drops of 2% KOH solution add the tip of a stem, and strip the leaves off to show the axillary hairs. Add a portion of the lower stem to examine the tomentum and gemmae, if any. Mount perichaetial leaves and capsules on a different slide—the KOH will help remove the operculum if you wait a bit or heat the slide. Make cross sections of the leaves and stem; adding a tiny drop of glycerin helps keep sections from floating around on the slide. Scrape the single-edge razor blade to remove adherent sections after sectioning, then add polyvinyl alcohol-glycerin (70% clear polyvinyl alcohol glue: 30% glycerol) mountant and add coverslip. Do not ring coverslip on slide with, say, fingernail polish, in any attempt to slow evaporation by luting because nothing works. At best, keep the cabinet you keep your slides in sealed so the glycerol saturates the air and impedes evaporation. If the slides degrade due to evaporation, you can steam the cover slips off and remount.

KOH 2% solution is valuable to help remove adherent opercula; to lend color of either yellow or red to tissues, which may be taxonomically significant; and to enhance the index of refraction of temporary water mounts to provide greater clarity, particularly when viewing details like laminal papillae. From experience, one can wipe clean the KOH slide preparation on one's clothing without damage.

Because the Streptotrichaceae species have been part of Pottiaceae until now, other comments on morphology and microscopy methods may be found in the treatment of Zander (1993).

Stains — In botanical studies, ruthenium red has been used for differentially staining gums, pectins and mucilage. It is a non-specific cationic dye that binds to negative charges at anionic sites, and will stain acidic polysaccharides such as those that make up pectin. It works well in basic solution; a pH greater than 7.0 is best. A stock solution precipitates out after two or three months. Rhizoids and tomentum of *Leptodontiella apiculata* and other Streptotrichaceae species were differentially stained pink by ruthenium red in distilled water, and strongly stained blue in 2% KOH. It is probable that the thick cell walls of tomentum include mucilage, pectin or gums. The combination of hyaline windows in the leaf base, and a packing of gum-rich tomentum inside the sheathing leaf base, doubtless acts like a sponge efficient at both morphological and chemical levels. The habitat of the taxa studied presently being misty and dewy areas of high insolation, e.g. tree branches and páramo soils, may prove strongly selective in ability to absorb and retain moisture.

In toluidine blue O and distilled water, rhizoids and tomentum of *Leptodontium* are violet purple while the distal 4/5 of the leaf are light blue, but in 2% KOH followed by mounting in glycerol or polyvinyl alcohol mountant, the rhizoids are deep violet and the distal 4/5 of the leaf are green to turquoise (perhaps polyphenolics, see Keating 2014: 88); the leaf base is unstained, probably because the basal cells are pure cellulose. A protocol (Zander 2106b) tested with *L. viticulosoides* was mounting in toluidine blue O, daub off excess water, all ruthenium red, daub, add KOH, add PVOH-glycerol mounting medium will give green distal laminae, yellow or green basal region, and tomentum that is stained blue internally (toluidine blue O) and cell walls pink (ruthenium red). The tomentum in only toluidine blue O has colorless walls and dark blue cell contents.

In sum, distal and proximal leaf portions, and rhizoids or tomentum are distinguishable by aqueous stains, and their chemical constituents are probably adaptive. The stains above are indicative but not specific, and further study is warranted.

Chapter 6
OUTGROUP ANALYSIS

Morphological information — The annotated morphologically based genus-level cladogram (Fig. 6-1), based on that offered by Zander (1993: 43), may be recouched in modern terms by assigning Bayesian posterior probabilities based on numbers of traits supporting each node. Nodes have no empiric reality, because the number of trait changes for each node do probabilistically support the clustering of the terminal taxa, it is valid to translate the numbers of traits into Bayesian posterior probabilities as described in the Methods chapter. This is okay as long as the range of values of an exponential distribution of informational bits and their equivalent probabilities matches the reasonable numbers of transforming traits for each cluster of taxa.

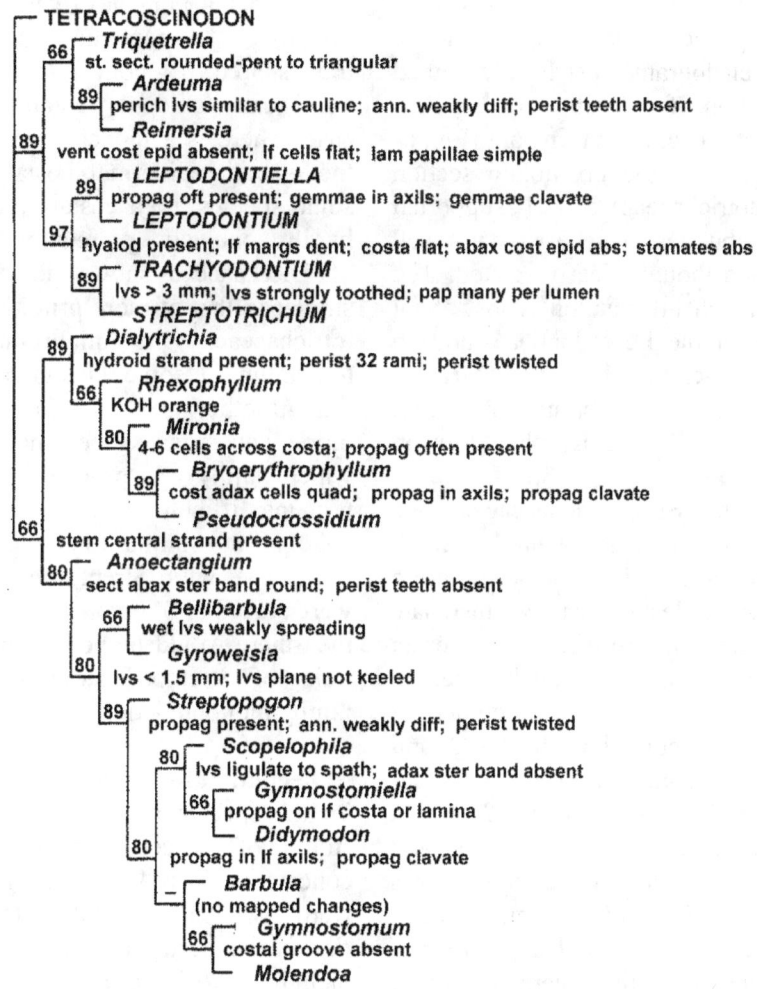

Figure 6-1. Annotated cladogram number 11 from Zander (1993: 43), being Leptodontieae and Barbuleae with *Tetracoscinodon* as outgroup. Advanced trait transformations that serve to cluster terminal taxa are placed just right of the nodes. Percentages for Bayesian posterior probabilities are given to left of nodes, and are based on numbers of *shared* traits for that node. The new Streptotrichaceae (*Leptodontiella, Leptodontium, Streptotrichum, Trachyodontium*) as represented in Fig. 6-1 is well supported at 0.97 BPP, while the potential outgroup (*Ardeuma, Reimersia,*

Triquetrella) is poorly supported as a group at 0.66 BPP, though clearly related at 0.89 BPP. *Ardeuma* is the new name for all species of *Hymenostylium* except the type, *H. xanthocarpumi*.

Too many traits imply a wrong or improper relationship given expectations of gradualist evolution. Too few is not so much of a problem. One trait is a dubious character on which to base a species, but is acceptable for a group of species sharing that trait in that each additional species in the group adds to the empirical value of that single trait. Given that the heteroplasy (radiation of molecular races) contributed by molecular analyses does not affect morphological data sets (but homoplasy—convergence of morphology— does), the morphological cladogram must be considered important in evaluation of outgroups for the here proposed Streptotrichaceae. As it turns out (Fig. 6-1), the Leptodontieae (in the taxonomic section renamed the Streptotrichaceae) is well supported by five shared traits, but nearby genera are not well supported as a group though clearly related. The conclusion is that much information is lost about ancient progenitors of the Leptodontieae, and an ingroup acting as functional outgroup, *Streptotrichum* Herz., is more appropriate than, say, *Triquetrella* Müll. Hal. See paragraph below on Taxonomic Abyss.

All genera of Streptotrichaceae have gametophytes lacking a stem central strand; tomentum often ending in series of short-cylindric cells; commonly dentate leaves that are lanceolate with a sheathing base; reniform (in section) costa with guide cells and two stereid bands but lacking a hydroid strand and adaxial and adaxial epidermal cells; perichaetial leaves convolute-sheathing; and peristome with teeth usually in groups of four rami. *Leptodontiella*, *Streptotrichum*, and *Trachyodontium* are robust and complex in character and share multiple traits, but have no close relatives other than *Leptodontium*. Assuming species generate other species through small steps of a few traits per descendant, these three genera may be considered relics of an ancient speciose group now mostly extinct. The variation in peristomes implies former experimentation in peristome development. These genera may be considered a functional outgroup because no clear species of other genera are particularly similar yet without considerable morphological reduction (e.g. *Ardeuma*, *Reimersia*, *Triquetrella*, etc.) that confounds trait polarization.

After a review of the tomentum of haplolepideous mosses, the Dicranaceae can be identified as closest to Streptotrichaceae in tomentum morphology. For instance, *Dicranum spidiceum* has thick-walled knotty tomentum of thick walls, with 1–2 cross walls in the knots, as well as elongate-celled thin-walled knotless tomentum on the same stem. Study of other species in that genus indicated, however, that these are largely distal branches of very thick rhizoids, not directly of stem tomentum. An unequivocal second trait shared by both Dicranaceae and Streptotrichaceae would be welcome, but such is not the case. There is a tendency in Dicranaceae to lack the adaxial epidermal costal cells, and also, in some species, both costal epidermises, but this is hardly conclusive of relationships. The peristome of Dicranaceae shows no odd experimentation similar to that of more primitive member of Streptotrichaceae. One might attribute the knotty tomentum of some Dicranaceae to convergence, but all other taxa examined have tomentum of more pedestrian appearance. The question of which family was progenitor of Streptotrichaceae remains difficult with Pottiaceae the frontrunner in spite of no similar tomentum. Of course, if *all* Pottiaceae with Streptotrichaceae-like tomentum were placed in Streptotrichaceae, as was done here, then there would be no connection left. The other traits of Streptotrichaceae, however, make this family monophyletic.

Molecular information — Molecular studies that include species of Streptotrichaceae are few. In the study of Kučera et al. (2013) based on concatenated rps4 and trnMtrnV datasets, there were two large clades found. *Leptodontium flexifolium* occurred at the base of a large clade including *Barbula*, *Didymodon*, *Aloina*, *Pseudocrossidium*, *Erythrophyllopsis*, *Microbryum*, *Tortula*, *Sytrichia* and *Hennediella*, among other genera. An equally large clade included *Streblotrichum*, *Anoectangium*, *Gyroweisia*, *Molendoa*, *Hymenostylium* (*H. recurivrostrum* now *Ardeuma*), *Eucladium*, *Trichostomum*, and *Weissia*. Given the morphological similarities (Zander 1993), there proved a considerable distance between what is now *Ardeuma recurvirostrum* and *Leptodontium* *flexifolium*. What is significant is the position of *Leptodontium flexifolium* near the base of the Pottiaceae, within two nodes of *Scopelophila*, the basalmost of Pottiaceae short of the deep Pottiaceae taxa that have been segregated as the Timmiellaceae, *Timmiella* and *Luisierella*. A segregation of Streptotrichaceae from basal members of Pottiaceae is a hypothesis thus not rejected in this study. A study of rps4 alone in the same paper put *Lepdodontium flexifolium, L. luteum,* and *Triquetrella tristica* together near the base of the Pottioideae but five nodes from *Scopelophila*.

In the molecular study of Cox et al. (2010), in a combined rps4/nad5/nuc26S data set, *Leptodontium luteum* occupied the same position at base of the Pottiaceae, supporting the results of the concatenated rps4 and trnMtrnV data of Kučera et al. (2013). Given this basal position, and the position of *Timmiella* quite farther towards the base of the Cox et al. (2010) cladogram, which included many families, one can hypothesize that Streptotrichaceae originated rather deep in the evolutionary past. On the other hand, the rps4 sequence study of Hedderson et al. (2004) places *Leptodontium flexifolium* rather deeply among other Pottiaceae genera. Likewise, the rsp4 data analysis of Werner et al. (2004) placed the same species deeply in the Pottiaceae. There is clearly a conflict between the results of various studies, and origin of the Streptotrichaceae requires more molecular study. Commonly, discrepancies between results of different molecular studies are not discussed in the literature in detail, in part because the same specimens are used as exemplars in different studies using information in GenBank, a public repository of nucleotide sequences. Thus, the discrepancies may be methodological.

The families in the Cox et al. (2010) study that are between *Timmiella* and *Scopelophila* are many, and include Calymperaceae, Ditrichaceae, Fissidentaceae, Grimmiaceae, Hypodontiaeae, Leucobryaceae, and Ptychomitriaceae. Given the similarities of *Timmiella* and Pottiaceae sensu stricto (a molecular aggregate), it is not inconceivable that Streptotrichaceae originated among the other families from an Ur group with the traits of Pottiaceae, here including Timmiellaceae (Inoue & Tsubota 2014) as a synonym. That Pottiaceae is indeed primitive is evidenced by the similarity to Polytrichaceae of anatomical details, particularly of costal sections. Polytrichaceae is also able to morphologically converge with Pottiaceae easily, and thus has less developmental phyletic constraint (Futuyma 2010; Schwenk 1998) than more advanced groups. Evidence is the uncanny resemblance of the polytrichaceous *Delongia glacialis* N.E. Bell, Kariyawasam, Hedd. & Hyvönento to *Pterygoneurum*. Of course, other families in this basal group have the same developmental flexibility, e.g. *Holomitrium* (Dicranaceae) and *Leptodontium* (Streptotrichaceae), and *Ditrichum ambiguum* Best with the twisted peristome of Pottiaceae.

In sum, molecular analysis is little help in clarifying the evolutionary position of studied species of this group. which is apparently near the base of the Pottiaceae molecular cladogram.

Serial outgroup analysis — Gradual, stepwise evolution as a rule-of-thumb is matched by an expectation of gradual, stepwise extinctions. Thus, taxa with no or very few near relatives are more ancient than groups with many near relatives, given somewhat equal rates of extinction. Therefore, *Trachyodontium* is ancient, but *Streptotrichum* and *Leptodontiella* are less ancient and grade into extant species. Given that *Timmiella* and *Luisierella* have strongly bulging laminal cells, and so do basal Trichostomoideae and Hyophileae, then flattened laminal cells should be advanced, as in *Tortula, Hennediella, Bryoerythrophyllum,* and *Oxystegus.* BUT *Leptodontiella, Trachyodontium* and *Streptotrichum* have flat laminal cells. *Streptotrichum* and *Trachyodontium* have tiny, simple to 2-fid scattered papillae. Papillae of *Leptodontiella* are very similar but 2--3 times larger, with the appearance of crowded scabs, reminiscent of those of *Molendoa.* Given that flat laminal cells are advanced, then the unique bulging with salient cells of *Stephanoleptodontium* are even more advanced. Thus, Streptotrichaceae has the unique traits of 4-rami peristomes and crown-like papillae over bulging cells derived from small, simple to bifid papillae over flat laminal cells.

Timmiella has no tomentum, rhizoids present and of all elongate cells. Although Timmiellaceae (of two genera, *Timmiella* and *Luisierella*) is characterized morphologically by S-twisted peristomes, the pottiaceous species *Streptocalypta santosii* also has an S-twisted (clockwise) peristome, as illustrated by Zander (1993: 95).

Sometimes macroevolutionary systematics is initially the same as cladistic analysis in which one or more taxa taxonomically near the study group is used to root a cladogram, polarizing the trait transformations and minimizing redundancy of information by identifying identical (homologous) traits. Commonly related taxa are then grouped and more primitive taxa are near the base of the cladogram. *but* nodes on a cladogram only represent shared traits, and cannot distinguish a shared ancestor by name or complete distinguishing traits. Skipping the cladogram analysis is thus okay unless there is much confusion of species. There is no software, however, at present to serially order taxa by increasing accumulation of advanced traits.

Outgroup analysis in macroevolution differs from that in cladistics in that one tries to select a recent direct ancestor. Because dissilient genera gradually go extinct, a specialized descendant of an extinct ancestor may bias analysis, therefore hypothetical core ancestors may be less informative but are also less wrongly informative. If identifying a recent ancestor is difficult or impossible, outgroup analysis may be done in the cladistic sense (emphasizing shared traits over serial descent) and can be as simple as selecting a closely generalist species in a related group. Or it can involve inference of a *hypothetical* taxon such as (1) such a generalist taxon but with any tacked on advanced traits removed, or (2) "triangulation" of a taxon by grouping all species with some distinctive trait and eliminating all traits not shared by in that group. For instance, the morphologically composite taxon with all traits shared by all or most taxa in the *Leptodontium* group that have rough, papillose calyptrae. This is the reverse of plotting traits on a cladogram under the assumption that no trait existed before the first node in which it occurred in the cladogram.

After distinguishing a primitive generalist species by outgroup analysis, other species in the group are tacked on serially, occasionally branching, with greater numbers of advanced traits superimposed on the basic set of traits of the species last in line. If two species share an advanced trait and also have different new advanced traits, an unknown shared ancestor may be hypothesized.

As an exemple, *Didymodon acutus* (Brid.) K. Saito is the theoretical ancestral taxon of the genus *Didymodon,* and it has a twisted peristome while the peristome is reduced in many other species of the genus. That other species of the genus nest neatly in *D. acutus* as paraphyletic macroevolutionary descendants is confirmatory of the peristome being a clue to reduction and elaboration during radiation of a dissilient genus. The same is true for peristomes and other traits in other segregates of *Didymodon* s. lat. (Zander 2014a,b,c).

Additional comments on various possibly related taxa — Easily mistaken for a *Leptodontium* is *Barbula eubryum* Müll. Hal., of central and southern Africa, and the Comoros, which may or may not have a central strand and is usually quite densely short-tomentose. It differs in the large rhizoidal tubers sometimes present, undulate laminal margins, coarse laminal papillae, medial basal cells torn across their longitudinal walls and hyaline excepting thick-walled orange transverse walls, and costa with small but distinctly differentiated adaxial epidermal cells. Curiously, *L. pungens* often has hyaline cells at the mid-insertion of the leaf base that are quite like those of *B. eurbryum.* One might suggest this is a rarely occurring "tell" for an evolutionary relationship, with *B. eubryum* as ancestral taxon. On the other hand, the papillae of *B. eubryum* are simple to bifid and centrally raised on the salient of the bulging laminal cell walls in a fashion quite like that of *Leptodontium longicaule* but slightly more coarse. The tomentum of *Barbula eubryum* is long-stalked, ending in a "broom" of smaller rhizoids, while that of *Leptodontium* arises directly from the stem or laterally on large dark rhizoids as tufts. The leaves of *B. eubryum* are much like those of *L. subintegrifolium.* It is apparently a relict of a long extinct group basal to both Streptotrichaceae and Pottiaceae. A new genus name is provided in the taxonomic section of this book.

Rhexophyllum subnigrum (Mitt.) Hilp. has much the same leaf shape (sheathing base, carinate limb with dentate margins) as has *Leptodontium,* the lamina red in KOH, presence of a stem central strand and epidermal cells in the costa indicate that that this taxon is convergent. It is probably another relict, like *Barbula eubryum.*

Triquetrella paradoxa (I. G. Stone & G. A. M. Scott) Hedd. & R. H. Zander (= *Leptodontium paradoxum* I. G. Stone & G. A. M. Scott) is midway in relationship (a missing link?) between the genera *Leptodontium* and *Triquetrella* (Hedderson & Zander 2007) but is more like *Triquetrella* because of the entire leaf margin and differentiated abaxial epidermis of the costa. The former feature is rare in *Leptodontium* s.lat. and the latter is unknown in that genus. *Triquetrella paradoxa* also has a distinctly triangular stem section, which is rare in *Leptodontium* s.lat., while such sections usually have "tails" that are remnants of the long and broad decurrencies typical of *Triquetrella,* but rare in *Leptodontium.* The stem of *T. paradoxa* also lacks quadrate cells on the tomentum, which are common in *Leptodontium.* However, the short-excurrent costa is reminiscent of that of *Barbula eubryum.* The laminal cells are punctate and similar to those of *L. viticulosoides,* and the long and broad decurrencies are like those of *L. scaberrimum,* while the lack of a hyalodermis is similar to both. In the molecular analysis of Kucera et al. (2013) *Triquetrella* is just proximal on the cladogram to *Leptodontium flexifolium* and *Leptodontium luteum,* that is, (*Triquetrella* (*Leptodontium flexifolium, L. luteum*). *Triquetrella paradoxa* is similar to *Leptodontium* in the stiff, widely spreading leaves, yellow to orange in KOH, costa with no adaxial epidermis, and lack of a stem central strand. It is similar to *Leptodontium* sect. *Verecunda* in the scattered bifid papillae, and to sect. *Leptodontium* in the thick-walled upper laminal cells with ovate to rounded-quadrate or angular lumens, and no stem hyalodermis. It differs from *Leptodontium* in the stem triangular or asymmetrically pentangular, leaves without teeth, costa terete and semicircular in section with enlarged abaxial epidermal cells, and basal cells of leaf not strongly differentiated. Few *Leptodontium* species occur at such low elevations (ca. 400–600 m). Sporophytes are unknown for this species. *Triquetrella paradoxa* is found in Australasia, as is the New Zealand endemic *Leptodontium interruptum,* and both are morphologically somewhat isolated species.

Reimersia might be a good ancestor for *Streptotrichum* and *Leptodontium viticulosoides,* but differs in the lack of marginal dentition or tomentum or laminal papillae. Also it has elongate distal laminal cells, a weakly differentiated abaxial

epidermis on costa, weakly differentiated perichaetial leaves, and lacks a peristome. It has the same trigonous nature of cells, no hyalodermis, equal-sized central-cylinder cells.

Taxa with flat leaf cells are probably advanced, given that *Timmiella* and *Luisierella,* of early Pottiaceae morphology, have bulging laminal cells as do those basal on the morphological cladogram. Taxa with flat distal laminal cells include *Leptodontium flexifolium, Bryoerythrophyllum recurvirostrum, Tortula* spp., and *Hennediells* spp. These are probably the most advanced in their transgeneric lineages.

Paraleptodontium is a good species and possibly a good genus. It may be in part an outgroup for *Leptodontium* (bright yellow color) except for strong marginal border which is advanced but nearly identical to that of *L. flexifolium.* But check *Didymodon erosodenticulatus* (C. Müll.) Saito (= *Prionidium e.*). *Paraleptodontium* has no central strand and is yellow in KOH, but it has a ventral epidermis on costa. It has the areolation of *L. flexifolium,* namely thin walls, quadrate cells and lumens, flat superficial walls, scattered 2-fid papillae. It is, however, most closely related to *Oxystegus.*

Hierarchical cluster analysis and the taxonomic abyss — Technically, a cladogram groups taxa on the basis of advanced shared traits alone. Phenetic algorithms, such as UPGMA and Neighbor-Joining, group by overall similarity, and treat all traits as evidence of grouping. The result is a phenogram, not a cladogram. The results of phylogenetic methods, such as maximum like-lihood and Bayesian techniques, use both advanced shared traits and autapomorphies as evidence to create phylograms that are cladograms that also show branch length, a measure of trait differences between different parts of a maximally parsi-monious cladogram and occasionally between two species. These are here considered all various sorts of dendrograms. Phylograms, as is usual in modern phylogenetic studies, are in the present work all called cladograms.

In this book and elsewhere I commonly lump cladistics and phylogenetics when the significance of their results are much the same, but cladistics is hierarchical cluster analysis, differing from standard phenetics by using minimized trait transformations on a non-ultrametric dendrogram.

Phylogenetics adds an evolutionary dimension in the length of the branches and designation of a cladogram (or phylogram) node as an unknown "shared ancestor." Both are hierarchical cluster analysis because the nodes of trees do not model species or other taxa in evolutionary order, but instead serve only to demarcate groups (never less than two taxa) sharing advanced traits (reversals are allowed).

Evaluation of the deeper connections of a *caulogram* can come to a halt signaled by a basal taxon nameable only at a high taxonomic rank, when, in spite of best efforts to eliminate redundancy, trait distances between taxa pass beyond 8 bits and efforts at modeling gradualism in speciation come to naught. At this point, macroevolutionary analysis reveals a disconcerting abyss of ignorance of evolutionary processes explaining the apparently disconnected serial relationships of the most primitive members of extant lineages, or of unique, evolutionarily isolated taxa. A cladogram. on the other hand, presents to the reader such ignorance of ancient evolutionary relationships simply as dichotomous branches no different than those nesting taxa that are closely related (e.g., see complex and deep cladogram of Cox et al. 2010). The present emphasis on distance measured by informational bits as modeled by caulograms, and also detailed by analytic keys, brings importance to the study of fossils and subfossils, and to the search for presently unknown and doubtless rare missing links. Expectation of gradual change between species, however, allows an evolutionary "uniformitarianism" in modeling both extant and some extinct relationships. Details of the abyss remain murky and are not modeled by cladograms.

Any cladogram has fully connected branches. A rooted cladogram has one deep stem. The implication is that a cladogram can evaluate evolutionary relationships between all involved taxa. In no case do cladograms trail off into multiple roots embedded in a taxonomic abyss. This is another way cladistics is actually cluster analysis. All scientists do want to find connectedness and structure in nature. But, impressing a dichotomous cladistic tree on one's modeling of real structure in nature is a major disservice to science.

A review of the more ancient fossils of mosses demonstrates a great disparity in morphology between those fossils and modern taxa. We might assume initial conditions and chaotic early evolution created a myriad morphotypes. We assume bottle-necking over time has reduced diversity of Bauplans to those presently extant. Yet, even those now extant have very odd morphology, for instance *Sphagnum, Takakia, Andreaobryum, Polytrichum, Hypodontium, Tetraphis, Diphyscium,* etc. If these are surviving representatives of the kinds of variation in the past, then there must have been a vast medley of plant forms that may have generated extant groups. The present study suggests that the proposed new family Streptotrichaceae (see taxonomic section) arose from an oddly peristomated epiphyte group generally similar to Pottiaceae but difficult to pin down as far as serial relationships go. Morphological and molecular cladograms of relationships including extant members of Streptotrichaceae are thus probably poorly informative and misleading. The idea that there must be an ancestor and it can be revealed through shared traits should be replaced by the simple acknowledgement that a cluster analysis has only one stem.

Chapter 7
THE DATA SET

The full data set for all relevant traits for species in this study are given in the descriptions of the genera and species. A subset of traits shared by taxa was selected for a cladogram of maximum parsimony plus Neighbor Joining tree, and UPGMA cluster diagram. These charts, particularly the cladogram, are intended to help minimize parallelism prior to analysis with an analytic key, which minimizes reversals. Together, the studies of shared descent and of direct descent maximize information on order and direction of evolution. Cladistic data set characters and states for each character follow:

Cladistic data set

1. Stem length: 1 = < 3 cm, 2 = > 3 cm.
2. Stem section: 1 = rounded-pentagonal, 2 = triangular.
3. Stem hyalodermis: 1 = absent, 2 = present, 3 = weak.
4. Tomentum: 1 = arbusculate, 2 = stringy, 3 = short micronemata, 4 = thin or absent.
5. Tomentum: 1 = no short cells ending tomentum, 2 = short cells present.
6. Leaf size: 1 = about equal, 2 = increasing or rosulate.
7. Leaf shape: 1 = long-triangular to linear-lanceolate, 2 = lanceolate, 3 = ligulate or ovate.
8. Leaf length: 1 = < 3 mm, 2 = > 3 mm.
9. Leaf marginal ornamentation: 1 = entire or crenulate, 2 = denticulate to dentate.
10. Leaf apex shape: 1 = acuminate to acute, 2 = broadly acute, 3 = narrowly obtuse.
11. Leaf base: 1 = sheathing, 2 = weakly differentiated.
12. Costa: 1 = percurrent or excurrent, 2 = ending before the apex.
13. Costa number of cell rows across midleaf: 1 = 2(–4), 2 = 4 or more.
14. Costa transverse section: 1 = round to semicircular, 2 = flattened to reniform.
15. Costa abaxial stereid band layers: 1 = 1–2, 2 = 2 or more.
16. Costa guide cells fully included at midleaf: 1 = 2, 2 = 4 or more.
17. Cell walls distally, thickness: 1 = thin to evenly thickened, 2 = thickened at corners.
18. Distal laminal border differentiation: 1 = no, 2 = yes.
19. Laminal cell width distally: 1 = 9–11(–13) µm, 2 = 11 or more µm.
20. Laminal cell superficial (exposed) walls distally: 1 = flat to weakly convex, 2 = bulging.
21. Laminal papillae: 1 = simple, punctate or 2-fid, 2 = multiplex, crowded, 3 = coroniform, centered.
22. Laminal basal cell color in KOH: 1 = yellow, 2 = hyaline.
23. Laminal basal cells with miniwindows juxtacostally: 1 = absent, 2 = present.
24. Laminal basal cells with longitudinal stripes of darker color: 1 = absent, 2 = present.
25. Specialized asexual reproduction: 1 = absent, 2 = gemmae.
26. Capsule stomates: 1 = absent, 2 = present.
27. Capsule annulus rows of cells: 1 = 2–4, 2 = 4–6.
28. Peristome length: 1 = < 300 µm, 2 = > 300 µm.
29. Peristome ornamentation: 1 = smooth or striate, 2 = spiculose.
30. Calyptra surface: 1 = smooth, 2 = antrosely papillose.

For the study of shared traits to minimize parallelism, a Nexus file was devised using the data set above. Species names are six letter compressed versions of the names, easily interpreted by the reader.

```
#NEXUS;
begin data;
   dimensions ntax=27 nchar=30;
   format symbols="1234";
   matrix
[        1         2         3]
[2345678901234567890123456789 0]
Strepr
2124212221112222?2211112111222
Laggre
2114212221122121111111112222211
Larauc
2124212221122121111111111122222?
Lbrach
2124112221222222121231112222211
```

```
Lcapit
21242122211222221222232122?121?
Leryth
11221122231221221212212211?????
Lexcel
21112122212222122111111112122211
Lfilic
21222122221211221112311122121?
Lflexi
11141231222211212211111221111
Lgemmi
11141222212111112111111112?????
Linter
11241121211112121111211221122??112
Llatif
21232122221212121222232111?????
Lapicu
11222121212222122111121111211212
Llongi
212421222112221211223111122221?
Lluteu
21241121231222121211111222211
Lproli
111411312121222212211111122111?
Lpunge
2122212221112122122212212212221?
Lscabe
2211112223221122211231111?????
Lstelc
1124112121212111111211112?????
Lstelf
11142131222212111112211112?1211
Lstolo
11242132222222222222112112?????
Lsubin
2122212213122122121221221?????
Lsyntr
211411222312221211223111?221?
Trachy
21222122211122221221112122222
Ltrico
21212122211222241211121121211
Lvitic
21111122212222122121111111122211
Lwalli
22222122211121221211212212221?
    ;
endblock;
outgroup 1;
set maxtrees=200000;
log file = leptlog.txt;
UPGMA;
NJ;
hs addseq = random nreps = 20;
showtrees;
describetrees /plot=phylogram brlens=yes;
savetr /brlens=yes;
contree;
boots nreps=2000 search = fast;
```

Chapter 8
CLADISTICS

A cladogram is a highly sophisticated means of grouping by shared advanced traits. The cladogram helps minimize parallelisms in models of linear descent by maximally concatenating species with homologous traits. This eliminates branches of taxa with much the same traits. Minimizing the number of traits needed to create a linear descent model of evolution minimizes redundancy. The less redundancy, the more Shannon entropy there is, which is information content. Traits not possible to eliminate as redundant contribute to informational bits that support order and direction of evolution. The cladogram, because it deals only with shared traits must be re-interpreted in terms of linear descent using an analytic key. Because the end product is not a cladogram, we do not need to find the shortest cladogram of shared descent. We want, instead, the caulogram of direct descent that minimizes parallelism and reversals.

A set of 30 traits were selected as parsimony-informative. A cladistic data set (detailed in a previous chapter) was created for the 24 *Leptodontium* species plus *Leptodontiella apiculata*, *Streptotrichum ramicola*, and *Trachyodontium zanderi*. *Streptotrichum ramicola* was selected as the outgroup in view of its long, four-parted, spiculose peristome, and simple papillae scattered over flat lumens, much like the otherwise rather different *Trachyodontium* and *Leptodontiella* species.

An analysis using PAUP* (Swofford 2003) was done using parsimony. The 30 characters were unordered, and heuristic search was used with 20 random replicates. The strict consensus of 35 equally parsimonious trees (length 116 steps) was in part poorly resolved, apparently from convergence of perhaps evolutionarily less important traits.

No weighting of characters was used in this study. There is a situation in which weighting of characters for analytic importance is worthwhile. That is when there is a question of which traits are best for tracking evolutionary lineages. Recently evolved or strongly adaptive traits are less valuable than relatively neutral traits that tag along through many speciation events. A trait that occurs in many habitats is better (more neutral) at tracking than one that is found in one or few habitats. A trait found in too many habitats, however, can be useless in that the character field becomes saturated, and one lineage becomes less distinguishable from another.

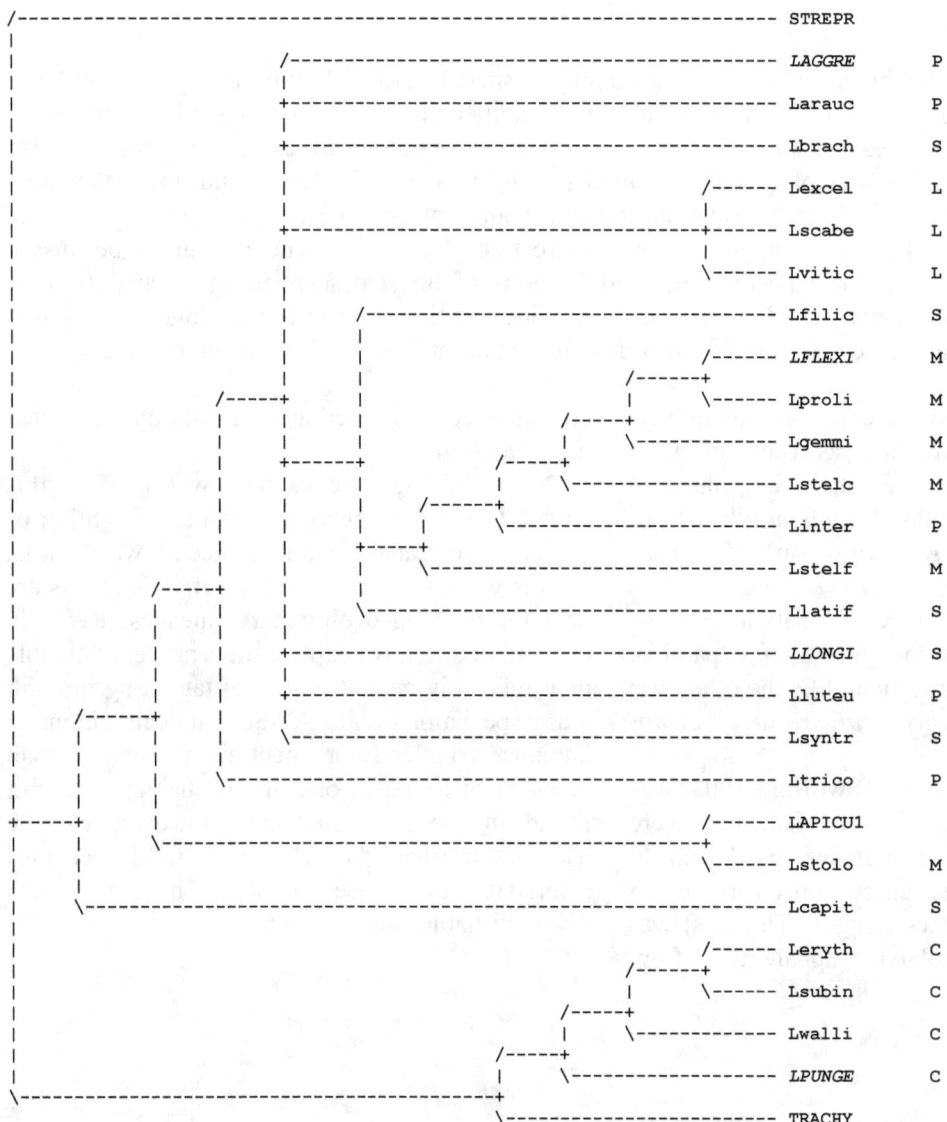

Figure 8-1. Strict consensus tree of 35 equally parsimonious trees from weighted parsimony analysis. Genera other than *Leptodontium* as previously conceived (Zander 1972) are in capital letters. Species thought to be core generative species are in caps and Italics. Species identified as primitive in the present study are labeled group P. Otherwise, identifiable groups of two or more species approximating those in the *Leptodontium* revision of Zander (1973) are distinguished by letter C, M, and L, and further contained in brackets. Group labels have the first letter of genus names awarded later in this book. Each genus is not a clade but is anchored by a putative generative core using the dissilient species concept (Zander 2013). Interpret the cladogram as species of other genera (*Leptodontiella, Streptotrichum, Trachyodontium*) or primitive Group P variously giving rise to other Groups.

The nonparametric bootstrap values in the cladogram of Fig. 8-1 were low or less than 0.50, but true support is not from shared traits, but will be (see Methods chapter) from Bayesian analysis of numbers of new traits distinguishing a descendant species from its progenitor. This is done in part with an analytic key (seen next chapter).

The species selected during the taxonomic study as primitive (Group P) in the genus *Leptodontium*, and also the three related genera, however, were somewhat scattered, due to being attracted by groups they separately generated, either because they were the probable progenitor or through the bias of long-branch attraction. Group C (*Leptodontium* sect. *Leptodontium*) and Group M (*L.* sect. *Verecunda*) are well resolved. The putative progenitor of Group M, *L. flexifolium,* is at the apex of the series, while the progenitor of Group C is at the base. The evolutionary order is apparently affected by a self-nesting ladder (Zander 2013: 53) in Group M.

Species of Group S (*Leptodontium* sect. *Coronopapillata*) were somewhat scattered but usually associated with a possible progenitor species in Group P. (Note here that true monophyly is restricted to sometimes branching series of species or genera and that a single dissilient genus may generate more than one descendant lineage without being a part of that descendant lineage.) A long-branch attraction problem should be soluble with further analysis in an analytic key and perhaps with molecular analysis as the distance between possible positions on this morphological cladogram is large enough to preclude bias from extinct molecular paraphyly (Zander 2013: 51, Zander 2014b).

Parsimony and Neighbor Joining act like maximum parsimony to help minimize parallelisms, the first by approximate minimum evolution (Saitou & Nei 1987) and the second by overall similarity. Apparently maximum parsimony is best of the three at creating shortest trees, but, as Saito and Nei (1987) reminded us, the true tree is not necessarily the shortest. A UPGMA tree of overall similarity may be useful in determining order after a core generative species is selected on the basis of generalization (absence of specializations) and association with geologically old habitats and broad or ancient geographic ranges.

The Neighbor Joining and UPGMA trees from the above study are instructive and are given below. They grouped the taxa much as in the parsimony tree, and primitive or related genera are approximately in the same positions. The two latter trees are fully resolved because of the nature of their algorithms, and both are not expected to be maximally parsimonious. At this point, it may be valuable to point out that the writing of software that will minimize both parallelisms and reversals in the context of direct descent of known or inferred taxa would be a valuable service.

The NJ and UPGMA trees (Figs. 8-2 and 8-3) seem to clarify the position of *Leptodontiella apiculata* as near Group M, and of Group L being the progenitor of *Leptodontium aggregatum.*

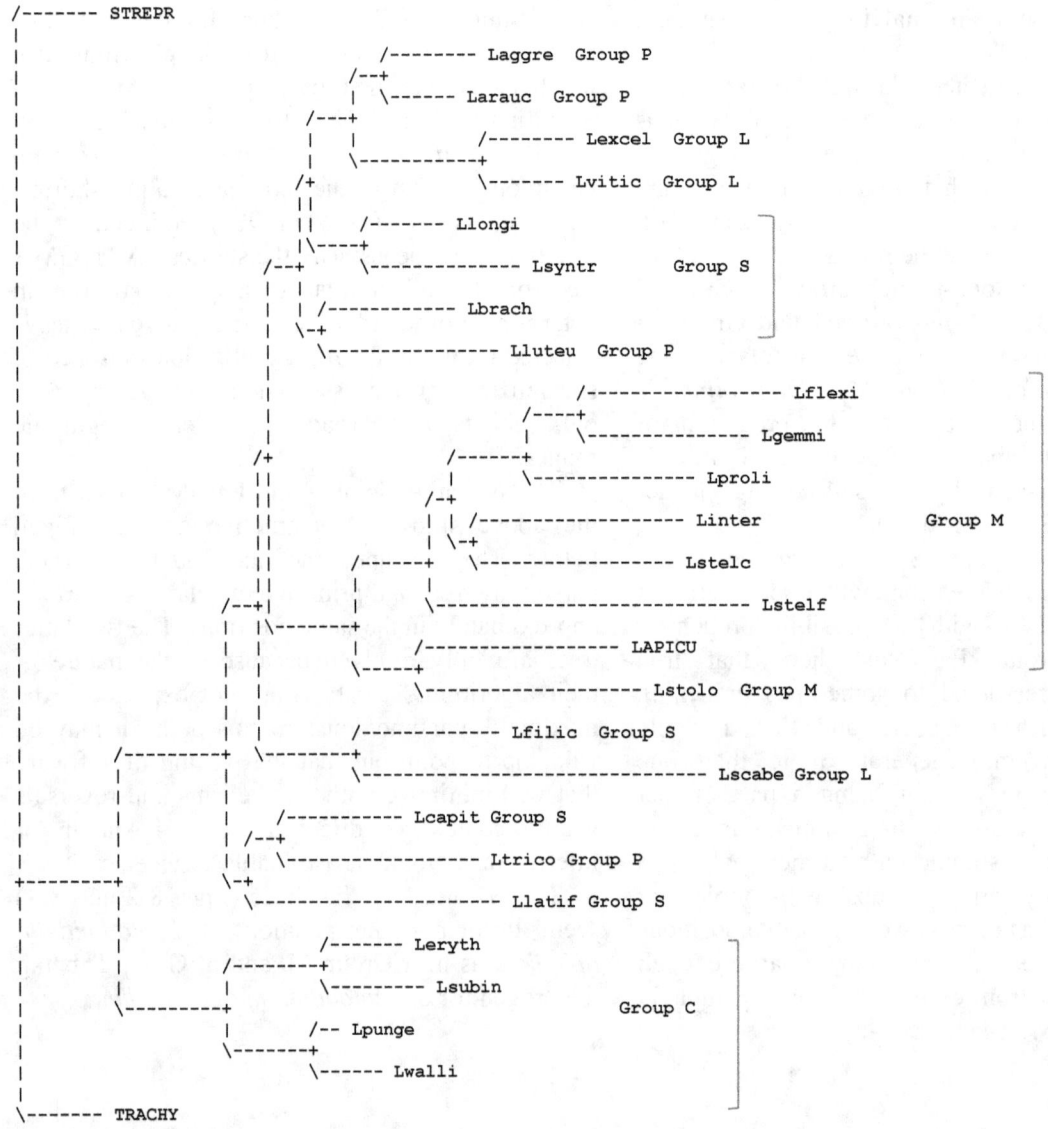

Figure 8-2. Neighbor Joining tree from parsimony analysis. Brackets indicate major groups. P = most primitive in analytic key evaluation.

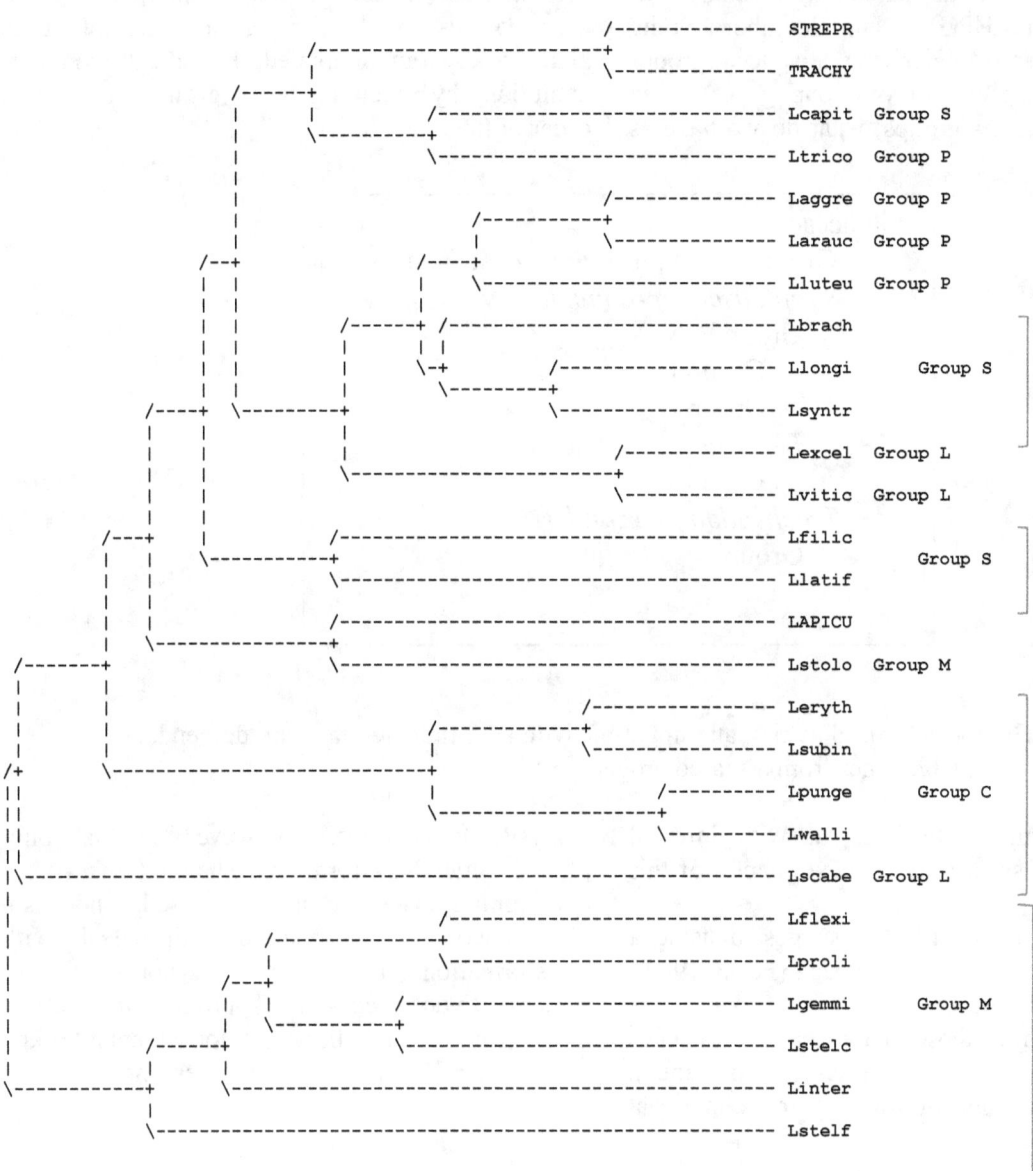

Figure 8-3. UPGMA tree of the parsimony analysis. Brackets indicate major groups.

So … if we combine the parsimony cladogram as best possible with UPGMA and Neighbor Joining trees, with a view to deriving advanced groups from primitive species or outgroup genera, then condense the species groups, what do we have as best attempt at elimination of parallelism? To describe this, we can try a first attempt at an analytic key (an annotated theoretical model of evolution) by indenting taxon groups in possible order of linear evolution.

Pottiaceae
    Ur Group (first family-level descendant)
        Streptotrichum ramicola
           Group P
             Group L
             Group S
           Leptodontiella apiculata
             Group M
           Trachyodontium zanderi
             Group C

Figure 8-4. Groups organized as first attempt at analytic key. Indented taxa are descendants. Taxa equally indented are branches from a shared progenitor taxon.

This model will be fleshed out with inferred relationships of species in the next chapter of this book.

Damien Aubert (2017) has suggested using a successive approximations approach (Farris 1969), which weights higher traits that prove less homoplastic, that is, those that appear fewest times in the cladogram. This is a way to provide a measure of the value of tracking traits in serial analysis. It is limited, however, in that only phylogenetically informative shared traits (those determining sister groups) are used, and also, phylogenetically informative traits deal with transformations of sister groups, not species to species. Even successive approximation results must be used as ancillary to a formal analytic key evaluating relationships of serial descent.

Chapter 9
ANALYTIC KEY AND CAULOGRAM

Quick overview —Macroevolutionary analysis by direct descent is fairly simple. In the ideal case, one selects one species from a group and finds another that is very similar. Using outgroup technique and evaluation of rarity of traits, one of these two species, if close enough to expect derivation by gradual steps of traits in evolution, can be hypothesized the descendant of the other. Try it with a group you know well.

Continue making a string of progenitor-descendant species and stop when a progenitor is quite generalist in character, and label this a core progenitor. This minimizes reversals and redundancy of traits. Then do this again, and see if a string of species will also attach to that core progenitor, usually at the core progenitor, or sometimes to a descendant. Create as many star-like core progenitor-descendant many-branching stars as possible (the descendant lineages will be short strings or even unicates). Then find the optimum attachment (continue to minimize reversals and redundancy of traits) for these fragments of lineages to create a branching pattern.

The attachments will often be core progenitor to core progenitor. If species in a lineage or star do not seem to represent gradual stepwise evolutionary transformation, it is permissible to postulate a missing link that fills out the stepwise transformation of a serial lineage or provides a core progenitor to loosely associated descendant lineages that may be somewhat specialized. Then make an analytic key (see below) that shows Bayesian support for the lineages, and from that devise a caulogram summarizing an evolutionary tree of direct descent. This is clearly forcing data to fit theory, which is good. Theory is the essence of science, and is tested by efficacy of prediction.

An *analytic key* focuses on differences between species, while similarities are signaled by increasing indentation (Zander 2013: 82—as a "natural key," 2014c, 2016a). A caulogram summarizes the evolutionary tree inferred from the analytic key, and minimizes both numbers of state changes and necessary postulated unknown shared ancestors. The transformations in the analytic key given below are cognate with transformations in other genera of the moss family Pottiaceae (Zander 1993), while taxa are characterized by a minimum variation of two traits, which are needed to encompass an inferred evolutionary function or process, particularly as adaptive or conservative-tracking at various levels of taxon hierarchy, or which group specimens about a variational mean presumably held together by some evolutionary process particular to the group.

Given for each evolutionarily informative trait is one informational bit supportive of descendancy. These estimates are based on an overview of *Leptodontium* s.str., the here recognized segregate genera of *Leptodontium* s.str., and certain genera considered outgroup or near relatives (probable progenitor siblings) to *Leptpdontium*, namely *Trachyodontium, Streptotrichum,* and *Leptodontiella.*

For each of the groups in the clades identified in the cladogram of weighted parsimony in the last chapter, one species can be inferred as a core generative species for most of the other species in the clade. Thus, using the dissilient genus concept (Zander 2013), monophyly is not determined by the continuity of the nodes in the clade but by the continuity of that core generative species, so the whole clade is not the basis for taxonomic names at the genus level. *Leptodontium* s.lat. is in this book split into several genera, each with a core generative species. These names are formally proposed later in the taxonomic chapters, but used here informally for clarity in understanding the analytic key.

The evolutionary tree — Based on work with several groups (Zander 2008, 2009, 2013, 2014a,b.c), caulograms may be expected to have a number of clusters of progenitor taxa and the descendant taxa of each. Each cluster will have one to several descendant lineages, each lineage being 1 to 4 taxa in length. Deepness of a lineage series is a kind of weighting, because they all fit together as gradual steps in evolution, so the longer the lineage, the better.

A mathematical binary tree is not a cladogram or a caulogram. In binary trees, each node is named (or at least numbered), but in cladograms only the terminal ends of the tree are named and each internal node is labeled only as an immediate shared ancestor of two siblings. In a caulogram, the branching is not necessarily dichotomous, but may be also simply seriate or multichotomous, and internal nodes are usually named in the same manner as terminal taxa.

Summary of technique — The Groups can be filled out with extant or inferred species in series of direct descent. Linearly, the series of species increase in specialization. Branching must be considered when reversals would be needed if the linear progression is to be continued, and reversals are minimized by such branching. A specialization associated with direct descent is awarded one informational bit. A reversal is taxed with one negative bit. Bits within a species are added to gauge support in terms of Bayesian posterior probability (BPP) for order of linear descent, and bits of all species in a lineage or dissilient genus are added to gauge support for direction of evolution. If some new traits are each considered part of a composite adaptive trait for a species, then the IRCI formula (Zander 2013: 59) may be applied to increase BPP given that in this case if one trait is informative of the evolutionary order then so are the others. Summed bits are directly translatable into Bayesian posterior probabilities. Given that gradual evolution by a few major traits is the desideratum, too few or too many new traits

are problematic; in the first case a species is expected to have a minimum of two traits to ensure linkage by isolation, and in the second case, too many traits require too many reversals or an assumption of and unknown progenitor or progenitor lineage not in the group studied. The analytic key is a theoretical model of evolution in the Groups, and modeled relationships may be diagrammed in a caulogram of linear and branching sets of taxa, known or inferred (i.e., predicted).

The analytic key models direct descent by indentation, with numbers reflecting progenitor(s) and number of descendants. The traits are divided into "tracking" traits that are found in both progenitor and descendants (and are therefore evolutionarily neutral or becoming neutral for the group), and "new" traits that may be adaptive or at least are advanced and informative of order and direction of evolution.

Analytic key
The model obtained from the cladogram of the last chapter is below, and completion requires addition of species. Each line is unique and may be termed an "entry" rather than a couplet as in classical dichotomous keys. A preliminary general organization of evolutionary relationships through direct decent can be outlined though an indented key. The symbol ">" stands for "generates," and indentations replace the angle bracket in the completed analytic key. Descendants are indented below progenitors.

Pottiaceae
> Ur Group (first family-level descendant)
    > *Streptotrichum ramicola*
        > Group P
            > Group L
            > Group S
        > *Leptodontiella apiculata*
            > Group M
    > *Trachyodontium zanderi*
        > Group C

The letters of the Groups are simply the first letters of the names of the segregate genera I will formally propose later in the book. As a first attempt at order and direction of evolution, species of Group P and of Group C are added to the above

chart, then the other more deeply embedded species are added in proper place. This simplification and overall view is based on expert familiarity with the species as first estimate, then re-evaluation (a kind of reciprocal illumination) with respect to the completed analytic key. Other genera and core species are boldfaced to emphasize the dissilient nature of the evolutionary order. Indented entries are inferred descendants of the same progenitor, and have the same entry identification number.

1. Pottiaceae
  2. Ur Group (first family-level descendant)
    3. ***Streptotrichum ramicola***
      4a. Group P **Core – *L. araucarieti***
        5a. Group P *L. aggregatum* and *L. tricolor*
          6a. Group L Unknown
            7a. Group L *L. viticulosoides*
            7b. Group L Unknown
              8a. Group L. *L. scaberrimum*
              8b. Group L. *L. excelsum*
          6b. Group P *L. luteum*
          6c. Group S **Core – *L. longicaule***
            9a. Group S *L. syntrichioides*.
            9b. Group S *L. capituligerum*
              10a. Group S *L. latifolium*
              10b. Group M. *L. stoloniferum*
            9c. Group S *L. brachyphyllum*
              11. Group S *L. filicola*
      4b. Group A *L. interruptum*
      4c. ***Leptodontiella apiculata***
        12. Group M **Core –** Unknown
          13a. Group R *L. stellatifolium*
          13b. Group M *L. flexifolium*
            14a. Group M *L. umbrosum* (= *L. proliferum*)
            14b. Group M *L. stellaticuspis*
            14c. Group M *L. gemmascens*
      4d. *Trachyodontium* **Core**
        15a. ***Trachyodontium zanderi***
        15b. Group C **Core – *L. pungens***
          16. Group C *L. wallisii*
            17a. Group C *L. erythroneuron*
            17b. Group C *L. subintegrfolium*

The Group P of putative primitive species (because similar to *Leptodontiella*, *Streptotrichum* and *Trachyodontium* in the critical laminal papillae morphology) is restricted to Entry 4a, and are involved in generation of the Group L and Group S lineages, although two species (*L. luteum* and *L. tricolor*) are apparently rather specialized dead ends.

One species, *L. interruptum*, fits nowhere well because of the trait antrorsely papillose calyptra, which it shares with *Leptodontiella*, *Streptotrichum*, and *Trachyodontium*, and thus may be interpreted as a remnant of an ancient lineage now with reduced traits that forced a fit via cladogram with Group M; but we can now isolate it as a monotypic Group A. The same applies to Group R containing the odd *L. stellatifolium*. *Trachyodontium* is simply too specialized to be a progenitor; this is because there would be at least three reversals of significant traits (see discussion of this genus). There are three "Unknown" progenitors inserted to minimize reversals and thus grease the slide of gradual evolution. Although *L. viticulosoides* is worldwide in distribution, it is autoicous and anisosporous, and a reversal to the dioicous, isosporous condition of *L. excelsum* seems extreme. On the other hand, *Leptodontium excelsum* is restricted to the New World but widespread there, and doubtfully the ancestor of the other two in Group L, while *Lepdontium scaberrimum* is highly specialized.

The next task is to add traits to each species or species group, those known or those unknown and inferred as missing links. At this time nomenclatural changes may be made to group species in dissilient genera when possible or at least segregate distinctive species and groups.

Table 9-1. Fully annotated analytic key for Streptotrichaceae, with Pottiaceae as outgroup. Vertical lines connect the immediate descendants from one progenitor.

Taxon	Bits	BPP
1. Pottiaceae — Progenitor. Calyptra usually smooth, or if papillose then simple-punctate not antrorse; laminal cells commonly bulging; papillae crowded and multiplex/thickened, rarely absent or simple; tomentum if present ending in elongate cells; perichaetial leaves not differentiated or uncommonly enlarged; peristome of 32 twisted (unless reduced) rami in pairs, basal membrane usually present; annulus of 2–3 rows of sometimes vesiculose cells, occasionally revoluble. Generative species for Streptotrichaceae not clear, perhaps *Barbula eubryum, Streblotrichum* sp., or *Ardeuma* sp.	Outgroup	
2. Streptotrichaceae — Tomentum thin but ending in short-cylindric cells (1); perichaetial leaves strongly differentiated, sheathing the seta (1); peristome of 32 straight rami primitively grouped in 4's (1), teeth primitively spiculose (1); annulus of 2–4 rows of weakly vesiculose cells, not revoluble (1); primitively antrorsely papillose calyptra (1).	6	0.98
3. Core *Streptotrichum ramicola* — Flat or weakly convex surfaced laminal cells (1); papillae very small, simple to bifid, crowded (1); leaves sheathing basally (1); peristome basal membrane short (1). .	4	0.94
4a. Core *Williamsiella araucarieti* — Tomentum arbusculate (1); leaves broadly channeled distally (1); costa ending before the apex (1), abaxial stereid band layers 1–2 (1); distal laminal border not differentiated (1); distal laminal cell width 9–11 µm (1); asexual reproduction by gemmae borne on stem (1); capsule stomates present (1); annulus of 4–6 rows (1); peristome of 16 pairs of teeth, spirally striate and low spiculose (1), basal membrane absent (1); calyptra ornamentation unknown but probably smooth.	11	0.999
5a. *Williamsiella tricolor* — Tomentum arbusculate or arising from stem in lines, deep red (1); distal lamina bordered by 1 row of epapillose cells (1); basal laminal cells strongly differentiated, inflated, hyaline (1); basal marginal cells of leaf forming a strong but narrow border (1).	4	0.94
5b. Core *Williamsiella aggregata* —Tomentum absent (1); leaves recurved distally (1); distal laminal cell walls often thickened at corners (1); basal cell stripes absent (1); distal laminal cells very small, ca. 7 µm wide (1); peristome teeth spirally striate or smooth (1).	6	0.98
6a. Core *Leptodontium* Unknown — Tomentum arbusculate and usually without short-cylindric cells (1); stem lacking hyalodermis (1); leaves serrate, not distantly dentate (1), laminal cell lumens commonly strongly angled (1), larger (1).	5	0.97
7a. *Leptodontium viticulosoides* — Autoicous (1); short-cylindric cells ending tomentum very rare (1); spores of two size classes (1).	3	0.89
7b. *Leptodontium scaberrimum* — Leaf marginal decurrencies elongate, red, as wings on stem (1); laminal cells enlarged (1) and with coroniform papillae (1) (by IRCI = 3 bits); leaf apex narrowly blunt as a unique ligule (1); leaf base lacking stripes (1).	5	0.97
7c. *Leptodontium excelsum* — Leaves often with rhizoid initials near apex (1); leaves polymorphic (1); limited to New World (1).	3	0.89

6b. *Williamsiella lutea* — Large plants, stems to 20 cm, leaves 4–8 mm long, costa 8–10 rows of cells across at midleaf (1); leaf margins often dentate to near base (1).   2   0.80

6c. Core *Stephanoleptodontium longicaule* — Distal laminal cells with centered (coroniform) group or circle of spiculose simple or bifid papillae (1) and lumens bulging (1) (by IRCI 3 bits).   3   0.89

    8a. *Stephanoleptodontium syntrichioides* — Stem hyalodermis absent (1), gigantism: leaves larger, 4–7 mm (1), distal laminal cells larger, 13–17 µm (1); New World only (1).   4   0.94

    8b. *Stephanoleptodontium capituligerum* — Dense tomentum with short-cylindric cells (1); leaves with sharp apices (1) and well-demarcated hyaline fenestrations of thin-walled, hyaline basal cells (1).   3   0.89

        9a. *Stephanoleptodontium latifolium* — Stem hyalodermis cells reduced in size to that of next innermost layer of stereid cells, often apparently absent (1); leaves strongly decurrent (1), costa blackened and thickened at base (1).   3   0.89

        9b. *Stephanoleptodontium stoloniferum* — With leafless branches in axils of distal leaves (1) and gemmae restricted to leafless branchlets (1) (by IRCI 3 bits); gemmae elongate-elliptic (1).   4   0.94

    8c. *Stephanoleptodontium brachyphyllum* — Leaves shorter, 2–3 mm long (1), ovate to short-lanceolate (1), distal laminal cells smaller, 9–11 µm wide (1).   3   0.89

        10. *Stephanoleptodontium filicola* — Leaves distally plane to weakly keeled or broadly channeled (1), leaves dimorphic (1): when unspecialized, leaves ovate and bluntly acute with costa ending 6–8 cells before apex (1), specialized in strongly gemmiferous portions of the stem as shortly acuminate-lanceolate, costa subpercurrent, apex blunt, marginal teeth mostly near apex (1); specialized leaves appressed over the stem-borne gemmae when wet and open-catenulate when dry (1).   5   0.99

4b. *Austroleptodontium interruptum* — Stems short (1); tomentum thin (1); leaves small, 1.8–2.2 µm long (1), costa excurrent as an apiculus (1); distal laminal cells small, 9–12 µm wide (1); axillary gemmae present (1); peristome of 16 paired rami, smooth (1), basal membrane absent (1).   8   0.996

4c. *Leptodontiella apiculata* — Leaf base with hyaline area (1); stripes on basal cells occasional (1); calyptra antrorsely papillose (1).   3   0.89

    11. Core *Microleptodontium* Unknown. — Stems short (1); hyalodermis absent or very weakly differentiated (1); leaves ligulate (1), less than 3 mm (1), costa narrow (1); gemmae common (1); calyptra smooth (1).   7   0.99

        12a. *Rubroleptodontium stellatifolium* — Leaf margins entire (1); laminal cells strongly bulging on free surfaces (1), papillae multiplex (1); basal cells red (1); brood bodies on rhizoids sometimes present (1).   5   0.97

        12b. *Microleptodontium flexifolium* — Leaf base weakly   3   0.89

differentiated (1); laminal papillae simple, hollow, scattered over the flat or weakly concave surfaces of the lumens (1); gemmae abundant on stem (1).

13a. *Microleptodontium umbrosum* — Excurrent costa (1) and gemmae restricted to excurrent costa (1) (by IRCI 3 bits); leaves dimorphic (IRCI two traits).	4	0.94
13b. *Microleptodontium stellaticuspis* — Leaf ending in unique terminal dentate cup (1) and gemmae and rhizoids borne in cup (1) (by IRCI 3 bits).	3	0.89
13c. *Microleptodontium gemmascens* — Excurrent costa (1) and gemmae restricted to excurrent costa (1) (by IRCI 3 bits); restricted to British Isles (1).	4	0.94
4d. Core *Trachyodontium* Unknown — Basal laminal cells with longitudinal brown to red stripes (1); short-cylindric cells often ending tomentum (1); gemmae absent (1).	3	0.89
14a. *Trachyodontium zanderi* — Leaves with cartilaginous border of elongate cells (1); peristome further divided into 64 rami (1); operculum of short cells with rounded lumens (1).	4	0.94
14b. Core *Crassileptodontium pungens* — Basal cells with miniwindows except in specimens with very thick-walled basal cells (1); distal laminal cells with crowded, solid, knot-like papillae (flower-like) (1); basal portion of costa reddish (1); peristome of 16 paired rami (1), smooth or slightly striated (1).	6	0.97
15. *Crassileptodontium wallisii* — Leaf base high, to 1/3 leaf (1) and basal marginal cells inflated (1) (by IRCI 3 bits).	3	0.89
16a. *Crassileptodontium erythroneuron* — Costa red throughout (1); marginal leaf teeth lacking or much reduced (1), leaves straight distally (1); leaf apex concave, blunt (1).	4	0.94
16b. *Crassileptodontium subintegrifolium* — Laminal teeth absent (1); costa ending 1–10 cells before apex (1), concolorous (1).	3	0.89

The analytic key reflects 130 trait transformations (including the IRCI interpretation for *Crassileptodontium wallisii*), equivalent to 130 bits of information on evolutionary descent. There are no negative bits representing major reversals. Although the cladogram of maximum parsimony of this group was 116 steps in length, comparison is difficult because the latter only includes shared traits, while the analytic key includes both shared and unique traits. If we subtract from the total trait transformations the 56 trait transformations that end in one terminal taxon (equivalent to autapomorphies in cladistics), then there are only 74 traits shared by a progenitor with one or more descendants involved in the analysis. The low number of shared trait transformations apparently reflects the elimination of a large number of phylogenetic methodologically required unknown, unknowable, putative shared ancestors. The three Unknown inferred ancestors are placeholders for extinct or unsampled species that may yet be discovered, and are, as such, scientifically predictive.

Examination of the data set yields an immediate difficulty. The basal genera *Leptodontiella*, *Streptotrichum*, and *Trachycarpidium*, and the two species *Williamsiella aggregatum* and *Rubroleptodontium stellatifolium* all lack stomates, yet at least the first three share the primitive and unique trait (vis-à-vis the Pottiaceae outgroup) of elongate, spiculose peristome teeth arranged in eight groups of four rami on a basal membrane. This peristome form is reduced and striate or smooth in other species of the family.

We then have the Dollo problem that we must choose between re-evolution of the complex peristome teeth or of the stomates. In this case the expression of the stomates are considered governed by some epigenetic factor that is itself repressed in more advanced taxa of the Streptotrichaceae, particularly as *P. aggregatum*, lacking stomates or these rudimentary, is rather clearly derived from *P. araucarieti*, which has stomates and also has weakly spiculose and spirally striate peristome teeth. A cladogram with *P. araucarieti* as (functional) outgroup simply moves the root with no illumination of a different nesting more optimal for minimizing parallelisms and reversals. (Software, as yet uncoded, might help with this.)

It is possible that the rather odd differences in the peristomes of the basal genera mentioned above may imply different lineages, these three taxa remain in the same relationships as in the original cladogram, with *Streptotrichum* and *Trachyodontium* as sister groups, and *Leptodontiella* embedded in the cladogram at the base of *Microleptodontium* species. In addition, the peristomes of *Trachyodontium* and *Leptodontiella* are unique in numbers of rami, and this may simply reflect the long isolation of two examples of unusual neutral traits or experiments in adaptation in an uncommon arboreal environment (shrubs and tree branches at high elevations).

In the Pottiaceae, the only genus lacking stomates is *Aschisma*, which also have laminal papillae centered on a bulging lumen, and two exposed stereid bands in the costa, but the bulging papillose laminal papillae are characteristic of the general *Weissia* assemblage of genera, the stereid bands are about the same size and are substereid, there is no sign of short-cylindric cells ending the few rhizoids present, and the capsule is extremely reduced to cleistocarpy with seta nearly absent. Although "stomates present" was considered pleisiomorphic (Zander 1993: 24), *Leptodontiella*, *Streptotrichum* and *Trachyodontium* are clustered (Zander 1993: 46) with *Leptodontium* deeply in the Pottiaceae near the *Barbula* assemblage, while *Aschisma* is placed far from these, at the base of the *Tortula* clade. I conclude that "stomates absent or rudimentary" is an advanced, specialized trait found today as revenant of a mostly extinct lineage of much reduced morphology.

The caulogram — The analytic key may be summarized by a diagram of direct descent, the caulogram (Fig. 9-1). Lineages are indicated by squared balloons and connecting lines. Given are taxon names and Bayesian posterior probabilities of the evolutionary order that one taxon is descendant of the one given before it. Dissilient genera are easily made out as a core generative species plus descendants; monotypic very distinct genera are not dealt with by the dissilient species concept, and classical justifications must be used for their distinction. There are four postulated Unknown "missing links" tagged with asterisks (*). The genus *Williamsiella* is as low in the caulogram as is the monotypic genera *Austro-*

leptodontium, Streptotrichum and *Trachyo-dontium,* but is young enough that its three descendant species remain extant, while one of these descendants is itself generative of two dissilient genera. The generalization that core generative species generate the next core generative species, made with study of the moss genus *Didymodon* s.str. in mind (Zander 2016a: 330), clearly does not apply here. The direction of evolution is away from Pottiaceae towards elaboration (e.g., *C. erythroneuron, L. scaber-rimum, P. luteum*) or reduction (e.g., *A. inter-ruptum, M. gemmiferum, S. filicola*). The Strepto-trichaceae is supported by 130 informational bits, which include postulated missing links. The 28 species of the family are supported individually by 113 bits in the range (2–)3–5(–11) bits, where the average is 4.04 bits.

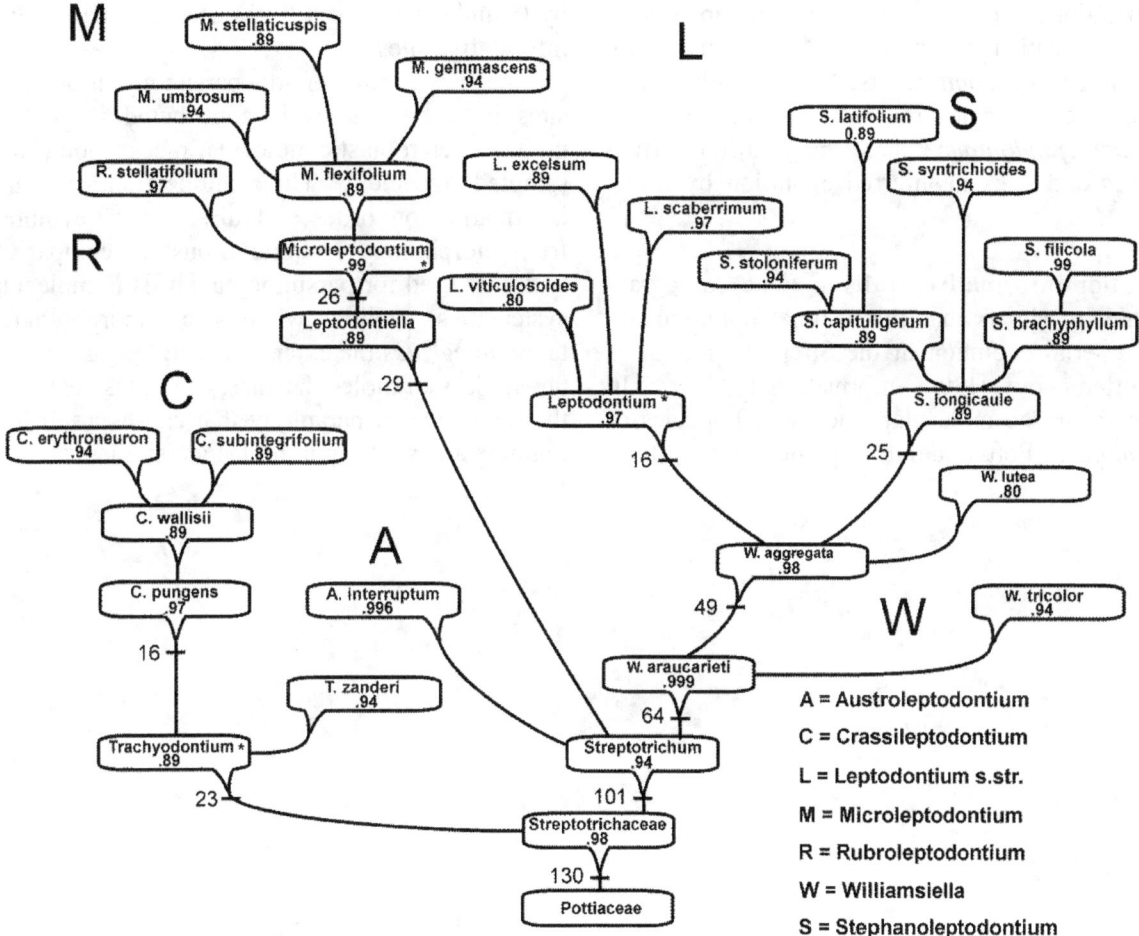

Figure 9-1. The caulogram of Streptotrichaceae. This summarizes justification for the evolutionary coherence of Streptotrichaceae, and for splitting *Leptodontium* into seven genera (in side bar), with their general positions noted by capital letters. Bayesian posterior probabilities give support for order of direct descent, that is, of one taxon being the descendant of the one before it in the lineage. Numbers on lines give numbers of bits supporting major lineages, adding all distal bits of individual species together. Three Unknown missing links are necessary to minimize reversals and redundancy of parallelisms. Unknowns are identified with asterisks (*) and may be interpreted as extinct taxa or predictions of taxa yet undiscovered.

The *Microleptodontium* postualted core species generates one rather old taxon, *Rubroleptodontium stellatifolium,* and another descendant that is probably recent as it has four direct descendants of its own. The genus *Leptodontium* s.str. requires three separate unknown origins for the species, two of which are evidently rather ancient because of their odd traits and one recently evolved, being common in the New World but known nowhere else. *Crassileptodontium* is somewhat isolated by combinations of rare traits and it apparently connected with the remainder of the family only through *Trachyodontium,* itself sufficiently specialized to require a negative informational bit. Only *Stephanoleptodontium* has no disconcerting oddities and gaps in inferred evolution by direct descent.

Evolution-informative traits — The caulogram graphically shows that this estimated maximally nested serial evolution of the Streptotrichaceae is supported by 130 informational bits. In comparison, the molecular study of 30 species of *Didymodon* (Pottiaceae) by Werner et al (2005) had 91 parsimony-informative sites. The study of 129 exemplars of haplolepideous moss families by Hedderson et al. (2004) used 269 parsimony-informative sites. The study of *Barbula* (Pottiaceae) by Kučera et al. (2013) separately analyzed three sequences, each with about 90 exemplars, using rps4, rps4+trnM-trnV and ITS, with 196, 157/260, and 826 parsimony-informative sites, respectively. The analysis of 46 exemplars, largely of *Tortella* and *Pleurochaete* (Pottiaceae) by Grundmann et al. (2006) used 223 parsimony-informative sites.

The large number of parsimony-informative sites in some studies is contributed in part by multiple heteroplastic molecular races among well-sampled species. The implication is that information on order and direction of evolution from morphology is quite robust in comparison with that used for parsimony analysis in molecular systematics studies, while each morphological taxon *integrates* the extant, unsampled, and extinct heteroplasy of molecular race,s and thus deals with the phylogenetic paraphyly that causes artefactual lumping and splitting in molecular studies.

Chapter 10
RETROSPECTIVE ON MOLECULAR ANALYSIS

The evolutionary tree — Based on work with several groups (Zander 2008, 2009, 2013, 2014a,b.c), caulograms may be expected to have a number of clusters of progenitor taxa and the descendant taxa of each. Each cluster will have one to several descendant lineages, each lineage being usually 1 to 3 or 4 taxa in length, excepting a new lineage involving generation of a new cluster of taxa.

A mathematical binary tree is not a cladogram or a caulogram. In binary trees, each node is named (or at least numbered), but in cladograms only the terminal ends of the tree are named and each internal node is labeled only as an immediate shared ancestor of two siblings. In a caulogram, the branching is not necessarily dichotomous; in cladistics a multichotomy is taken as a lack of resolution. A caulogram may be simply seriate or truly multichotomous, and internal nodes are usually named in the same manner as terminal taxa.

So … what is wrong with molecular clado-grams, other than being more akin to cluster analysis (with likewise have no names for nodes) than to caulograms? Is there not more data in molecular analysis? What about the high or even certain Bayesian support measures?

Molecular systematics — Certain salient problems bedevil interpretation of molecular cladograms:

First, the morphotaxa listed at the termini of the cladogram include all molecular races and strains yet only one or just a few strains may be analyzed. ("Morphotaxon" is a shorthand meaning taxa defined by expressed traits.) The taxon names do not include extinct strains that may be themselves paraphyletic. Although tracking traits may have great support they may not track the whole morphotaxon, which may appear in more than one place in the molecular cladogram. The heteroplasticity of divergent molecular races is never evaluated as a hypothesis against the homoplasticity of morphological convergence, but is instead shunted to the side as "incomplete lineage sorting" or "attainment of reciprocal monophyly" (Hudson & Coyne 2002). Divergent infraspecific genetic elements were demonstrated in White Pine and other species by Moreno-Letelier and Barraclough (2015), who termed them "mosaic patterns of genetic differentiation." Those divergent molecular races may separately generate one or more new species each. A molecular cladogram node does not eliminate other taxa from relevance, but a node in a morphological cladogram does, i.e., molecular clades are not closed causal groups. Although one may argue that postulating extinct, presently invisible, molecular races as the cause of incongruence while morphological convergence is in general demonstrable and obvious, molecular races that separately generate other species are demonstrable in molecular studies with many samples of each species. A scientist *must* find a reasonable solution to incongruence that deals with all available data, both morphological and molecular.

This problem is part of a larger statistical difficulty called cherry picking of data or "selective windowing" (Levitin 2016: 160). One aspect of a solution is adequate sampling on the molecular side, easier now than in the past. It is usually admitted that molecular analyses are bedeviled by small sample problems, initially because of the costa of DNA analysis early on. Determining what is an adequate sample is sometimes facilely passed off as easy, where "30" is the magic number for simple distributions with one maximum. With molecular analysis, however, adequate samples are often considered those that exhaust variability. This might be fine if this were commonly accomplished, but such is seldom the case. Using more sophisticated methods to determine sample size can be difficult. For instance, something that needs correct estimation is the statistical "base rate" for heteroplasy (generation of multiple molecular races that generate their own lineages).

Second, as with morphological cladistics, the nodes are subsets of the data set that match no known taxon. The nodes are characterized only by the set of molecular traits common to sister-group terminal taxa distal on the cladogram. Continuity between nodes in a molecular cladogram is at least as far from clarity in modeling evolutionary descent as is the same situation when using morphological data in a cladogram. Support for phylogenetic clades may be high or statistically certain, but if the model is not of direct evolutionary descent, then the support is not evidence for evolutionary descent (see Gunn et al. 2016).

Third, evolution proceeds through the phenotype. Molecular analysis can track at least molecular races and probabilistically cluster by advanced shared traits. But, because progenitors show a great degree of stasis, direct descendants can be morphologically identified as such. This is because in analysis by direct descent, advanced traits can be identified and taxa arranged serially in a sometimes branching caulogram, or evolutionary tree of direct descent. This is not the case with molecular data, because both descendant and progenitor continue to mutate at about the same rates. With molecular data, only shared traits are informative, but with morphological data both shared and unique traits are informative.

Fourth, because only shared traits are informative of evolutionary relationships in cladistics, there can be no information on derivation of one sister group from another, particularly when molecular races are separated from one another on a cladogram. Thus, any two morphotaxa terminating a clade cannot be polarized as progenitor and descendant on the basis of shared traits. Additionally, the two taxa at the base of a cladogram cannot be polarized as progenitor and descendant on the basis of shared traits. The groups separated by a node inside a cladogram (two or more nodes from a terminal taxon) are also not polarized as progenitor or descendant groups, although for recently evolved taxa (with no or little expected heteroplasy) the order of branches to terminal taxa on a molecular tree may be informative of the order of speciation.

Fifth, it is sometimes stated that there are more molecular traits than morphological traits, and therefore combining all traits in a joint analysis justifiably emphasizes the molecular. Now, hang on there! We have an apples and oranges situation here. Morphological traits define the whole species, though variable and the description applies to infraspecifically variable species. Morphological traits are, however, not those of molecular races of species. Surely, if variation in morphology can be

evaluated as characteristic of species, then heteroplasy (generation of widely divergent molecular races that that generate paraphyletically separate evolutionary lineages) should be saluted as expected variation in classical taxon concepts in molecular systematics. This might end the seemingly unstoppable recognition of unjustified cryptic species, genera and families (e.g., see the apologistic of Adams et al. 2014), and end the random lumping of the well-established taxa that is characteristic of molecular mechanical taxonomy. Strict phylogenetic monophyly imposed on paraphyletic patterns of molecular cladograms is the source of the kerfuffle (Flegr 2013) over the correct name of the model system once called *Drossophila.*

Sixth, molecular sequences are not necessarily immune to apparent convergence. Studies that show improbable but real convergence of independent molecular data between species include those of Projecto-Garcia et al. (2014), Castoe et al. (2009), Parker et al. (2013), Christin et al. (2007), and various studies of the RubisCO gene that shows convergence. Multiple sequence analyses are intended to deal with these anomalies.

Some molecular studies are comparative paragons of analysis in that they include many exemplars of the same species, the inclusion of which tests to some extent for paraphyly (a constant evolutionary property of a progenitor species or group) in older species, and also morphological analysis resulting in distinctions that can stand alone and are not merely traits that happen to match the molecular clustering. Two that are of particular interest in the Pottiaceae (mosses of harsh environments, Zander 1993) are the study of *Barbula* and related taxa by Kučera et al. (2013), and of *Oxystegus* by Köckinger et al. (2010). Each of these show divergent molecular races that look in the cladogram like a broom with the dust bunnies of separately derived other species embedded in it.

This is *heteroplasticity,* the molecular obverse of the homoplasy associated with morphological convergence. Directional selection (powered by stabilizing and purifying selection in a gradually changing environment) can encourage homoplastic convergence of several morphospecies into phenocopies occupying a single niche. In the absence of strong selection, molecular races of a single taxon may diverge massively as heteroplasticity of the molecular taxonomic units. Usually there are several conservative traits that allow proper taxonomic disposition of convergent morphospecies, but the only conservative traits that may be used to properly meld divergent races are those of their morphospecies.

Molecular taxa include those that may be difficult to distinguish from morphological taxa but have distinctive molecular signatures (Renner 2016). The present study considers purely molecular taxa simply as molecular races of morphological taxa that are due to isolation of the races leading to gradual regional accumulation of DNA distinctions. These distinctions can be used for tracking microevolution without concern for new adaptive features that may signal macroevolution (taxon-taxon transformations). Molecular tracking traits should not define a taxon, or if they are made to do so, what good are they? They describe no evolutionary process, as possible selection and adaptation is ignored. (This may be why the reader may find in the phylogenetic literature assertions that most evolution necessarily consists of gradual cladogenic splitting of anagenetically gradually transforming species. In such case, an ancestral species gradually and supposedly transforms into two daughter species while anagenetically disappearing, a process termed "pseudoextinction.")

Differential phyletic constraint in expressed traits of morphology may be the reason for molecular paraphyly. If we define phyletic con-straint as the restriction on possible morphological transformations of a lineage caused by a certain amount of adaptive selection, then the lack of such a restriction on molecular transformations means that the tail end of the probability distribution for DNA base changes may introduce larger possible patristic distances than morphological changes on the same cladogram. An ancestral taxon with a distinctive morphology may, over time, have several rather extreme DNA signatures in molecular races, some or most of which races are extinct. Thus, from one ancestral taxon, many molecularly distant directly descendant taxa may be extant.

Enroth (2015) wrote: "From my own experience I can say that molecule-based groups are sometimes morphologically nearly undefinable, but geographically consistent, so there must be lessons there to be learned of moss evolution." Geographic distinctions without much morphological change is consistent with a theory that such is infraspecific transformation, e.g., local biotype and neutral DNA differentiation through genetic drift. In macroevolutionary systematics, we can define a species empirically as a set of individuals that are grouped by two or more expressed traits sustained together by some often unknown evolutionary process, like drift after isolation or selection pressure, and not explainable as chance distribution of independent or quasi-independent traits in one species.

Molecular races

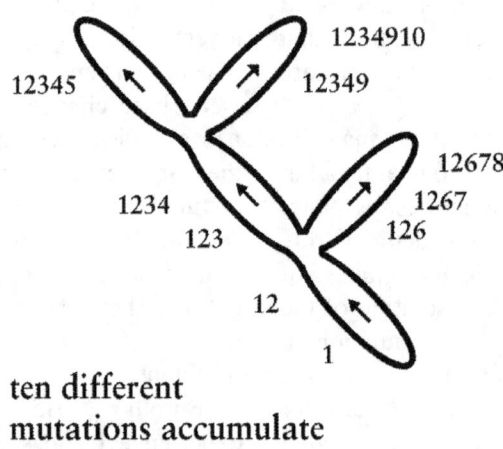

ten different
mutations accumulate
over time

Figure 10-1. Heteroplasy diagram. Molecular races, identified by numbered (1 through 10) new mutations, sometimes occurring serially in time and sometimes splitting in an evolutionary tree (a caulogram). As elsewhere, this is a contrived tree for demonstration purposes. Arrows indicate direction of gradual accumulation of mutations through time. The analysis is microevolutionary in that the same result occurs whether some or all molecular races are attached to different morpho-species. A macroevolutionary analysis deals instead with transformations between taxa.

Let us take a basic, simple example of a molecular *evolutionary* tree. Figure 10-1 is a contrived caulogram of molecular transformations of tracking traits in a hypothetical evolving gene pool. Suppose we have an outgroup with trait 1 and no others of the above that lets us hypothesize what the direction of evolution is. The best tracking traits are those that do not change because of selection, i.e., they are evolutionarily neutral or nearly neutral. Ten neutral mutations numbered 1 through 10 gradually accumulate in the populations, and various combinations are here termed molecular races. Note that trait 1 is found in all molecular strains; so is trait 2 which newly mutates just after trait 1. Next there is an isolation (but not necessarily speciation) event. The first two branches are distinguished in that traits 3 and 4 appear (sequentially) in the left branch, while 6, 7 and 8 appear in the right. The next isolation event places trait 5 in the left branch and 9 and 10 in the right. Because there are many spots for mutations to occur in a DNA sequence there is little overwriting and old traits usually remain as the populations transform themselves through

mutation. Those old traits can be used to track direction and order of new molecular trait accumulation through mutation. This is an exciting dimension of modern molecular analysis, but there are problems using the method in systematics.

The major problem is that not every one of the molecular races in any one molecular lineage survives and is presently extant. This yields cryptic paraphyly, a source of bias. That is, while homoplasy is common feature of morphological evolution, molecular heteroplasy due to paraphyletic races that individually speciate. Divergent molecular races that may speciate separately (Fig. 10-2) is a problem usually ignored by phylogeneticists as a problem of incomplete lineage sorting that will settle into something straight-forward (sort itself out) as time (geologic) goes by. Molecular races may be present and sampled or not, or mostly extinct. The data set in most molecular analyses should be expected to be incomplete. What would a *cladogram* of this caulogram look like if just molecular races 1, 1267, 12345 and 1234910 survived? Say X is the outgroup also with trait 1.

Thus: 1,1(1267(12345, 1234910)).

What about if 1, 12678, 1234, 12345 survived?

Thus: 1,1(12678(1234, 12345)).

How about 1, 123, 1234 and 12345?

Thus: 1,1(123(1234, 12345)).

So, cladograms cannot represent serial evolution with only data on shared descent. They can only imply in a general manner serial evolution through clustering of taxa with shared traits. What if all of the races survived to the present and were analyzed cladistically—what would the cladogram look like? The reader might develop such a cladogram as an exercise. Clearly the internal portion of the caulogram labeled 123 and 1234 is not the same as an internal branch in a cladogram. More discussion of the problems of molecular cladograms, including self-nesting ladders, is given by Zander (2013: 31ff.). Thus, to avoid false nodes on clado-grams, phylogenetic analysis requires extinction of all ancestral taxa in morphological analysis and all ancestral molecular races in molecular analysis. This requirement is doubtfully achieved in practice. In sum, cladists prefer considering morphological homoplasy due to parallelism rather than equally possible molecular heteroplasy due to persistent molecular races.

So what good is molecular systematics? — If taxa are young, there may well be little extinct paraphyly, or it may be limited to very short phylogenetic distances. In this case, a clade may simply be nested set of molecular races that match well morphospecies. This is excellent though expected. If taxa are older and confounding extinct paraphyly is possible, then (1) phylogenetic (patristic) distance if larger than expected for extinct paraphyly (as judged from extant paraphyly for the group) is informative, and (2) heterophyly (paraphyly and short patristic distance phylo-genetic polyphyly) can bracket another taxon with implications that the paraphyletic taxon is ancestral. If exemplars of a single taxon (species, genus, family) are discovered to be distant on a molecular cladogram, for them to be described as new, they need to stand on their own as morphotaxa. Even the best molecular studies,

however, if not corroborated by a morphological analysis that can stand on its own, have the two basic problems that sister groups (for instance, two terminal taxa) cannot be ordered, and self-nesting ladders (Zander 2013) may reverse all or parts of the order of a lineage. (The term "morphospecies" in this book means "distinguished by fixation of expressed traits involved in the different processes of evolution, including selection and adaptive radiation, in addition to neutral processes like drift.")

Taxic waste — Entirely or mostly cryptic mole-cular species, genera and families are rather useless in the study of evolution, in large part doubtless being the recrement (dross, refuse) of past evolutionary experimentation that found no niche into which they might evolutionary adapt via phenotype transformation. It is possible that molecular systematists might name all extant races as different "molecularly cryptic" species whether they are morphologically somewhat different or not. Corroboration, on the other hand, must come from molecularly coordinated morphological taxa that can stand on their own.

Naturally, a cladogram and a caulogram are not directly comparable, yet a morphological cladogram can be converted to a caulogram by naming nodes when possible, and collapsing into a nested, sometimes branching, serial set of taxa, with only a few unknown shared ancestral taxa tacked in. Modifying the interpretation of molecular trees to reflect a somewhat less parsimonious result often conciliates morpho-logical and molecular analyses. But, explaining away contrary morphological results by invoking convergence and delayed lineage sorting in the context of enforced optimality of analytic dichotomization (the desideratum of a resolved cladogram) and pseudoextinction leads only to a myriad cryptic taxa or morphologically indescribable lumpings of taxa, i.e., taxic waste. Each taxoid founded on molecular races without accompanying morphological traits that stand alone to corroborate evolutionary distinctiveness confounds, confuses, and misleads biodiversity study in a time of major extinction.

Repeatability in science is repeatability of evidence supporting an inferred process in nature. This means repeated resampling of all relevant material, i.e., worldwide sampling as is done in morphologically based monographs. It does not mean analyzing different DNA sequences of the same one or two samples, or using different software, or different studies using in part the same accessions in GenBank. Repeatability is also not the only criterion for correct scientific analysis. The resulting same *cladogram* obtained every time from different phylogenetically informative (shared descent) data sets processed by various boutique phylogenetic software may be wrong if other methods analyzing the same evolutionary process with different evolutionarily informative (direct, serial descent) data can create different *evolutionary* scenarios.

New methods of species delimitation have been advanced by molecular systematists (reviewed by Flot 2015). These coalescent-based techniques (Leaché et al. 2014) commonly require rather narrow assumptions, such as the biological species concept. Flot warns non-taxonomist end users that they may be challenged by having to check their study populations for molecular cryptic diversity as a prerequisite for other studies. A naïve (ignoring the requirement of total data relevance) use of such methods may further dis-integrate the morphological and molecular scientific boundaries.

Summary of why morphological analysis is more informative than molecular analysis —

Morphological analysis — In theory, morphological traits (and other expressed traits) are subject to selection. Morphological taxa are commonly characterized by *punctuated equilibrium*, where races begin alike after isolation and one of them undergoes differential selection in a different environment. One taxon, which we can call the progenitor, remains in morphological stasis through continuing stabilizing selection in a weakly changing environment, and the other transforms into a different species adapted to a different environment and then goes into stasis itself with difference balancing selection. Comparison of kinds of adaptive traits can reveal direction of evolution between ancestor and descendant taxa.

Molecular analysis — Differentiation of isolated molecular races are genuinely characterized as *pseudoextinction*, in which two races through isolation differentiate by accumulated mutations in tracking traits not or weakly subject to selection. Because both races continue to mutate, they diverge from the common ancestor which as a molecular race ceases to exist. This difference requires that molecular races be analyzed by shared traits alone, because any two sister taxa cannot be distinguished as progenitor and descendant by comparing differences in nonselective, nonadaptive tracking traits.

Therefore, cladograms only show groups of taxa based on shared traits optimized by a minimum set of character transformations. If a cladogram is pectinate (a comb), then the two terminal sister taxa cannot be polarized as progenitor and descendant by shared traits alone, and neither can the two basalmost taxa. If a cladogram is much branched, then it may be that none of the taxa are in closely related pairs that can be distinguished as ancestor and descendant. In addition, (1) morphologically distinguished taxa in a caulogram are arranged such that critical trait transformations are between two individual taxa, but in a cladogram, important differences are between groups as parts of clades; and (2) extinct molecular races can lead to molecular paraphyly and polyphyly, while extant molecular races can lead to misinterpretation as cryptic species or genera or families.

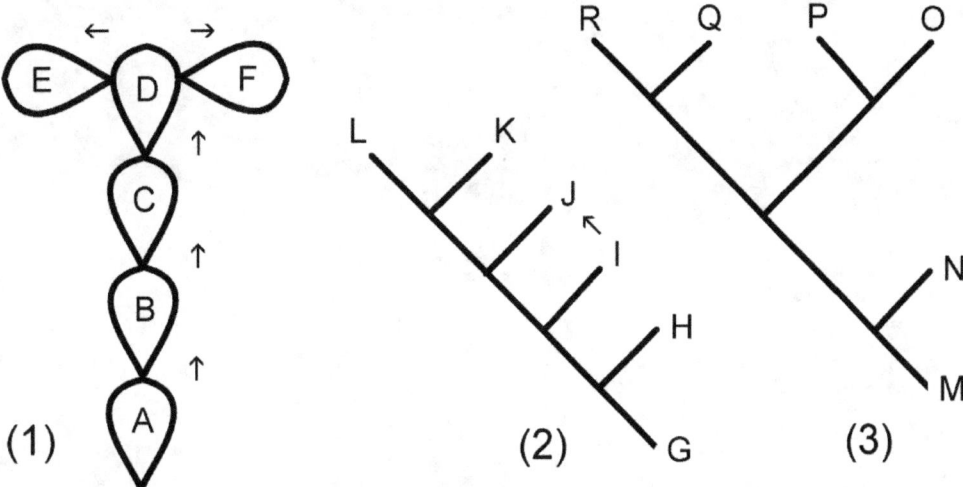

Figure 10-3. Comparing direct descent modeled on (1) morphologically based caulogram versus molecularly based (2) pectinate and (3) fully branched cladograms. Direct descent relationships of all pairs of morphotaxa are modeled in (1). Only one molecular race pair is related by direct descent (see arrow in cladogram 2), with G as outgroup for polarization. No pairs of molecular races are modeled in direct descent in cladogram (3).

Chapter 11
RETROSPECTIVE ON THE CLADOGRAM

Cladograms can produce quite different views of relationships than do caulograms.

Morphological cladograms — Let us say that taxon A is the direct ancestor of taxa B and C, and X is an outgroup similar to A. An outgroup can be a taxon that is a nearby direct ancestor or a branch of some nearby ancestor of the group studied. These taxa have the macroevolutionary formula

$$X, A > (B, C).$$

That is, ancestral taxon A is inferred to have generated B and then also C with the additional information from X.

Case 1. Consider the cladistic morphological data set where only shared traits are informative of relationship. In this case, the species share no traits.

C 00111
B 11000
A 00000
X 00000

This has taxon X for outgroup, it being quite similar to A (other traits not shown). A and B have each at least two constant independent traits sufficient to distinguish each as a species but C has 1 more trait than B. The data show no advanced *shared* traits and is cladistically uninformative. Given that the trait 0 is pleisiomorphic, there are no synapomorphies (0 cannot be a synapomorphy if it is plesiomorphic), and the cladogram of maximum parsimony is a simple multifurcation X, (A, B, C) from a central ancestor shared by A, B and C, diagrammed like the Greek letter psi, Ψ. The species might be evolutionarily nested if we had data that would allow serial nesting, say C deriving from B with 2 reversals, or B from C with three reversals, or more parsimoniously both B and C derived from A.

Case 2. Suppose B and C as derivatives of A *randomly* share one new morphological trait.

C 0111
B 1100
A 0000
X 0000

B and C are terminal on the cladogram, with 1 trait (the second) as synapomorphy. The cladogram is a dogbone: >—<, with A and X on one end and B and C at the other end. Taxa B and C show a shared relationship (because of one shared trait) on the other.

In Case 2 the cladogram is misleading in that an unknown shared ancestor is implied for B and C and that ancestor is not identified or identifiable from the cladistic data as taxon A if A were probably the ancestor of both as determined from other information.

Case 3. This last is more obvious when many morphological traits are involved, as in the case of B sharing two advanced traits with C:

C 00000000000000000000000000000111
B 00000000000000000000000000000110
A 00000000000000000000000000000000
X 00000000000000000000000000000000

where the similarities of B and C are minimal compared to their similarities to A. Again we get a dogbone cladogram >—< with A and X at one end and with B and C at the other end. B and C have an implied shared ancestor, but cladistically that ancestor cannot be identified as A. In fact, given that B has a zero branch length, and that zero branch lengths are often considered to imply ancestral status, B may be wrongly considered the direct ancestor of C.

Molecular cladograms — The situation with molecular cladograms is somewhat similar. Let us say that A as ancestral taxon similar to outgroup X generates sequentially in time B, C, D, and E. That is, formulaically X, $A > (^1B, {}^2C, {}^3D, {}^4E)$. The superscripts indicate theoretical order of speciation, when such can be inferred (or theorized).

Suppose that the time between speciation events is long enough for an investigated molecular sequence to gradually mutate. Thus, A generates B, then both mutate (differently) and A′ (a molecular race) generates C, then A″ generates D, and A‴ generates E. All are different molecularly but A, A′, A″, and A‴ are all molecularly heteroplastic offshoots of one morphological taxon.

Case 4. Suppose a sample includes only the latest geographic race of A, namely A‴. Consider a data set of this:

A‴   1111
E   1111
D   1110
C   1100
B   1000
X   0000

We get X, (B, (C, (D, (A‴, E)))) as a cladogram.

Case 5. If A were a surviving population of the original A, then the data set would be:

E 1111
D 1110
C 1100
B 1000
A 0000
X 0000

And the molecular cladogram would be X,(A, (B, (C, (D, E)))).

Case 6. If two populations of A were present, one being A and the other A‴, then the data set would be:

A‴   1111 where A‴ is here
E   1111
D   1110
C   1100
B   1000
A   0000 and A is also here
X   0000

The cladogram would be X, (A, (B, (C, (D, (A‴, E))))), with A and A‴ at opposite ends and paraphyletic. It would appear that A and A‴ are different taxa but they are simply two molecular races of one morphological ancestral taxon. If all molecular races of A survive and were sampled, they would all be separated on the cladogram. Thus, depending on which races survive extinction, an ancestral taxon can turn up anywhere and everywhere in a cladogram among its derivative taxa.

If most molecular analyses may have incomplete data sets (because extinct or unsampled races may be paraphyletic), then just how incomplete should we expect such data sets? A measure is the largest extant number of molecular races in cladograms of related taxa with a great deal of species-level sampling worldwide. The largest amount of infraspecific paraphyly in some one species is an indication of the amount of extinct paraphyly in more ancient related species.

The only value of molecular cladograms is distance between taxa greater than that which might be due to extinct paraphyly, and whenever heterophyly brackets one taxon with two examples of another taxon in which case a progenitor-descendant relationships is implied.

Trepidation — Species of the Streptotrichaceae have been little studied by molecular systematists. Given the problems caused by strict phylogenetic monophyly and the wrong assumption that a molecular race is the same as a species, future such study is worrisome. It is clear that molecular races exist and comprise the heteroplastic twin of homoplasy in morphological systematics, and that they may go extinct. *Heteroplasy* results when one species generates two or more species out of two or more molecular races of the same progenitor species. The progenitor species may be recognized as molecularly paraphyletic if the races are extinct. Extinct or unsampled molecular strains can destroy much evolutionary modeling as reflected in useless or misleading name changes when a molecular cladogram is used to generate a classification that shows some (only apparent) paraphyly or polyphyly due to molecular heteroplasy not morphological homoplasy.

When molecular cladograms end in a brush at several samples of one terminal species, that is, when all molecular lines originate at one point, this does not mean that all the specimens of the one species have identical sequences, though often they do. It means that the sequences may be different but no one matches another to make a unique sister group apart from the others. A "broom," however, with many branches originating at different points (therefore becoming a number of sister groups), does indicate that pairs of specimens or pairs of groups of specimens are sister groups, which can be evidence of sufficient isolation to imply beginning speciation.

Consider the cladogram of Alonso et al. (2016) shown in part in Fig. 11-1. This shows cladistic relationships of several genera of the moss family Pottiaceae. This study is of importance in that many specimens were sequenced (nuclear ITS and certain plastid markers) for *each* of several species. Molecular differences were demonstrated within these species whenever the cladogram did not represent the exemplars of each species as sister groups. Many species showed Bayesian support such that their molecular differences were so significant (so different that they represent isolation events) that the specimens probably represent stable molecular races. Branches that originated from the same node may or may not differ molecularly, but there were apparently no demonstrable shared traits among them.

For *Oxystegus minor,* the cladogram, using nested parentheses to represent sister groups, showed great Bayesian support (in bold face) for the molecular set (the 0.99 BPP distinguishes the *O. minor* set by separating its sister group), and statistical certainty that the two U.K. specimens have shared molecular signatures different from that of the Austrian specimen.

For *Oxystegus tenuirostris,* the cladogram demonstrated much well-supported nesting and the inclusion of *Ox daldinianus* and *O, recurvifolius,* two species, deeply within the *O. tenuirostris* group of molecular races.

For *Oxystegus daldinianus* (nrITS), the clado-gram shows considerable genetic differences among molecular races, and two species, *O. recurvifolius* and *O. tenuirostris* are embedded in the cladogram of *O. daldinianus.*

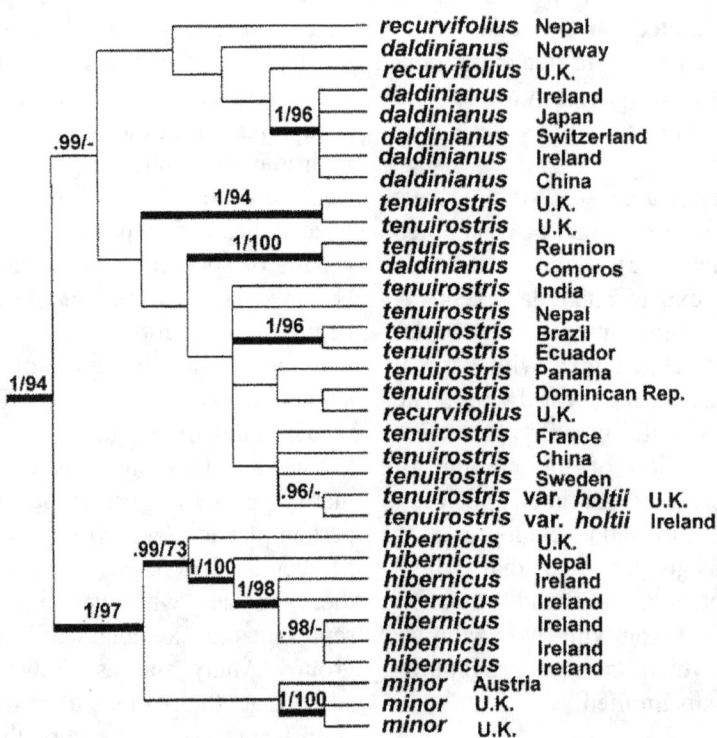

Figure 11-1. Demonstration of heteroplasy in molecular cladogram re-drawn from part of that of Alonso et al. (2016). The species of *Oxystegus* are shown on the right with the provenance of each sample. The cladogram on the left gives Bayesian posterior probabilities and maximum likelihood bootstrap values for nodes with strong support. Well-defined molecular races or strains are those with Bayesian support for some sharing differential traits within the species.

Many other species of Pottiaceae in this paper and in others (e.g., Cano 2009) demonstrate that extant molecular races as heteroplasy exist within them. The lack of representation of paraphyletic molecular races through non-sampling or extinction in a molecular cladogram can contribute confusion among cladistic relationships of extant species.

Thus, if taxa in the present paper are newly speciated, and molecular races are possibly not yet widespread, the order and pattern of speciation is probably secure from extinct paraphyly, and molecular results need to be considered. For example:

Microleptodontium flexifolium > (*M. gemmascens, M. umbrosum, M. stellaticuspis*)

This group should have no molecular contradiction attributable to extinct paraphyly because the three descendants are very closely related, clearly specialized offshoots of *M. flexifolium*. However, consider the caulogram formula:

Stephanoleptodontium longicaule > (*S. capituligerum, S. brachyphyllum, S. syntrichioides*)

These species are rather deep in the Streptotrichaceae caulogram (Fig. 9-1), and their relationships, though most parsimonious by direct descent based on morphological considerations, may be helpfully better informed by some future molecular study.

In addition, the Streptotrichaceae is probably derived from elements of the Pottiaceae, and is itself liable to synonymy because its recognition would make the Pottiaceae paraphyletic. Therefore, it is with trepidation that I await molecular phylogenetic analysis of the Streptotrichaceae, with its associated confounding principle of strict monophyly that rejects the idea that one taxon might give rise to another of the same or higher rank.

PART 2: TAXONOMY
Chapter 12
CLASSIFICATION OF STREPTOTRICHACEAE

STREPTOTRICHACEAE R. H. Zander, **family nov.** Type genus *Streptotrichum* Herz.

Leptodontieae Herz., Geogr. d. Moose 101. 1926.
Leptodontioideae (Herz.) Hilp., Beih. Bot. Centralbl. 50: 679, 1933.

Plants mats or tufts, yellowish or reddish green, (1–)2–4(–12) cm in length, with or without leaf buttresses. **Stem** transverse section rounded-pentagonal, central cylinder of equal-sized cells, central strand absent, sclerodermis usually present, hyalodermis present or occasionally absent or much reduced in cell size. **Axillary hairs** elongate, basal 1–3 cells firm-walled or brownish. **Tomentum** variously absent, weak, arbusculate, straggling-elongate and short-branching, in tufts on the stem in leaf axils or just below leaves or sometimes arising in longitudinal rows, cells ending tomentum elongate or a series of short-cylindric cells or some combination. **Leaves** when dry usually appressed and contorted or twisted, when wet wide-spreading to squarrose, about equal in size along stem or occasionally comose, ligulate to lanceolate, rarely linear-lanceolate, (1–)3–4(–9) mm, broadly channeled with a narrow groove along costa distally; **leaf margins** recurved in proximal 1/2–4/5 of leaf, occasionally revolute along sheathing base, usually not or narrowly decurrent, leaf margins serrate to dentate, rarely denticulate or entire; **leaf apex** acute, occasionally rounded acute or narrowly acute, apex adaxial surface flat to more commonly keeled; **leaf base** usually sheathing, rectangular or broadly ovate or elliptic, basal cells differentiated either throughout sheathing base or in proximal half, usually straight across leaf but often with an area of hyaline cells juxtacostally (a "miniwindow"), with or without longitudinal brown stripes, walls straight or porose; **costa** concolorous to brown or rarely blackened at the base, ending before the apex to percurrent or rarely excurrent, adaxial cells superficially elongate, costa (2–)4–6(–10) rows across at midleaf, costa transverse section rounded, semicircular or flattened and reniform, adaxial epidermis absent, adaxial stereid band present, adaxial stereid band in size smaller than the abaxial, adaxial band in 1–2 layers, abaxial band in 2(–4) layers, fully included guide cells 2–4(–6), guide cells in 1, rarely 2–3 near base, hydroid strand absent, abaxial band in section semicircular to reniform, adaxial outgrowths absent; **distal laminal cells** rounded-short-cylindric to short-rectangular, distal laminal border not differentiated or with 1–2(–4) rows of less papillose, thicker walled cells, or rarely 2-stratose narrow cartilaginous border, thickness throughout leaf of cells near leaf margins about same as medially or occasionally 1/2 as thick, distal laminal cells ca. 9–15 µm wide, 1(–2):1, cell walls evenly thickened or thickened at the corners, occasionally trigonous or porose, superficially flat, weakly convex or bulging; **distal laminal papillae** simple to bifid, often multiplex or spiculose, rarely absent. **Specialized asexual reproduction** absent, or as obovoid or clavate gemmae on stems or leaves, occasionally on specialized branches or leaf cups. **Sexual condition** usually dioicous, occasionally monoicous (autoicous). **Perigonia** gemmate, usually terminal, often in clusters. **Perichaetia** terminal, leaves strongly differentiated, 4–15 mm. **Seta** 0.8–1.7(–3) cm. **Capsule** peristomate, cylindric, straight or weakly curved, exothecial cell walls usually evenly thickened, short- to long-rectangular; stomates at base of capsule, occasionally rudimentary or absent; annulus differentiated as 4–6 rows of vesiculose cells, these deciduous in pieces; **peristome teeth** of straight filaments, delicate, often broken, 32 (rarely more, to 40) in groups of four, or variously reduced, (100–)200–600(–1000) µm, smooth or papillose, occasionally spiculose, usually 9–15 articulations; basal membrane absent or very short and hidden by annulus; **operculum** conic-rostrate or short-conic, 0.8–1.5 mm, cells in straight rows. **Calyptra** cucullate, covering most of capsule, smooth or rough distally with antrorse prorulae, rarely also with simple papillae. **Spores** spheric, 12–20(–36) µm in diameter, occasionally in 2 size classes, colorless to light brown, lightly papillose. **KOH laminal color reaction** yellow to yellow-orange.

Family-level descendant or co-descendant of immediate progenitors of Pottiaceae, along with Cinclidotaceae, Ephemeraceae, and Splachnobryaceae.

Genera 10.

A summary of the most important distinguishing traits of the Streptotrichaceae is as follows: **Stem** lacking central strand. **Tomentum** often with short-cylindric cells ending strands (see Fig. 13-4, 3 and 13-8, 3). **Leaves** usually dentate distally, squarrose or strongly recurved-reflexed when large; **costa** lacking adaxial and abaxial epidermis, adaxial and abaxial surface cells elongate; **laminal cells** usually with differentially thickened walls. **Perichaetial leaves** strongly differentiated. **Capsules** lacking stomates in ancient surviving taxa. **Peristome** of basically 16 teeth each split into in fours (Fig. 13-10, 11) though often much reduced, teeth spiculose in relatively primitive genera and otherwise reduced to spiral grooving or smooth. **Calyptra** large and primitively rough with antrorse prorulae (Fig. 13-1, 14). **Evolutionary** radiation often by asexual species, these usually with gemmae.

The species were extracted from the Pottiaceae, which usually has stomates, does not have antrorsely prorulate calyptrae, and peristomes are not highly dissected into numerous rami in primitive taxa of the family.

The ten genera here included in the Streptotrichaceae are *Austroleptodontium, Crassileptodontium, Leptodontiella, Leptodontium, Microleptodontium, Williamsiella, Rubroleptodontium, Stephanoleptodontium, Streptotrichum* and *Trachyodontium*. The five genera that are new among these are formally proposed and described in the taxonomic section of this book. The genera have a distribution that is tropical and subtropical worldwide, with a concentration of numbers of species in the Andes.

Classical key to the genera
1. Calyptra papillose with antrorse prorulae (forward pointing papillae at ends of cells.
  2. Leaves circinnate when wet; stomates present .... 1. *Austroleptodontium*
  2. Leaves recurved, wide-spreading or reflexed to squarrose when wet, stomates absent.
    3. Leaves with a thickened, cartilaginous border ....10. *Trachyodontium*
    3. Leaves with unistratose margins.
      4. Leaves with juxtacostal windows of short, hyaline, enlarged cells; peristome transversely striate; gametophyte similar to those of *Microleptodontium* .... 3. *Leptodontiella*
      4. Leaves with narrow, translucent basal cells; peristome well-developed, branching-spiculose; gametophyte nearly identical to that of *Williamsiella araucarieti*............... 9. *Streptotrichumn*
1. Calyptra smooth.
    5. Plants small, leaves usually less than 3 mm long; stem hyalodermis absent or only very weakly developed.
      6. Leaves with red basal cells, stomates absent.. 7. *Rubroleptodontium*
      6. Leaves with yellow or hyaline basal cells, stomates present .... 5. *Microleptodontium*
    5. Plants larger, leaves usually more than 3 mm long, stem hyalodermis usually present or rarely absent.
      7. Leaves with miniwindows (hyaline, delicate juxtacostal area); costa somewhat thickened, gemmae absent..2. *Crassileptodontium*
      7. Leaf base with translucent cells or large windows extending from costa to near margin; costa thin; gemmae sometimes present.
        8. Laminal papillae low, scattered and simple to bifid over flat or weakly convex lumens .... 6. *Williamsiella*
        8. Laminal papillae simple and punctate or hollow and scattered, or spiculose and centered over the lumens.
          9. Tomentum arbusculate, rarely ending in short-cylindric cells; leaf margins closely dentate to irregularly serrate, medial cell lumens generally angular, leaf base often

with longitudinal yellow-brown stripes; stem lacking hyalodermis; laminal papillae punctate or if spiculose then leaves with large decurrencies; gemmae absent .... 4. *Leptodontium*

9. Tomentum lacking or stringy or emerging in lines on the stem, commonly ending in short-cylindric cells; leaf margins dentate, usually distantly, medial cell lumens rounded-quadrate, leaf base lacking longitudinal stripes; stem with hyalodermis, or if hyalodermis lacking then leaves lacking large decurrencies; gemmae often present .... 8. *Stephanoleptodontium*

One of the most interesting traits among Streptotrichaceae is the tomentum ending in a series of short-cylindric cells. This trait is found in many species, including the most primitive extant. In other families, I have found this trait only in *Tuerckheimia guatemalensis* Broth. (Pottiaceae). The illustration (Zander 1993: 93) is clearly the same as that of Streptotrichaceae, but was interpreted in 1993 as "heterotrichous protonema." The specimen was old and had no chlorophyll, so there remains some question as to whether it has genuine streptotrichaceous tomentum or if the tomentum of Streptotrichaceae phenocopies heterotrichous protonema, though lacking chlorophyll.

The longitudinal "stripes" of brownish basal cells that are characteristic of many species are probably associated with the axillary hairs, which are of rather elongate serial cells, of 4–15 cells in this family. The basal cell stripe color may be of the same physiological origin as that of the brownish basal cells of the axillary hairs.

Given the apparent origin of the family from epiphytic species now mostly extinct, as discussed in the chapter on outgroup analysis, the stomates now present in advanced species are apparently not re-evolved but released from long epigenetic repression.

Related genera in Pottiaceae, possibly at the base of the Streptotrichaceae caulogram include *Ardeuma, Reimersia,* and *Triquetrella.* Stomates are absent in the monotypic genera *Leptodontiella, Rubroleptodontium, Streptotrichum,* and *Trachyodontium,* and also one species of *Williamsiella, P. aggregatum,* though generally present in Pottiaceae.

The internal anticlinal walls modified as inner peristomes in *Streptotrichum* are spirally grooved as are the teeth of *Leptodontium*. The peristome teeth of *Leptodontium* are quite like the interior walls of *Streptotrichum*. The "properistome" or preperistome, which is a linear excrescence on the base of the teeth of *Leptodontium,* particularly of *L. viticulosoides* or *L. excelsum,* is probably the remains of the teeth as fully expressed in *Streptotrichum,* although they are not spiculose but generally smooth in *Leptodontium*. According to Zander (1972: 219) "[The peristome] is best developed in certain specimens of *L. viticulosoides* ... from the Indian Himalayas. Other species in which a preperistome is sometimes present are *L. flexifolium, L. stellatifolium, L. pungens, L. capituligerum, L. araucarieti,* and *L. longicaule."* *Trachyodontium* has no preperistome but the teeth are spiculose and thus primitive, being interpretable as further reduced from those of *Streptotrichum.*

The family name —Available infrafamilial names include the Leptodontieae Herz., Geogr. d. Moose 101. 1926 and Leptodontioideae (Herz.) Hilp., Beih. Bot. Centralbl. 50: 679. 1933. The Code of Nomenclature indicated:

10.6. The type of a name of a family or of any subdivision of a family is the same as that of the generic name on which it is based (see Art. 18.1). For purposes of designation or citation of a type, the generic name alone suffices. The type of a name of a family or subfamily not based on a generic name is the same as that of the corresponding alternative name (Art. 18.5 and 19.8).

19.4. The name of any subdivision of a family that includes the type of the adopted, legitimate name of the family to which it is assigned is to be based on the generic name equivalent to that type (Art. 10.6; but see Art. 19.8).

19A.1. When a family is changed to the rank of a subdivision of a family, or the inverse change occurs, and no legitimate name is available in the new rank, the name should be retained, with only the termination (-aceae, -oideae, -eae, -inae) altered.

Although Recommendation 19A.1 suggests that a name combining "Leptodonti-" and "-aceae" should be coined because of the existence of subfamily Leptodontioideae, it is considered here that such a name would be confounded with Leptodontaceae Schimp., a very similar, correct and heterotypic name for quite a different moss group and which is now in common use. Kanchi Ghandi (pers. comm.), expert in nomenclature, agreed that advancing the tribal name Leptodontieae to family status would introduce a nomen confusum.

Chapter 13
GENERA OF STREPTOTRICHACEAE

1. AUSTROLEPTODONTIUM R. H. Zander, gen. nov.

Type species: *Austroleptodontium interruptum* (Mitt.) R. H. Zander.

Hyalodermis caulina praesens. Tomentum plerumque deest. Folia circinnata, lanceolata, ca. 2 mm longa, indecurrentia; margines folii unistratosae, remote exigueque denticulatae; apex folii anguste acutus; costa ut apiculus multi-cellularis excurrens stratis stereidarum inter se e serie uno compositis; cellulae medianae laminales aequaliter incrassatae, luminibus rotundato-quadratis, parietibus superficialibus planis vel convexis, papillis simplicibus vel 2-fidis, per lumen 2–4 vel in massam applanatam multiplicem confertis; basis folii vaginans, e cellulis rectangulis translucentibus composita, fenestrellis (miniwindows) nullis. Reproductio asexualis propria per gemmas. Status sexualis dioicus. Peristomium e 16 ramis brevibus, laevibus constans. Calyptra antrorsum prorulosa.

Stem hyalodermis present. **Tomentum** usually absent. **Leaves** circinnate, lanceolate, ca. 2 mm in length, not decurrent; **leaf margins** unistratose, distantly weakly denticulate; **leaf apex** narrowly acute; **costa** excurrent as a several-celled apiculus, stereid bands each in 1 layer; **medial laminal cells** evenly thickened, lumens rounded-quadrate, superficial walls flat to convex, papillae simple to 2-fid, 2–4 per lumen or crowded into a multiplex flattened mass; **leaf base** sheathing, of rectangular translucent cells, miniwindows absent. **Specialized asexual reproduction** by gemmae. **Sexual condition** dioicous. **Capsule** peristome of 16, short, smooth rami. **Calyptra** antrorsely prorulose.

Species 1.

1. Austroleptodontium interruptum (Mitt.) R. H. Zander, comb. nov.

Didymodon interruptus Mitt. in J. D. Hooker, Handb. N. Zeal. Fl. 421. 1867, basionym. Type: New Zealand, Kerr, NY—lectotype selected here; Sinclair, NY—syntype. (The Kerr specimen should be lectotype as "in Herb Mitten" is annotated on it by Hooker).

Leptodontium interruptum (Mitt.) Broth., Nat. Pflanzenfam. 1(3): 401. 1902.

Trichostomum interruptum (Mitt.) Besch., Compt. Rend. Hebd. Séances Acad. Sci. 81: 724. 1875.

Illustrations: Sainsbury (1955: 157). **Figure 13-1.**

Plants in a dense turf. deep yellow. **Stem** length 1.5–2 cm, section rounded-pentagonal, often somewhat flattened, central strand absent, sclerodermis present, hyalodermis present. **Axillary hairs** ca. 10–12 cells in length, all hyaline or basal 1–2 very pale brownish. **Tomentum** absent or occasional as groups of short reddish brown rhizoids, rarely short rhizoids end in a series of short-cylindric cells. **Leaves** strongly incurved and circinnate when dry, squarrose-recurved when moist, stem apices pin-wheeling counterclockwise when dry, lanceolate, 1.8–2.2 mm in length, distal lamina carinate; **leaf margins** recurved in lower 1/2–2/3, distantly denticulate; **leaf apex** narrowly acute, apex keeled; **leaf base** sheathing rectangular, differentiated as a rounded M, basal cells 6–8 μm wide, 5–7:1; **costa** concolorous, excurrent as a several-celled apiculus, adaxial costal cells elongate, rows of cells across costa at midleaf viewed in section (2-)–4, width at base (50–)–60–65(–80) μm, section reniform in shape, adaxial stereid band in 1(–2) layers, abaxial band in (1–)2 layers, fully included guide cells 4; **medial laminal cells** rounded-hexagonal to rounded-quadrate or short-rectangular, leaf border not differentiated, marginal cell thickness in section same compared to medial cells, medial cell width (6–)9–12 μm, 1:1, cell walls evenly thickened and lumens rounded-quadrate or rounded-hexagonal, superficial cells walls convex to flattened; medial laminal papillae simple to 2-fid, 2–4 per lumen or crowded into a multiplex flattened mass. **Specialized asexual reproduction** by obovoid gemmae in leaf axils, 45–55 μm, mostly 2 transverse septa, 5–7-celled. **Sexual condition** dioicous. **Perichaetial leaf length** ca.

2–2.3 mm. **Seta** 0.6–0.7(–1) cm in length. **Capsule** narrowly elliptical, 1–1.2(–1.5) mm, exothecial cells walls evenly thickened, short-rectangular, 2–3:1, stomates phaneroporous, at base of urn; annulus of 4–6 rows of vesiculose cells, persistent; **peristome** of 16 short rami, 200–250 μm, inserted under the mouth, smooth to lightly striated, irregularly cracked or split into two rami, rami arranged in groups of 4, basal membrane absent, **operculum** erect, short-conic to conic, 0.4–0.6(–0.7) mm, cells straight, 2:1, thick-walled. **Calyptra** cucullate, ca. 2.3 mm long, antrosely low-prorulose throughout, more distinctly so near apex. **Spores** spheric to somewhat angular, 11–13 μm in diameter, colorless to light brown, papillose. **KOH laminal color reaction** yellow-orange.

Old World (New Zealand) specimens of *Austroleptodontium interruptum* differ from those of *L. subintegrifolium,* a similar New World species, by leaves with a lower sheathing base, apex denticulate, very strongly recurved when wet, and costal section commonly with only 1 layer of adaxial stereid cells. The high leaf base and less strong leaf recurvature of *L. subintegrifolium,* at least, seem to allow a clear distinction at the species level that is consistent in recently examined specimens from South America additional to those cited in the 1972 revision. Specimens of *A. interruptum* from New Zealand are identical to each other and may be clonal given rarity of sporophytes.

Sainsbury (1955: 158) found that sporophytes are "extremely rare." Jessica Beever and Antony Kusabs kindly facilitated a loan of North Island, T. Moss, Aug. 17, 1986 (WELT M008870), which was abundantly fruiting and exhibited one calyptra, which was still embedded in the perichaetial leaves and was clearly antrorsely prorulose. The perichaetial leaves were often dentate, perhaps a primitive trait retained in an organ less exposed to the environment.

The less thickened adaxial stereid band in *A. interruptum* is probably a primitive trait compared to similar species in *Crassileptodontium.* Given that both *A. interruptum* and *Williamsiella aggregata* seem to have primitive traits in their respective sections of the genus, it is possible that southeast Asia or some Gondwanaland derivative flora might be the center of origin of these representatives of both genera. It is difficult to determine a center of origin for one species, but if two or more related taxa seem to agree, the element of chance regional migration is lessened. The stenomorphic populations of New Zealand on the other hand imply a possible founder origin from South America, but the primitive trait of papillose calyptra argues against a progenitor in the genus *Crassileptodonum,* similar to *C. subintetrifolium.*

Some *A. interruptum* specimens have papillae that appear to be centered, like those of *Stephanoleptodontium.* These grade into the broad flower-like gnarly papillae quite like those of. *Crassileptodontium.* According to Catcheside (1980: 189), *A. interruptum* is known for New Zealand and S. Africa (but is not found in Magill's 1981 South African bryoflora). Bob Magill (pers. comm.) has no knowledge of this species in South no knowledge of this species in South Africa. Sainsbury (1955) reported that the distribution is Australia, Amsterdam Island, New Australia, Amsterdam Island, and Zealand. Clearly more investigation is warranted.

Zygodon species have much the appearance of *A. interruptum* but can be distinguished by the clearly delimited, small, simple, hemispherical laminal papillae and plane distal lamina, lacking the distinctive undulation near the apex of *A. interruptum.*

Distribution: New Zealand; low to moderate elevations; soil, lawns, sand.

Representative specimens examined:
New Zealand: North Island, Three Kings Islands,, Bald Hill, 172°10'E, 34°10'S, cliff top,, G. W. Ramsay 164550, November 21, 1970, MO; North Island, Rotorua, M. Fleischer B 42, May 5, 1903, MO; North Island, Hawkes Bay, near Wairoa, dry hillside, E.A.H. (MacFadden Herb. no. 7801), "Dec. 30," CA; Wellington, 39.02 Cook Strait, summit of road from Makara to Karori, short grassy turf, 308 m, Tom Moss s.n., 17 Aug. 1986, WELT M008870; South Island, Westport, Westport Airport, rough lawn, 10 m, A. J. Fife 11532, March 19, 2001, MO; South Island, Allan's Reach, Otago Peninsula, Dunedin, sand, short grass, W. B. Schofield 49831, July 9, 1972, CA, CAS, MO.

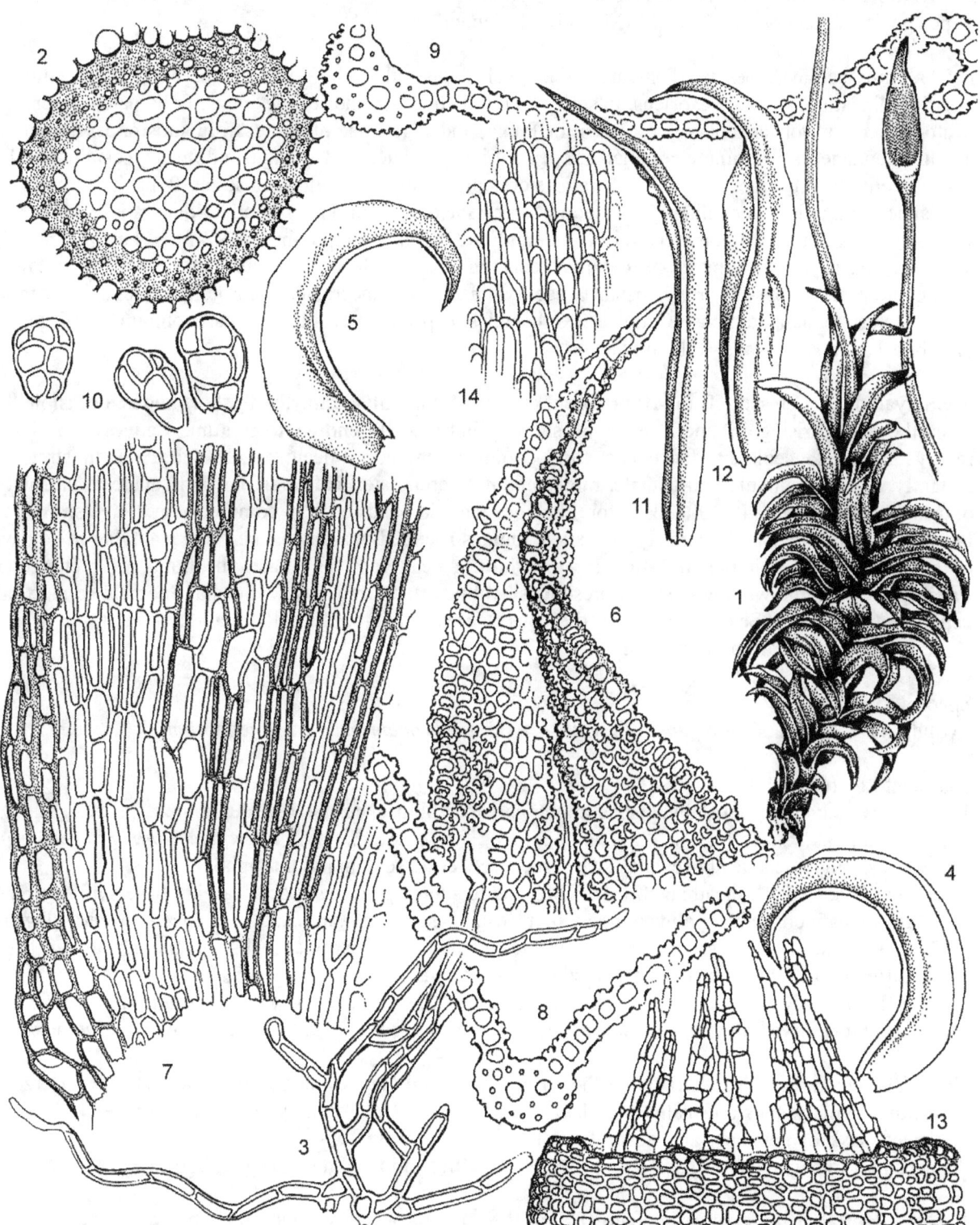

Figure 13-1. *Austroleptodontium interruptum.* 1. Habit, wet. 2. Stem section. 3. Tomentum. 4–5. Leaves. 6. Leaf apex. 7. Leaf base. 8–9. Leaf sections. 10. Gemmae. 11–12. Perichaetial leaves. 13. Peristome. 14. Antrorsely prorulose apical cells of calyptra.

2. CRASSILEPTODONTIUM R H. Zander, gen. nov.

Type species: *Leptodontium pungens* (Mitt.) R. H. Zander.

Hyalodermis caulina praesens. Tomentum deest vel magis vulgo chordosum nodis brevibus praeditum, interdum lineatum, ex caule exoriens, in series e cellulis brevi-cylindricis desinens. Folia patenti-recurva, ovato-lanceolata vel lanceolata, ca. 3–5 mm longa, indecurrentia; margines folii unistratosae, remote tenuiter dentatae, rarius minute crenulatae; apex folii anguste acutus vel anguste obtusus; costa usque 1–4(–10) cellulas sub apicem evanida, stratis stereidarum inter se vulgo in series 2 positis; cellulae medianae laminales aequaliter modiceque incrassatae, luminibus rotundato-quadratis, parietibus superficialibus planis vel convexis vel protuberantibus, papillis simplicibus vel 2-fidis, per lumen 2–4, dispersis; basis folii vaginans, e cellulis rectangulatis translucentibus, fenestrellis (miniwindows) plerumque praesentibus uti prope costam areae hyalinae, saepe laceratae. Reproductio asexualis propria deest Status sexualis dioicus. Cellulae exotheciales capsulares quadratae; peristomium e 16 ramis brevibus, laevibus constans. Calyptra laevis.

Stem hyalodermis present. **Tomentum** absent or more commonly stringy, with short knobs, occasionally from the stem arising in lines, series of short-cylindric cells ending tomentum. **Leaves** spreading-recurved, ovate-lanceolate to lanceolate, ca. 3–5 mm in length, not decurrent; **leaf margins** unistratose, distantly weakly denticulate to dentate, rarely minutely crenulate; **leaf apex** narrowly acute to narrowly obtuse; **costa** ending 1–4(–10) cells before the apex, stereid bands each in commonly in 2 layers; **medial laminal cells** evenly and moderately thickened, lumens rounded-quadrate, superficial walls flat to convex or bulging, papillae simple to 2-fid, 2–4 per lumen, scattered; **leaf base** sheathing, of rectangular translucent cells, miniwindows usually present as hyaline often torn areas near costa. **Specialized asexual reproduction** absent. **Sexual condition** dioicous. **Capsule** exothecial cells quadrate; peristome of 16, short, smooth rami. **Calyptra** smooth.

Species 4.
Evolutionary formula: *C. pungens* > *C. wallisii* > (*C. erythroneuron*, *C. subintegifolium*)

Classical key to species

1. Leaves weakly dentate distally, apex straight, narrowly obtuse, costa red throughout .... 1. *Crassicostata erythroneuron*
1. Leaves strongly dentate above, apex sharp, recurved, costa seldom red throughout.
    2. Leaf base usually sheathing 1/3–1/2 leaf length, submarginal cells of sheathing base similar to the distal leaf cells, with hyaline window juxtacostal basal cells ....2. *Crassicostata pungens*
    2. Leaf base usually sheathing 1/2 the leaf length, submarginal cells of the sheathing base differentiated, wider or longer than other laminal cells.
        3. Leaf nearly circinnate ....4. *Crassicostata subintegrifolium*
        3. Leaf recurved to reflexed above a narrowly rectangular base .... 3. *Crassicostata wallisii*

Crassileptodontium is characterized by the stringy tomentum, axillary hairs unusually long, leaf costa thickened, lask of gemmae, and quadrate cells of the capsule exothecial cell walls.

1. Crassileptodontium erythroneuron (Herz.) R. H. Zander, comb. nov.

Leptodontium erythroneuron Herz., Beih. Bot. Centralbl. 26(2): 62. 1909, basionym. Type:

Bolivia, Cochabamba, San Benito, Herzog, 1908, JE—holotype, S-PA—isotype.

Illustrations: Zander (1972: 260).

Plants in dense mats, light brown above, reddish brown below. **Stem** length 2–3 cm long, section rounded-pentagonal, central strand absent, sclerodermis of 1–2 cell layers, hyalodermis present. **Axillary hairs** to ca. 24 cells in length, basal 2–3 cells with somewhat thicker walls but all cells hyaline. **Tomentum** absent or stringy, with short knobby branches on a long rhizoid, red, ca. 3 short, short-cylindric cells ending many of the knobs on the tomentum. **Leaves** appressed, slightly incurved, scarcely twisted when dry, spreading-recurved when wet, lanceolate, 2.5–4 mm in length, distal lamina carinate; **leaf margins** recurved to revolute in proximal 1/2–2/3, weakly dentate in distal 1/4–1/2; **leaf apex** acute or narrowly obtuse, apex channeled; **leaf base** moderately sheathing, elongate cells filling high-sheathing base, stripes uncommon but distinct when present, basal cells long-rectangular, 9–11, ca. 5:1, walls slightly thickened laterally, scarcely porose; **costa** red throughout, ending 2–4 cells before apex, adaxial costal cells elongate, rows of cells across costa at midleaf viewed in section 2–4, width at base 90–100 μm, section semicircular in shape, adaxial stereid band in 2(–3) layers, abaxial band in 2(–3) layers, fully included guide cells 4–6; **medial laminal cells** subquadrate, leaf border of 1(–2) rows of epapillose cells, marginal cell thickness in section 1/2 as thick in one row marginally compared to medial cells, medial cell width 9–11 μm, 1:1, cell walls moderately and evenly thickened, superficial cell walls convex; **medial laminal papillae** crowded, bifid or much fused, 4–6 per lumen. **Specialized asexual reproduction** absent. **Sexual condition** dioicous. **Perigonia** terminal. **Perichaetial leaf length** to 5 mm. **Seta** 1.5–1.6 cm in length. **Capsule** cylindric, 1.5–2 mm, exothecial cells short-rectangular, ca. 2:1, annulus not seen; **peristome** not seen, **operculum** not seen. **Spores** spheric, 22–42 μm in diameter, yellowish brown, papillose. **KOH laminal color reaction** yellow-orange with a red costa.

Distribution: South America (Colombia, Peru, Bolivia); over rock, high elevations.

Representative specimens examined (See Zander 1972.)

2. Crassileptodontium pungens (Mitt.) R. H. Zander, **comb. nov.**
Didymodon pungens Mitt., Jour. Linn. Soc., Bot. 7: 150. 1864, basionym. Type: Cameroon, Cameroons Mt., Mann, 1862, NY—holotype.
Leptodontium pungens (Mitt.) Kindb., Enum. Bryin. Exot. 63. 1888.

Leptodontium joannis-meyeri Müll. Hal., Flora 71: 412. 1888 fide de Sloover 1987.
For additional synonymy see Zander (1972).

Illustrations: Zander (1972: 257); Zander (1993: 133; 1994: 266). **Figure 13-2**.

Plants in dense mats, greenish yellow to reddish brown above, reddish brown below. **Stem** length 2–5 cm, section rounded-pentagonal, central strand absent, sclerodermis present, hyalodermis present. **Axillary hairs** of 10–16 hyaline cells, basal 2–3 more thick-walled. **Tomentum** stringy, long red rhizoids with short knobs, occasionally ending in series of short-cylindric cells. **Leaves** erect, little contorted or twisted when dry, spreading-recurved when wet, ovate-lanceolate to lanceolate with a narrowly acute apex, 3–4 mm in length, distal lamina deeply carinate; **leaf margins** recurved in proximal 1/2–2/3, dentate in distal 1/3–1/2; **leaf apex** narrowly acute, filled with scarcely papillose rhomboidal cells, apex keeled; **leaf base** sheathing elliptical, differentiated as a rounded M, alar cells often differentiated (except for a single line of transparent clear marginal cells) as a group of yellowish brown, short-rectangular cells with thick and porose walls, hyaline basal cells porose, stripes present, basal cells 9–11, 4–6:1, miniwindows present, or absent in thick-cell-walled plants; **costa** usually red in proximal 1/3, percurrent or ending 1–4 cells before apex, adaxial costal cells elongate, rows of cells across costa at midleaf viewed in section 4–6, width at base 60–75 μm, section semicircular in shape, adaxial stereid band in 1–2 layers, abaxial band in 2–3(–6) layers, fully included guide cells 4; **medial laminal cells** subquadrate, leaf marginal row of cells above scarcely papillose, marginal cell thickness in section about 1/2 compared to medial cells, medial cell width 11–15, 1:1, cell walls moderately thickened, superficial cells walls flat to weakly convex; **medial laminal papillae** crowded, multifid. 2–4 per lumen. **Specialized asexual reproduction** absent. **Sexual condition** dioicous. **Perichaetial leaf length** 7–8 mm. **Seta** 1–1.1 mm in length. **Capsule** cylindric, exothecial cells short-rectangular, 20–30 × 30–45 μm, stomates phaneropore, at base of capsule, annulus differentiated, of 4–5 rows of reddish brown cells; **peristome** linear, 15–20 μm wide at base, prostome sometimes present, short, 310–375 μm, indistinctly striated, with about 10–12 articulations, basal membrane absent, **operculum** conic-rostrate, about 0.8 mm in length, cells straight. **Calyptra** not seen. **Spores** rounded, (14–)17–20 μm in diameter, dark brown, very papillose. **KOH laminal color reaction** yellow.

Crassileptodontium pungens never has gemmae, the costa has usually two full layers of stereid cells adaxially (not one or one and a half), and the juxtacostal basal cells are generally hyaline at the leaf insertion, while alar cells (except for a marginal row of hyaline cells) and the intramarginal cells in several rows are bright red, orange or yellow. *Leptodontium joannis-meyeri* Müll. Hal., according to de Sloover (1987), belongs here.

Distribution: North America (Mexico); Central America (Guatemala); South America: Venezuela, Colombia, Ecuador, Peru, Bolivia, Argentina, Juan Fernandez Islands, Brazil); Africa (Cameroon, Kenya); soil, humus, lava, rock, at high elevations, particularly páramos.

Representative specimens examined (see also Zander 1972.)
Cameroon: Meoli Ndiva, Victoria Div., Western Prov., Cameroon Mt., summit of Ridge, pockets of soil among rocks, exposed, 3300 m, J. P. M. Brenan 4267, March 28, 1948, MO; Mt. Cameroon, dense forest, 1700 m, D. Balázs 76/b, Nov. 28, 1967, EGR; Mt. Kenya, ridge between S & N Naro Moru steams, Gamia Peaks, 3875 m, P. Kuchar B8781, October 21, 1979, MO. **Kenya:** Mt. Kenya, Teleki Valley, dry soil, 4180 m, J. Spence 2603a, June 25, 1984, EGR; Mt. Elgon, S slope, rocks, 4000 m, S. Rojkowski 64, Dec. 16, 1975, EGR; Aberdare Mts, trail to Mt. Satima, soil, 3540 m, J. Spence 2801b, July 17, 1984, EGR; **South Africa:** Cape, Worchester Div., Du Toits Peak, 1830 m, E. Esterhuysen 3319 CC, MO; Natal, Drakensberg, Sani Top, alpine heath-grassland, on soil over rock., ca. 2900 m, J. Van Rooy 3587, MO; Natal, Cathedral Peak Forest Station, Organ Pipes Spire, above research area, subalpine grassland, on boulder in shade, R. E. Magill 2829 CC, MO; KwaZulu-Natal, Cathedral Peak Forest Station, Organ Pipes Spire, above research area, subalpine grassland, on rock cliff, R.

E. Magill 5754, Nov. 2, 1978, MO; **Tanzania,** Kilimanjaro, Shira Needle, wet lava cliffs, 3660–3730 m, T. Pocs 90034/F, Feb. 17, 1990, EGR; Trail from Mweka to Kibo Peak on Mt. Kilimanjaro, A. J. Sharp et al. 7458, July 28, 1968, UC; **Uganda:** Mt. Elgon, Suam Valley, Hot Springs, rock, 3600 m, M. Chuah et al. 9220, Jan. 16, 1992, EGR.

3. Crassileptodontium wallisii (Müll. Hal.) R. H. Zander, **comb. nov.**
Trichostomum wallisii Müll. Hal., Linnaea 38: 603. 1874, basionym.
Leptodontium wallisii (Müll. Hal.) Kindb., Enum. Bryin. Exot. 63. 1888.

Leptodontium gambaragarae Negri, Ann. Bot. (Rome) 7: 163. 1908, FT microphoto.
Leptodontium sublaevifolium Broth., in G. W. J. Mildbraed, Wiss. Ergebn. Deut. Zent.-Afr. Exped., Bot. 1907–1908, 2: 146. 13 f. 14. 1910. Type: Congo, Afr. Vulkangebeit, Karisimbi, 3400 m, 9/1909, G. Mildebraed 2070, Sept. 1909 (H-BR, also de Sloover 1987).
For additional synonymy see Zander (1972).

Illustrations: Allen (2002: 123); Zander (1972: 260).

Plants in mats, greenish to yellowish brown above, brown below. **Stem** length 2–20 cm, section rounded-pentagonal, occasionally rounded-triangular, central strand absent, sclerodermis of 2–3 layers stereids, hyalodermis present. **Axillary hairs** of up to 10 cells in length, 2–3 basal cells brownish and thicker-walled. **Tomentum** dense, stringy, with short side branches or directly from stem in lines, red, abundant, ending in series of short-cylindric cells. **Leaves** erect to spreading recurved when dry, spreading-recurved to squarrose-recurved when wet, lanceolate to long-lanceolate, 4.5–5.5 mm in length, distal lamina carinate; **leaf margins** revolute and incurved submarginally in proximal 1/3–1/2, dentate in distal 3/4–1/2; **leaf apex** narrowly acute, sometimes filled with scarcely papillose rhomboidal cells, apex carinate; **leaf base** oblong, high-sheathing, little decurrent long-rectangular, often porose, basal submarginal cells of the sheathing base short-rectangular, basal cells 11–13 μm wide, 2–3:1; **costa** concolorous or red along differentiated leaf base, percurrent or ending 1–3 cells before apex, adaxial costal cells elongate, rows of cells across costa at midleaf viewed in section ca. 6, width at base 60–90 μm, section semicircular in shape, adaxial stereid band in 2 layers, abaxial band in 2–3 layers, fully included guide cells 4; **medial laminal cells** subquadrate, leaf border less papillose in 1–3 rows of elongate cells, marginal cell thickness in section about same compared to medial cells, medial cell width 8–11, 1:1, cell walls evenly and moderately thickened, superficial cells walls flat; **medial laminal papillae** crowded, simple to multifid, usually fused into 2–4 groups over each lumen. **Specialized asexual reproduction** absent. **Sexual condition** dioicous. **Perichaetial leaf length** 6–8 mm. **Seta** 0.8–1 cm in length. **Capsule** cylindric, ca. 2 mm, exothecial cells 1:1, moderately thick-walled, stomates phaneropore, at base of capsule, annulus differentiated, of 4–6 rows of reddish brown cells; **peristome** teeth 16, split into straight, paired filaments, variously anastomosing, smooth, 200–290 μm, smooth, light brown, of 4–8 articulations, basal membrane absent, **operculum** not seen. **Calyptra** not seen. **Spores** spheric, 10–12 μm in diameter, brown, papillose. **KOH laminal color reaction** yellow.

In South America, the stem epidermis is occasionally apparently absent, but this may be due to simple variation in size given the usually small size of the hyalodermal cells. In the type of *Crassileptodontium sublaevifolium* the hyalo-dermis is of small cells the same size as those of the stereid cells but the exterior walls are clearly thin.

The gametophytes of *C. wallisii* are similar to those of *Streptotrichum ramicola*, including the characteristic inflated marginal basal cells, but the latter differs in having no miniwindows, tomentum thin or absent, coarser papillae, these centered over somewhat larger laminal cells. The exothecial cells of *C. wallisii* are nearly isodiametric and the peristome teeth are rudimentary to short and nearly smooth, with no basal membrane. The exothecial cells of *S. ramicola* are rectangular and the peristome teeth much elongate from a short basal membrane, red, and spiculose.

The distal laminal cells of *L. wallisii* may be moderately thickened or very thickened and porose, but not trigonous as with *L. viticulosoides*.

Crassileptodontium wallisii rarely fruits (e.g., Colombia, Cleef 9657, MO; Congo, De Sloover 13.184, EGR).

Distribution: Central America (Costa Rica); South America (Colombia, Ecuador, Peru, Bolivia, Brazil); Africa (Republic of the Congo); on soil, high elevations.

Specimens examined (see also Zander 1972). **Colombia:** Cundinamarca, Cabeceras de la Quebrada Chuza, paramo de Chingaza, soil, 3650 m, A. M. Cleef 9657, April 28, 1973, MO. **Congo:** Massif des Birunga, 4250 m, J. De Sloover 13.184, Jan. 26, 1972, EGR. **Rwanda:** Parc des Volcans, E side of Mt. Karisimbi, elfin woodland, 3900–4200 m, W. G. D'Arcy 9072, April 16, 1975, MO; **Tanzania:** Kilimanjaro Mt., along UMBWE Route, soil, 3550–3800 m, T. Pocs 6792/M, 20 and 22 Sept. 1972, EGR. **Uganda:** Ruwenzori Mts., Kinandara Valley, soil, 4300 m, N. Less 39, Jan. 6, 1993, EGR.

4. Crassileptodontium subintegrifolium (Thér. ex Herz.) R. H. Zander, **comb. nov.**
Leptodontium subintegrifolium Thér. ex Herz., Feddes Repert. Spec. Nov. Regni. Veg. 45: 45. 1938, basionym. Type: Peru, Yerupajá, Kinzl, 1936, JE—holotype, S-PA—isotype.

Illustration: Zander (1972: 263).

Plants in dense mats, green to yellowish brown distally, brown proximally. **Stem** length 4–7 cm, section rounded-pentagonal, central strand absent, sclerodermis of 2–3 layers, hyalodermis present. **Axillary hairs** of 13–17 cells, all hyaline, basal 2–3 more thick-walled. **Tomentum** stringy, long red rhizoids with short knobs, knobs of 4–6 short-cylindric cells, these often tapering into elongate rectangular cells. **Leaves** incurved, slightly twisted when dry, spreading-recurved when wet, oblong-lanceolate, 3–4 mm in length, distal lamina carinate; **leaf margins** recurved in proximal 1/2–3/4, minutely crenulate distally by projecting cell walls and papillae, occasionally denticulate near apex; **leaf apex** narrowly acute, apex keeled; **leaf base** elliptic, sheathing, stripes often present, basal cells differentiated across leaf, basal cells 9–11 μm, 5–6:1, thin walls, or slightly thickened; **costa** concolorous, ending 1–10 cells before apex, adaxial costal cells elongate, rows of cells across costa at midleaf viewed in section 5–7, width at base 70–90 μm, section semicircular to reniform in shape, adaxial stereid band in 2 layers, abaxial band in 2 layers, fully included guide cells 4; **medial laminal cells** subquadrate, leaf border in 1 row less papillose, marginal cell thickness in section about 1/2 as thick in 1 row compared to medial cells, medial cell width 10–12 μm, 1:1, cell walls moderately and evenly thickened, superficial cells walls bulging; **medial laminal papillae** crowded, simple to multifid. **Specialized asexual reproduction** not seen. **Sexual condition** probably dioicous, sexual structures not seen. **KOH laminal color reaction** yellow.

Distribution: South America (Peru, Bolivia); rock, at high elevations.

Specimens examined (see also Zander 1972). **Peru:** Ancash, Cerro Yanarrajo, rock, 4800 m, W. Schultze-Motel P-66, Sept. 30, 1982, MO. **Bolivia:** La Paz, Murillo, 5 km NE of Milluni, N of La Paz, grassy slopes, 4800 m, M. Lewis 79-1693A, MO.

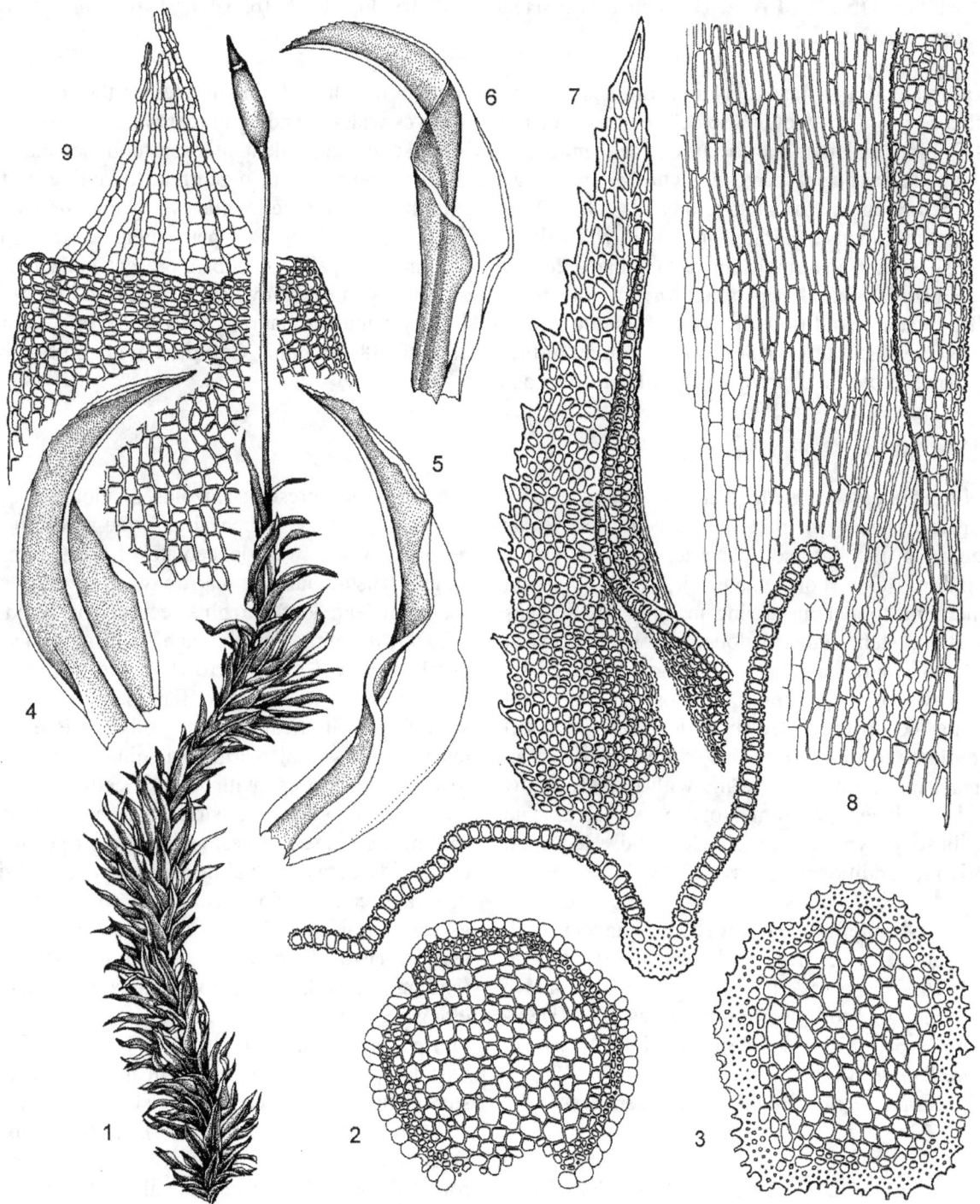

Figure 13-2. *Crassileptodontium pungens.* 1. Habit, wet. 2. Stem section, near apex. 3. Stem section, mature. 4–6. Cauline leaves. 7. Leaf apex. 8. Leaf base margin, showing miniwindow. 9. Peristome.

3. LEPTODONTIELLA R. H. Zander & E. Hegewald, Bryologist 79: 16. 1976. Type: *Leptodontium apiculatum* R. H. Zander.

Stem hyalodermis present. **Tomentum** absent or more commonly short, borne in lines on the stem, often with a series of short-cylindric cells ending tomentum. **Leaves** wide-spreading to recurved, lanceolate, ca. 1.8–2.2 mm in length, not decurrent; **leaf margins** unistratose, distantly dentate, each tooth separated by 5–10 cells; **leaf apex** narrowly acute, ending in a flat filiform point; **costa** percurrent to ending up to 6 cells before apex, abaxial stereid band each in 2 layers; **medial laminal cell walls** evenly and moderately thickened, lumens quadrate, superficial walls flat, papillae simple to bifid, 2–4 per lumen, scattered; **leaf base** sheathing, marginally of rectangular green cells and medial hyaline windows present juxtacostally. **Specialized asexual reproduction** possibly by stem-borne cylindric brownish gemmae of uniseriate cells. **Sexual condition** dioicous. **Capsule** exothecial cells rounded-quadrate; peristome of 32, short, transversely striate or smooth rami, often further split into 38–50 rami. **Calyptra** with both simple papillae and antrorsely projecting cell ends near apex.

Species 1.

Traits in *Leptodontiella* that are evolutionarily significant, being in part the basis for particular groups in the Streptotrichaceae, are the short-cylindric to short-rectangular cells ending tomentum, tomentum arising in longitudinal lines on stem; stem with hyalodermis; hyaline basal cells differentiated juxtacostally and bordered by 4–7 rows of cells similar to those of the distal lamina; no differentiated distal leaf margin; distal marginal teeth distant or rudimentary; flat superficial laminal cells walls with scattered, low bifid papillae, peristome of many teeth, and papillose calyptra. See Zander and Hegewald (1976) for additional discussion.

1. Leptodontiella apiculata (R. H. Zander) R. H. Zander & E. Hegewald, Bryologist 79: 16, 1976. Type: Peru, Lima, heights above Atacongo, Stork, Horton & Vargas 9278 (US—holotype; FH—isotype), distributed as *L. laxifolium* Broth.
Leptodontium apiculatum R. H. Zander, Bryologist 75: 238. 1972.

Illustrations: Zander (1972: 246); Zander (1993: 131); Zander & Hegewald (1976: 18). **Figure 13-3.**

Plants in thin, loose mats, greenish brown. **Stem** length 1–3 cm, section rounded-pentagonal, central strand absent, sclerodermis present, of 1–2 layers, hyalodermis present. **Axillary hairs** of 7–10 cells, basal 1–2 brownish or thicker-walled. **Tomentum** reddish, short, borne in lines on the stem, short-cylindric cells present; rhizoids of elongate cells. **Leaves** appressed when dry, wide-spreading to recurved when wet, lanceolate, 1.8–2.2 mm in length, distal lamina narrowly grooved down center of leaf; **leaf margins** recurved in proximal 1/2–2/3, distantly dentate each tooth separated by 5–10 cells; **leaf apex** narrowly acute, ending in a flat filiform point, apex flat; **leaf base** low-sheathing, elliptical, basal cells differentiated juxtacostally, basal marginal laminal cells in 4–7 rows, concolorous with distal cells, rounded-quadrate to elliptical, stripes absent on most cauline leaf bases, occasionally one stripe of 1–2 cells wide between hyaline window and marginal quadrate cells; stripes common on perichaetial leaves, basal cells 10–12 µm wide, 4–5:1, thin-walled; **costa** concolorous, essentially percurrent (ending 2–6 cells before apex), adaxial costal cells elongate, rows of cells across costa at midleaf viewed in section 3–4, width at base 50–60 µm, costal section flattened-reniform in shape, adaxial stereid band in 1 layer, abaxial band in 2 layers, fully included guide cells 3(–4); **medial laminal cells** irregularly subquadrate, occasionally rounded-triangular or hexagonal, leaf border not differentiated, marginal cell thickness in section about same compared to medial cells, medial cell width (7–)9–12(–15) µm, 1:1(–2), cell walls evenly thickened, superficial cells walls flattened; medial laminal papillae low, simple to bifid, crowded, scattered across lumens. **Specialized asexual reproduction** possibly by stem-borne cylindric brownish gemmae of uniseriate cells, 70–120 µm long, with a clear basal cell. **Sexual**

condition dioicous. **Perigonia** lateral. **Perichaetial leaf length** 5–5.5 mm. **Seta** 1.5–2.5 mm in length. **Capsule** short-cylindric, 1–1.5 mm, exothecial cells evenly thickened, 1:1, rounded-quadrate, stomates absent, or occasionally represented a by a small hole surrounded by a circle of rhomboid cells, annulus differentiated, of 3–4 rows of vesiculose cells; **peristome** of 16 bifid teeth cleft into 32 straight or weakly twisted filaments and usually crowded into eight groups of 3–5 anastomosing rami, further split in various capsules into as many as a total of 38–50 rami, 300–380 μm in length, transversely striate, of 10–12 articulations, basal membrane absent, **operculum** rostrate, inclined, cells straight. **Calyptra** cucullate, with scattered simple papillae, plus projecting antrorse cell ends near apex, calyptra length ca. 2 mm. **Spores** spherical, 15–20 μm in diameter, light brown, densely papillose. **KOH laminal color reaction** yellow.

Leptodontiella belongs with the Streptotrichaceae by the short-cylindric cells often terminating the tomentum; the enlarged, sheathing perichaetial leaves; the nearly straight, striate peristome teeth; the unistratose, carinate cauline leaves; and the costa with a ventrally exposed stereid band. *Leptodontiella* is distinctive in the very short seta, the short urn, 16 peristome teeth each cleft into three or four rami, and the propagula (if truly belonging to this species where seen only once) being cylindrical, uniseriate, and dark brownish green.

Distribution: restricted to Peru; rocks, tree limbs, high elevations.

Representative specimens examined (see also Zander 1972).
Peru: Arequipa, Caraveli, Lomas de Atiquipa, tree, 600--800 m, P. & E. Hegewald 8567, June 17, 1977, MO; Peru, Cajamarca, Contumazá, Cerro Cunantan, tree, 2900 m, P. & E. Hegewald 7300, Sept. 16, 19783, MO; Peru, Trujillo, La Libertad, Cerro Campana, over rocks, 800 m, A. Sagástegui 9221, Sept. 4, 1978, MO; Peru, Ancash, Yungay, Laguna Llanganuco, tree limb, 4035 m, E. &P. Hegewald 7525, Oct. 17, 1993, MO.

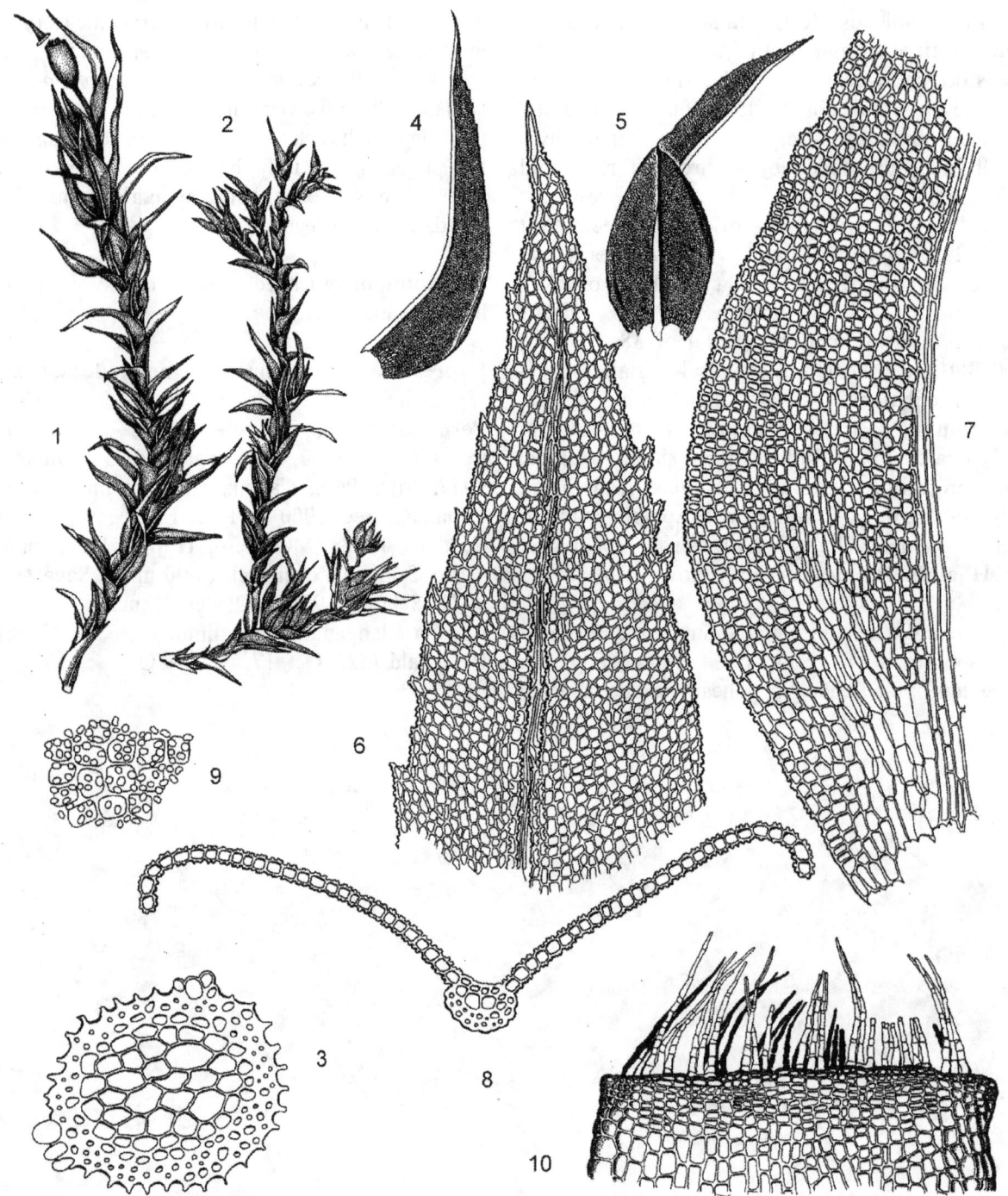

Figure 13-3. *Leptodontiella apiculata*. 1. Habits, wet. 2. Male plant. 3. Stem section. 4–5. Cauline leaves. 6. Leaf apex. 7. Leaf base. 8. Leaf section. 9. Medial laminal cells and papillae. 10 Peristome.

4. LEPTODONTIUM (Müll. Hal.) Hampe ex Lindb., Öfvers. Förh. Kongl. Svenska Vetensk.-Akad. 21: 227. 1864. Type: *Leptodontium squarrosum* (Hook.) Hampe (= *L. viticulosoides* (P. Beauv.) Wijk & Marg.)
Trichostomum sect. *Leptodontium* Müll. Hal., Synop. Musc. 1: 577. 1849.

Stem hyalodermis absent. **Tomentum** arbusculate, with short knobs, ending in tapering elongate cells, only occasionally are series of short-cylindric cells present on knobs. **Leaves** squarrose-recurved, ovate-lanceolate to long-lanceolate, sometimes cirrhate, ca. 2.5–4(–6) mm in length, not decurrent or occasionally strongly decurrent as a broad wing; **leaf margins** unistratose, distantly dentate to irregularly serrate in distal 1/3–1/2; **leaf apex** narrowly to occasionally broadly acute, rarely with a narrowly ligulate tip; **costa** ending 3–5 cells before the apex, adaxial stereid band of 1–2 layers, abaxial of 1–2 layers; **medial laminal cells** with walls moderately thickened and often with corners thickened or trigonous, lumens rounded-quadrate to rhomboidal, superficial walls flat to convex or bulging, papillae simple to bifid and 2–4 per lumen and scattered, or short-spiculose and arranged in a ring about the bulging lumen center; **leaf base** sheathing, of rectangular translucent cells, miniwindows absent, stripes absent or present. **Specialized asexual reproduction** absent. **Sexual condition** dioicous or monoicous. **Capsule** exothecial cells rectangular, ca. 2:1; peristome of 16, short, smooth or lightly striated rami. **Calyptra** smooth.

Species 3.

Evolutionary formula: ***Leptodontium* sp**. > (*L. viticulosoides, L. excelsum, L. scaberrimum*)

The tomenta of *Leptodontium* species are much the same, being arbuscular or rapidly multi-branching into a thick pad from a few slightly thicker, reddish stalk rhizoids originating in leaf axils. The tomentum of *L. excelsum* seems neotenic in that the ends of the tomentum only partly elongate into long tapering clear cells, as is more common in *L. viticulosoides*. Thus, *L. excelsum* has more reddish and shorter tomentum than does *L. viticulosoides*; it also has tomentum knobs that are somewhat more commonly differentiated with straight crosswalls to form a short-cylindric series. Although *Leptodontium scaberimum* has some of the traits of *Stephanoleptodontium,* particularly the coroniform papillae, it is placed here by the arbusculate tomentum, serrate distal leaf margins, and lack of gemmae. Bulging laminal cells topped with spiculose papillae are also found in genera of Pottiaceae, e.g. *Syntrichia. Barbula eubryum* of the Pottiaceae and *Dicranum scoparium* of the Dicranaceae also have arbusculate tomentum.

Classical key to species
1. Stem section rounded-triangular to rounded-pentangular, with broad, red (in KOH) leaf decurrencies running down the stem; leaf apex ending in a tiny, plane ligulate tip, obtuse or rounded-acute or with one rounded terminal cell; distal laminal papillae usually centered, 1–2 papillae per lumen each of 2–4 spiculose salients .... 3. *Leptodontium scaberrimum*
1. Stem section rounded-pentagonal, with narrow uncolored decurrencies running shortly down the stem or none; leaf apex ending in a keeled narrowly acute tip, terminating in a sharp cell; distal laminal papillae usually simple, occasionally bifid, scattered over the lumens.
   2. Leaves short- to long-lanceolate, but not cirrhate, apex broadly acute to narrowly acute, basal cells seldom striped with longitudinal rows of brown cells; distal laminal cells (8–)11–15 μm in width, quadrate except for short-rectangular cells occasional at apex, rhizoid initials (clear cells) or apical rhizoids absent, leaf marginal wall somewhat thickened; stem tomentum usually abundant, dense, whitish or red-brown, short-cylindric cells ending tomentum very rare; autoicous, cells in two size classes (variable in var. *exasperatum*), the larger chlorophyllose (when fresh), the smaller brown and often collapsed .... 2. *Leptodontium viticulosoides*

2. Leaves short- to very long-lanceolate and often cirrhate, apex narrowly acute, basal cells often striped with longitudinal rows of brown cells; distal laminal cells 7–11 μm in width, occasionally longitudinally elongate to 2–3:1, rhizoid initials (slightly enlarged, isolated, clear cells) or brown rhizoids often adaxially near costa at extreme leaf apex or absent, leaf marginal cell walls same thickness as other walls; stem tomentum absent or sparse, red, many filaments short and ending in a club of 1–3-short cells; dioicous, spores usually in one size class (rarely anisosporous), concolorous .... 3. *Leptodontium excelsum*

1. Leptodontium excelsum (Sull.) E. Britton, Bryologist 11: 66. 1908.

Syrrhopodon excelsus Sull., Musci Allegh. 170 [Schedae 41]. Type: U.S.A., North Carolina, Grandfather Mt., Sullivant, s.n. (Sullivant & Lesquereux, Musc. Allegh. 170), BM, NY, isotypes.

Leptodontium viticulosoides var. *sulphureum* (Lor.) R. H. Zander, Bryologist 86: 86. 1983. For additional synonymy, see Zander (1972), where treated as *Lepdontium viticulosoides* var. *panamense* (Lor.) R. H. Zander.

Illustrations: Allen (2002: 101, 120); Zander (1972: 246; 1994: 263).

Plants in turfs or loose mats, greenish to yellowish brown distally, brown proximally. **Stem** length 2–20 cm, section rounded-pentagonal, central strand absent, sclerodermis absent, hyalodermis present. **Axillary hairs** of ca. 6 cells, the basal 2 brown, or basal more firm-walled, or of 9–14 cells, all hyaline and thin-walled. **Tomentum** arbusculate, red to reddish white, knobby, most tomentum ending in clubs of 1–4 short cells, occasionally these lengthening into whitish elongate tapering cells, the short-cylindric cells present but occasionally difficult to demonstrate. **Leaves** erect to spreading, flexuose-twisted to contorted when dry, squarrose-recurved when wet, lanceolate to cirrhate, 3.3–6 mm in length, distal lamina carinate; **leaf margins** recurved in proximal 1/4, distantly dentate in distal 1/3–1/2; **leaf apex** narrowly acute, apex keeled, rhizoid initial cells (clear cells appearing like holes in the leaf) occasional; **leaf base** short-ovate, sheathing, differentiated across leaf, often with stripes, marginal cells not particularly differentiated, basal cells filling only 1/4–1/3 of ovate leaf base, basal cells 7–9 μm, 3–4:1; **costa** concolorous, ending 3–5 cells before apex, adaxial costal cells longitudinally elongate, rows of cells across costa

at midleaf viewed in section 4–6, width at base 60–70 μm, section reniform in shape, adaxial stereid band in 1–2 layers, abaxial band in 1–3 layers, fully included guide cells 4; **medial laminal cells** subquadrate to short-rectangular, leaf border not differentiated, marginal cell thickness in section same compared to medial cells, medial cell width 7–11, 1:1, cell walls little to moderately thickened at corners, superficial cell walls weakly convex; medial laminal papillae simple to bifid, 2–4 per lumen. **Specialized asexual reproduction** absent. **Sexual condition** dioicous. **Perigonia** terminal, as tight buds closely surrounded by larger leaves and thus difficult to find. **Perichaetial leaf length** 7–10 mm. **Seta** 0.6–1.2 cm in length. **Capsule** cylindric, 3–4 mm, exothecial cells thin-walled, 25–45 μm wide, 2:1, stomates phaneropore, at base of capsule, annulus of 4–6 rows of differentiated reddish brown cells; **peristome** linear, occasionally with prostome; in 8 sets of 4 rami, 150–300 μm, smooth, basal membrane absent, **operculum** rostrate, 0.6–0.9 mm, cells elongate distally, grading to rounded-quadrate at top of expanded base, smaller rounded-quadrate at mouth of capsule. **Calyptra** cucullate, smooth, calyptra length 4–5 mm. **Spores** homogeneous, rarely heterogeneous, spheric, 22–25 μm in diameter, light brown, lightly papillose. **KOH laminal color reaction** yellow.

Leptodontium excelsum (Sull.) E. Britt. (= *L. viticulosoides* var. *sulphureum* (Lor.) R.H. Zander) is found only in the New World and is characterized mainly by dioicy, spores in only one size class, and a tendency to long-lanceolate leaves with elongate distal laminal cells. The differences discussed by Zander (1972) between *L. viticulosoides* and *L. excelsum* were re-examined in the course of this study, but no traits are always present that distinguish the two species. *Leptodontium viticulosoides* generally has larger distal laminal cells, and somewhat bulging cells in the marginal laminal decurrencies, but this is not constant. Certain traits when they are present are diagnostic, however. Any of the following traits will distinguish *L. excelsum*: some tomentum that ends in short-cylindric cells, isolated clear cells (nematogenous) in distal lamina, isosporous capsules, dioicous sexual condition, cirrhate leaves, rectangular distal laminal cells, or very small distal laminal cells, ca. 7–10 μm.

Anisosporous capsules and autoicy will distinguish *L. viticulosoides*. Helpfully, *L. excelsum* does not occur in the Old World. One anomalous specimen from Mexico (Rezdowski 21843, MO) is clearly *L. excelsum* by knotty red tomentum, stripes on leaf bases, small distal laminal cells, and dioicy, but has heterogeneous spores. Although traits of sexuality are considered important in distinguishing species in bryology, *Leptodontium excelsum* (as determined by gametophytic traits) may have both false anisosporous and isosporous capsules in the same clump (Anderson & Zander 1986). A lethal factor is possible, as has been suggested for other bryophytes with the same sort of dimorphic spores—one dead, the other larger and alive (Mogensen, G. S. 1978), perhaps in association with heterozygy, although a simple recessive trait that both suppresses autoicy and false anisospory is possible. Given that sexual condition and spore size are traits not linked in some collections, a direct derivation of *L. excelsum* from *L. viticulosoides* seems less probable than postulation of a missing link with one or the other of these major traits (see caulogram).

Allen (2002) segregated *L. excelsum* from the present concept of *L. viticulosoides* to apply to small, somewhat flagellate plants with rhizoid (nematocyst, tmema) initials (isolated tiny windows of isolated slightly enlarged hyaline cells) in the leaf apex usually close to the costa. He also recognized *L. ulocalyx* (Müll. Hal.) Mitt., as a larger species lacking the rhizoid initials. The scattered occurrence throughout the range of *L. excelsum* and *L. ulocalyx* in both large and small plants, plus the lack of such initials in many leaves of plants that otherwise have them, indicates that *L. ulocalyx* is not a taxon valuable to recognize. The rhizoid initials are apparently a common trait of *L. excelsum*. In 2% KOH mounting fluid, the rhizoid initials are clearly identifiable as white cells in the deep yellow areolation.

Monoicous plants are associated with false anisospory and are identifiable as *L. viticulosoides* var. *viticulosoides* (or the more uncommon and local var. *exasperatum* if papillae are tall and columnar). Given that dioicy is associated with a complex of unusual traits that are found only in the New World including dioicy itself, all plants with the gametophytes of *L. viticulosoides* var. *viticulosoides* in the Old World may be named as that taxon when sporophytes or sexual structures are unavailable. When plants exhibit dioicy, or if sterile have very long-lanceolate to cirrhate leaves with short-rectangular distal laminal cells, and often but not always have scattered clear areas the size of a cell among the distal laminal cells that may be interpreted as rhizoid initials, then these plants may be identified as *L. excelsum* var. *excelsum* (or as *L. excelsum* var. *flagellaceum* if distinct clusters of flagellate branchlets are present).

Leptodontium excelsum most saliently differs from *L. viticulosoides* by the dioicous sexual condition, short-cylindric-celled tomentum tips present though uncommon, smaller laminal cells, and colored longitudinal stripes of 1–5 serial cells reaching up from the leaf base. *Leptodontium viticulosoides* does not have such colored stripes but the same sort of cells, similarly filled with protoplasm form uncolored stripes with the same morphology. *Streptotrichum ramicola* has colored stripes and a high-sheathing leaf base, and ca. 8 punctate simple papillae per distal laminal cell lumen, while *L. excelsum* and *L. viticulosoides* have only ca. 4 simple papillae per distal laminal cell lumen, and these are slightly larger. Although *L. excelsum* has short-cylindric cells often terminating stringy tomentum, while *L. viticulosoides* usually has elongate narrow cells terminating arbusculate tomentum, the two share more traits with each other than with other groups.

Leptodontium excelsum has been found to be successional on fir trees in the high Appalachian mountains of North America after damage to bark caused by Balsam Woolly Aphid (Zander 1980).

Distribution: North America (U.S.A., Mexico); Central America (Guatemala, Honduras, El Salvador, Nicaragua, Costa Rica, Panama); West Indies (Haiti, Dominican Republic, Jamaica); South America (Venezuela, Colombia, Ecuador,

Peru, Bolivia, Brazil); soil, trees, rock, humus, at moderate to high elevations.

Representative specimens examined (see Zander 1972).

1a. Leptodontium excelsum var. flagellaceum (E. B. Bartram) R. H. Zander, **comb. nov.**
Leptodontium sulphureum var. *flagellaceum* Bartr., Contr. U.S. Natl. Herb. 26: 75. 1928.
Leptodontium viticulosoides var. *flagellaceum* (Bartr.) R. H. Zander, Bryologist 75: 255. 1972
Leptodontium ulocalyx var. *flagellaceum* (E. B. Bartr.) B. Allen, Mo. Fl. Centr. Amer. 2: 119. 2002.

Illustrations: Allen (2002: 121); Zander (1972: 247; 1994: 263).

The variety *flagellaceum* is distinguished by clusters of thin branchlets with reduced, scale-like leaves arising from leaf axils. See discussion by Allen (2002: 119) for additional observations.

Distribution: North America (Mexico); Central America (Costa Rica, Guatemala, Honduras); on trees, at high elevations.

Specimens examined: see Zander (1972).

2. Leptodontium scaberrimum Broth., Symbolae Sinicae 4: 36. 1929. Type: Yünnan, bor.-occid.: Ad confines Tibeticas sub jugo Doker-la, bambusetis, 3600 m, Handel-Mazzetti 8174, 18.IX.1915, H.

Plants in dense mats, yellow-green distally, brown proximally. **Stem** length 4–6 cm, section triangular to rounded-pentagonal on same stem, central strand absent, sclerodermis present, hyalodermis absent. **Axillary hairs** with basal 1–2 cells brown, ca. 14 cells long. **Tomentum** arbusculate, very pale brown to white, all elongate or some ending in several short-cylindric cells. **Leaves** appressed-incurved with tips spreading and twisted once, margins infolded when dry, squarrose-recurved when wet, lanceolate, 2.5–3.4 mm in length, base ovate, not strongly differentiated, distal lamina carinate; **leaf margins** broadly recurved in proximal 2/3, strongly decurrent as a wing of many red cells retained on the stem after leaves are stripped off, sharply and irregularly closely dentate in distal 1/3–1/2; **leaf apex** narrowly obtuse, ending in a narrow ligule which may be weakly thickened in some leaves, either narrowly rounded at apex or ending in a blunt cell, leaf apex plane; **leaf base** ovate, weakly sheathing, gradually differentiated in proximal 1/4–1/3 of leaf base, smooth in 4–5 rows at insertion, red in 2–3 rows, no stripes, basal cells 10–12 μm wide, 3–5:1; **costa** concolorous to pale brown, ending 5–6 cells before the apex, adaxial costal cells elongate, rows of cells across costa at midleaf viewed in section 3–4, width at base 90–100 μm, section reniform in shape, adaxial stereid band in 1 layers, abaxial band in 2–3 layers, fully included guide cells 3–4; **medial laminal cells** rounded-hexagonal, leaf border not differentiated, marginal cell thickness in section about same compared to medial cells, medial cell width 9–13 μm, 1:1, cell walls thin to thickened at corners, lumens rounded-hexagonal, superficial cells walls bulging on both sides; medial laminal papillae spiculose, branching from a central position or column over the lumen with the appearance of a crown-like ring (the same as those of *L. capituligerum*, etc.). **Specialized asexual reproduction** absent. **Sexual condition** not seen. **Sporophyte** not seen. **KOH laminal color reaction** yellow.

Illustrations: Chen (1941: 319); Li et al. (2001: 199). **Figure 13-4.**

Leptodontium scaberrimum is restricted to China, and few collections have been seen. It is here associated with *L. viticulosoides* by the presence of arbusculate tomentum (red below, hyaline above), lack of a stem hyalodermis, the more or less trigonous distal laminal cells, the long decurrencies of leaf margins, and lack of gemmae. Distinguishing traits between *L. scaberrimum* and *L. viticulosoides* and *L. excelsum* are given in the classical key. The laminal cells are rather large and clear, much like those of *L. capituligerum*.

The leaf decurrencies are larger than those of *L. viticulosoides* and extend down the stem as a long triangle, often up to 6 cells wide where the leaf was torn off and up to 15 cells in length down the stem. The stem transverse section commonly shows these as "tails" depending from the section (similar to those of the genus *Triquetrella*). The cells of the decurrencies are likewise somewhat inflated, rectangular, 3–5:1. The tall narrow papillae basally branching 2–4 times are quite like those of *Stephanoleptodontium*. That, and the blunt ultimate tip of the leaf apex in all or at least some leaves, associates this taxon with *Stephanoleptodontium* but the triangular stem section and broad leaf decurrencies are distinctive. Horizontal gene transfer may be considered but *L. scaberrimum* does not (now) occur with species of *Stephanoleptodontium*. As with *L. viticulosoides* var. *exasperatum*, the papillae are centered over the lumens, but in *L. scaberrimum* papilla are usually branching bifid, and are narrower. Occasional transitional specimens of *L. viticulosoides* from China have stems with some winged decurrencies, but such specimens have the rounded-pentagular stem section, a sharp leaf apex, and the characteristic scattered papillae of *L. viticulosoides*.

Geheebia giganteus (Funck) Boulay may be mistaken for *L. scaberrimum* because of its large size, similar leaf shape, broadly decurrent laminal margins, and angular laminal cell lumens, but the former differs in the sinuous-walled laminal basal cells, lack of marginal teeth, and central strand present in the stem.

Distribution: China (Sichuan and Yunnan provinces); soil, rock, shrubbery, moderately high elevations.

Specimens examined (see also holotype given above):
China: Sichuan Prov., Jiulong Co., below Wuzu Hai (lake), wall of granite boulder, 3700 m, J. Shevock 36119, Aug. 4, 2010, MO; Sichuan Prov., Ba Tang Co. Zi La (Yi Dun) District, soil, 3500-3560 m, Si He 31608, July 8, 1983, MO; Sichuan Prov., Xiang-Cheng Co., 3600-3750 m, X.-J. Li 3082, Aug. 12, 1981, KUN; Yunnan Prov., soil under shrub, 2925 m, Ma W.-Z. 11-1856, Mar. 5, 2011, KUN.

3. Leptodontium viticulosoides (P. Beauv.) Wijk & Marg., Taxon 9: 51. 1960.
Neckera viticulosoides P. Beauv., Prodr. 78. 1805. Type: Réunion, Bory-St.-Vincent, s.n. (not seen).

Didymodon squarrosus Hook., Musci Exot. 2: 160, f. 1–8. 1819.

Leptodontium squarrosum (Hook.) Hampe, Öfvers. Förh. Kongl. Svenska Vetensk.-Akad. 21: 227. 1864.
Leptodontium squarrosum var. *hildebrandtii* Cardot, Hist. Phys. Madagascar, Mousses 206. 1915 fide Sloover (1987).
Leptodontium squarrosum var. *paludosum* (Renauld & Cardot) Cardot, Hist. Phys. Madagascar, Mousses 206. 1915 fide Sloover (1987.)
Holomitrium maclennanii Dixon, Smithsonian Misc. Collect. 72(3): 2. 1920.
Leptodontium maclennanii (Dixon) Broth., Nat. Pflanzenfam. (ed. 2) 10: 268. 1924.
Leptodontium acutum (C.H. Wright) Broth., Die natürlichen Pflanzenfamilien, Zweite Auflage 10: 268. 1924. *Holomitrium acutum* C.H. Wright, Journal of Botany, British and Foreign 30: 264. 1892. = *Leptodontium viticulosoides* fide Sloover (1987).
Leptodontium abbreviatum Dix., Notes Roy. Bot. Gard. Edinburgh 19: 284. 1938 fide Moss Fl. China, p. 200.
Leptodontium viticulosoides var. *abbreviatum* (Dix.) Wijk & Marg., Taxon 11: 221. 1962.
[*Leptodontium chenianum* X.J. Li & M. Zang, Int. Bryol. Symp. Chen's Cent. [Abstracts] 16. f. 1. 2005. The holotype, China, Yunnan, Wei Xi County, Ye Zhi, alt. 1900 m, near Lan-Cang River, in forest of Pinus and Citrus, Feb. 11, 2003, presumably at HKAS, is unavailable but from the description this species is probably *L. viticulosoides*.]
For additional synonymy see Zander (1972).

Illustrations: Aziz & Vohra (2008: 249); Chen (1941: 318–319); Gangulee (1972: 780, 783); Magill (1981: 190); Norris & Koponen (1989: 107); Zander (1972: 247, 257; 1993: 134).

Plants in turfs or loose mats, greenish to yellowish brown distally, brown proximally. **Stem** length 2–10 cm, section rounded-pentagonal, central strand absent, sclerodermis absent, hyalodermis present. **Axillary hairs** of ca. 6 cells, basal 2 brown or basal more firm-walled or of 9–14 cells, all hyaline and thin-walled. **Tomentum** arbusculate, rarely macronematous and ropy, very pale brown to white, usually terminating in elongate cells, short-cylindric terminal cells rare, knobs usually lacking. **Leaves** erect to spreading, flexuose-twisted to contorted when dry, squarrose-recurved when wet, ovate- to long-lanceolate, 3.3–3.5 mm in length, distal lamina carinate; **leaf margins** recurved in proximal 1/2–2/3, distantly dentate in distal 1/3–1/2; **leaf apex** acute or broadly acute, apex keeled, rhizoid initial cells (scattered clear cells in the apex) absent; **leaf base** ovate, sheathing, differentiated across leaf, no stripes, 2–4 rows marginal cells shorter and less colored, basal cells filling only 1/4–1/3 of ovate leaf base, basal cells 9–12 μm wide, 3–4:1; **costa** concolorous, ending 3–6 cells before apex, adaxial costal cells longitudinally elongate, rows of cells across costa at midleaf viewed in section 4–8, width at base 60–70 μm, section reniform in shape, adaxial stereid band in 1–2 layers, abaxial band in 1–3 layers, fully included guide cells 4; **medial laminal cells** subquadrate, leaf border not differentiated or outside lateral wall of marginal cells thickened, marginal cell thickness in section same compared to medial cells, medial cell width (8–)11–15 μm, 1–2(–3):1, cell walls thickened at corners to strongly trigonous, superficial cell walls weakly convex; medial laminal papillae low, simple to bifid, 1–4 per lumen. **Specialized asexual reproduction** absent. **Sexual condition** autoicous, seldom dioicous. **Perigonia** lateral on stem, gemmate, 0.7–1 mm in length; antheridia 300–350 μm long. **Perichaetial leaf length** (4–)7–12 mm. **Seta** (0.5–)1.6–2.2(–2.5) cm in length. **Capsule** cylindric, 3–4 mm in length, exothecial cells thin- to thick-walled, 25–40 μm wide, 2:1, stomates phaneropore, at base of capsule, annulus of 4–6 rows of differentiated reddish brown cells; **peristome** teeth linear, occasionally with trabeculate prostome up to half the height of the peristome tooth, in 8 sets of 4 rami, (75–)250–400 μm, lightly striated, basal membrane absent, **operculum** conic-rostrate, 0.9–1(–1.5) mm, cells straight. **Calyptra** cucullate, smooth, calyptra length ca. 6 mm. **Spores** heterogeneous in two size classes, the larger nearly spheric, the smaller lenticular to tetrahedral, 17–20 and 23–25 μm in diameter, light brown, lightly papillose. **KOH laminal color reaction** yellow.

Varieties 2, *L.* var. *viticulosoides* and *L.* var. *exasperatum.*

Leptodontium viticulosoides is easily identified by the serrate leaves, only weakly superficially convex distal laminal cells, which have thick internal cell walls and angular lumens, and the stem epidermis is not differentiated as a hyalodermis. The stem is tough and makes a distinctive snap when sectioned, possibly due to lack of a softening hyalodermis. This species is autoicous and therefore often abundantly fruiting. The perichaetial leaves are much longer than the cauline leaves, and those of this species are even longer than those of most species in the Streptotrichaceae, and so are immediately evident. The rather small perigonial buds are usually hidden among the thick tomentum, and are gemmate, 0.7–1 mm in length, single or 2(–3) in a cluster, often on a short, thick stalk. The perigonia may be present on both perichaetiate plants and also on separate stems far from the fruiting plants, all in the same mat in one collection (e.g., Rwanda, T. Pócs 6035, EGR).

In Toluidine Blue O stain, the leaves are green to bluish green, and the tomentum blue to pinkish blue. As this stain is used with vascular plant tissue to distinguish lignified tissue (staining green to bluish green) from primary walls (pinkish blue), it is possible that lignin-like cell wall components are present. Despite much study, lignin is not known for the bryophytes (Hébant 1977: 66) and the toluidine blue test is not entirely diagnostic.

Arbusculate tomentum means that the tomentum arises from a central thicker and more colored stalk or small set of stalks, and branches outwards from that like a shrub, with ultimate branches gradually tapering. The tomentum is usually red in KOH, but in nature it grades from grayish to red. Arbusculate tomentum is found in *Leptodontium* s.str. species, *Williamsiella araucarieti,* and also elsewhere in Pottiaceae, e.g. *Tortella tortuosa,* but not in other species or genera segregated from *Leptodontium* s.lat. The arbusculate tomentum has thick walls, and thinner internal cell walls angled at 40°, with, often, one branch emerging before the internal wall from the space left by the angle. This form of tomentum is similar to the arbusculate protonema emerging from hurt stems of other mosses, e.g. *Anoectangium euchloron* (Schwägr.) Mitt., but such protonema has thin cell walls, internal walls mostly transverse at 90°, and one or sometimes two branches emerging just before the internal cell wall. The tomentum of *L. viticulosoides* may have a few knobs laterally but the ends of the tomentum are almost always of tapering, elongate cells. Some specimens (e.g. Nyasaland, J. B. Davy 1506, BM), however, may show knobs with 3 to 4 short-cylindric cells in a series, as is characteristic of most species of the Streptotrichaceae. By "short-cylindric cells" is meant a stacked series of chunky disks seen from the side.

The spores in mostly empty capsules may appear entirely of one size class, but some few spores may be found that are in the large size class. The small size class spores are apparently empty, but the larger spores have protoplasmic contents. The var. *viticulosoides* is not monoicous because of chromosome doubling, as its spore chromosome number is 13 (Anderson & Zander 1986). A detailed study was made of a specimen: Malawi, Pawek 1938, March 27, 1976 (MO). The perigonia were clearly lateral on perichaetiate plants, gemmate, antheridia 400–440 μm long, and the paraphyses 400–440 μm long. Careful free-hand sectioning of perigonia on perichaetiate plants resulted in clear parenchymatous (non-rhizoid) connections between perigonial buds and stem. Even the youngest perigonia had a clear connection to the stem. These are autoicous buds (as per Zander 1972, and Anderson & Zander 1986). The spores were heterogeneous in size. The var. *viticulosoides* is therefore autoicous, though according to Zander 1972 sometimes dioicous (or perhaps rhizautoicous). Meiosis is completely normal according to Anderson and Zander (1986). On the other hand, a specimen from Mexico (México, Popocatépetl, Rzedowski 21843, MO) has long-lanceolate leaves, heterogeneous spores, dioicy, and upper laminal cells with no sign of rhizoidal initials. It could be that the "dioicy" that appear to be separate plants are actually long branches with perigonial bud clusters at the apices and now not distinguishable as connected basally. Hypermutability may be the case in *L. excelsum* involving an accelerating complex of additional traits, e.g., cirrhate leaves, long leaf cells, narrow decurrencies, and isospory.

The sporophytes of *L. viticulosoides* have peristomes of sixteen split rather rudimentary teeth that commonly group in four's (four quartets per sporophyte) with large pieces of the high annulus attached to each (thus four quartets each of four teeth plus an external shield of annular cells). There may be some adaptive function for this.

The basal cells of *L. viticulosoides* are distinctive in the red color (in KOH) across the basalmost 2–3 rows of cells. There is no sign of the hyaline juxtacostal area ("miniwindows") as is found in *Crassileptodontium*.

Leptodontiopsis fragilifolia Broth. (Orthotrich-aceae), of central Africa, and *L. orientalis* Dixon of Borneo are similar to *Leptodontium viticulosoides* in general aspect including the dense tomentum, and in anatomy by the lack of a stem central strand, hyalodermis absent, perichaetial leaves not sheathing the seta, cauline leaves entire or with weakly serrulate margins, with long, differentiated leaf base but this not sheathing the stem, yellow in KOH, lamina cells thick-walled and porose or trigonous, with walls of middle lamella evident, with simple and scattered small papillae, but the stem section is strongly convex-pentagonal, the leaf base lacks shorter bordering cells on the margins, and the costa anatomy shows only a layer of abaxial stereid cells enclosing two rows of guide cells.

Distribution: North America (Mexico); Central America (Guatemala); South America (Ecuador, Bolivia, Argentina, Brazil); Africa (Burundi, Cameroon, Congo, Kenya, Malawi, Tanzania, Uganda, Zambia, Zimbabwe, South Africa, Malagasy Republic); Indian Ocean Islands (Réunion, Comoro Archipelago); Asia (India, Nepal, Bhutan, China, Formosa); Southeast Asia (Indonesia, New Guinea); Australia; on soil, trees, bamboo, humus, rock, moderate to high elevations.

Representative specimens examined (see also Zander 1972)
Australia: New South Wales, Blue Mountains National Park, Mt. Wilson, small branch, 960 m, A. E. Newton 5283, October 19, 2000, MO. **Bhutan:** Griffith 24, Herb. East India Co, MO. **Burundi:** Prov. Buvuriz, Mt. Kenya Natl. Park, Percival's Bridge, W slopes of Mt. Kenyaee de la Syguvyaye, galerie forestiere de montagne, talus aride, 1850 m, M. Reekmans 8309A, June 21, 1979, MO; Prov. Bujumbura, Gakara, roadside, 1900 m, J. Lambinon 80/134, Apr. 5, 1980, EGR. **Cameroon:** Western Prov., Victoria Div., Western Prov., Moyange, moist gully, 2500 m, J. P. M. Brenan 4372, April 3, 1948, MO. **China:** Sichuan Prov., Ludingxian, 1900 m, Y. Jia 01869, Aug. 16, 1997, PE; Yunnan Prov., Fugong County, Lishadi

Xiang, N bank of North Fork Yamu River, forest, branch of *Castanopsis* tree, 2475 m, D. G. Long 34443, August 7, 2005, MO; Yunnan Prov., Chang-shan, on stem at ca. 6 m, ravine, 2800 m, Su Yong-ge 957, July 1, 1984, KUN; Yunnan Prov., Tengchong Co., W slopes of S portion of Gaoligongshan, base of tree, 2185 m, J. R. Shevock 28353, May 21, 2006, MO; Yunnan Prov., Yongde Co., Daxueshan National Nature Reserve, metamorphic rock, 2616 m, Ma W. Z. 13-5307, Sept. 26, 2013, KUN. **Comoro Archipelago:** Ngazidja (Grand Comore) Island, W slope of Kartala summit, braches, 1500 m, T. Pocs 9159/BZ, March 19-20, 1991, EGR. **Congo:** Mt. Minagongo, 2820 m, D. H. Linder 2165, February 16-17, 1927, MO; Massif du Kahuzi, km 22 de la piste Mukaba, trunk, 2320 m, J. De Sloover 12.628, Dec. 24, 1971, EGR; volcan Karisimbi, bamboo, J. Louis 5442, Aug. 1987, EGR; Prov. Kivu, Kahuzi-Biega Nat. Park, Mt. Biega, stem of Agauria, 2620, T. Pocs 7592, Sept. 3, 1991, EGR. **India:** Meghalaya, Khasia Hill, Elephant Falls, 1520 m, H. C. Gangulee 6177, October 1966, MO; Kumaon, bridle path to Dhakuri from Loharkhet, Rhododendron tree, 2440 m, Srivastava 22401, May 1966, MO. **Kenya** Mt. Kenya Natl. Park, near summit of Naro Moru National Park Road, W slope of Mt. Kenya, fallen from tree, 3200 m, M. R. Crosby 13317a, December 30, 1972, EGR, MO; Aberdare Natl. Park, near entrance, tree branches, 3100 m, P. Kuchar B8550, December 3, 1979, MO; Mt. Elgon Nat. Park, Erica bark, 3250–3350 m, T. Pocs et al. 9215/BS, Jan. 12, 1992, EGR. **Madagascar:** Ranomafana Natl. Park, tree branches, 1110 m, R. Magill 9465/EQ, Sept. 27, 1994, EGR; Centr. Madagascar, Massif de l'Ankaratra Reserve, Manjakatompo, tree branches, 2050-2130 m, T. Pocs 9481/DD, Sept. 30, 1994, EGR; near Ambositra, mont Vatomavy, siliceous rocks, 1500-1870 m, H. Humbert, s.n., July 23, 1928, CA. **Malawi:** Cholo Mt., rocks, 1400 m, L. Brass 17695, Sept. 20, 1946, EGR; Luchenya Plateau, Mlanje Mt., epiphytic in forest, 1850 m, L. Brass 16679, July 5, 1946, EGR; Zomba Plateau, Zomba Distr., rocks, 1820 m, L. Brass 16165, May 31, 1946, EGR; N. Prov. Nkhata Bay Distr., Vipya Plateau, 37 mi. SW of Mzuzu, vernal pool dome, 5500 ft., Jean Pawek 1938, March 27, 1976 (MO); Nyika Plateau, branches of trees, 2350 m, L. Brass

17305, 17307, Aug. 17, 1946, EGR. **Nepal:** East Nepal, between Basantapur and Dor, humus, 2450 m, Z. Iwatsuki 358, June 6, 1972, MO; Dongga, Kobresia mat, 4680 m, G. & S. Miehe 10658, September 4, 1986, MO. **Réunion:** Cirque de Cilaos, tree trunk, 1600 m, J. De Sloover 17.967, Jan. 1, 1974, EGR; Arrt. du Vent, Plaine des Chicots, tree, 1850-1980 m, M. R. Crosby & C. A. Crosby 8927b, MO. **Rwanda:** Parc des Volcans, E side of Mt. Karisimbi, elfin woodland, 4900-4200, 2800 m, W. G. D'Arcy 9071, 8467, Apri 16, 1975, Mar. 10, 1975, MO, EGR; Pref. de Gisenyi, valley of Sebeya, trunk of tree, 2000 m, J. De Sloover 18.725, July 30, 1974, EGR; Pref. de Gikongoro, forest of Nyungwe, Rwasenoko, tree branch, ground, 2500 m, T. Pocs 6035, Frahm 6106, Aug. 11, 1991, EGR. **Tanzania:** S. Highlands, Rungwe Dist., Mt. Rungwe, bark, 1800 m, T. Pocs 6777/A, Aug. 21, 1972, EGR; Arusha Nat'l Park, E slopes of Meru Crater, tree, 2530 m, M. Crosby 7228, s.d., EGR, MO; Arusha Nat'l Park, Meru Crater, on branch, 2500 m, A. J. Sharp et al. 7841, Aug. 2, 1968, CA; Nguru Mts., peak of Mafulumala, 1920 m, T. Pocs & H. Schlieben 6440/P, Aug. 20, 1971, EGR; Kiboriana Mts., Mpwapwa Peak, 1800 m, T. Pocs & L. Mezosi 6566/M, May 11, 1972, EGR; Morogoro Distr., Mindu Mtns, W of Morogoro, tree bark, 1150-1230 m, T. Pocs, Apr. 12, 1970, EGR; W-Usambara Mts., Mazumbai, bark, 1540 m, T. Pocs 89254/P, Dec. 7, 1989, EG R. **Uganda:** Ruwenzori Mts., Bajuku Valley, on rock at river, 3050 m, E. Esterhuysen 25319, January 1, 1956, MO. **South Africa:** Transvaal, pied des Mamobssimini (?), combe humide, H. A. Junod, 1906, MO; Transvaal, South Transvaal, Woodbush, T. Jenkins s.n., 1910, CA; Transvaal, Mariepskop, NE summit plateau, fynbos, epiphytic, 1900 m, P. Verster 797, June, 11, 1969, UC; Transvaal, Mariepskop, summit, stone, 1970 m, P. Vorster 1110, Oct. 2, 1969, CA; Transvaal, District Lydenburg, near Lydenburg, F. Wilms 2458, Dec. 1894, CA. **Zambia:** Muchinga Escarpment, Mukowonshi Mt, xerophytic bush, 1900 m, T. Pocs 6630, January 18, 1972, MO. **Zimbabwe:** Melsetter, Chimmanimani Mts, N of Mt. Club Hut, moist soil, 1700 m, D. S. Mitchell 242, January 2, 1952, MO; Eastern districts (north), between Penhalonga and Inyangani, Mt. Invanga, S. C. Seagrief 20, 1958, MO.

Key to varieties

1. Distal laminal papillae low, less than 5 μm, usually less than the thickness of the lamina ....
....*Leptodontium viticulosoides* var. *viticulosoides*
1. Distal laminal papillae high, 5–7(–15) μm, usually half the thickness of the lamina or greater
....*Leptodontium viticulosoides* var. *exasperatum*

3a. *Leptodontium viticulosoides* var. *exasperatum*
(Card.) R. H. Zander, Bryologist 75: 254. 1972.
Leptodontium exasperatum Card., Rev. Bryol. 36: 74. 1909. Type: Mexico, Vera Cruz, Orizaba, Müller, s.n., NY—lectotype.

Illustration: Allen (2002: 99); Zander (1972: 247).

One should not recognize a species (e.g., as *L. exasperatum*) on the basis of one trait. Derivation by a distinct character does not translate to distinction at the species level. Two traits are commonly required. A process-based explanation is that two apparently linked traits, if found in widely distributed populations, means that the genome is protected in some fashion. Although the individual traits may be each evaluated as adaptive, together, as a pair, they form a conservative trait complex that is a good basis for species identification. Any one (simple) trait may be only a mutation scattered within a species.

The papillae in *Leptodontium viticulosoides* s.lat. are low in leaves with trigones small or absent, but may be low or high in leaves with large trigones; thus, large trigones do not necessarily generate high papillae. Var. *exasperatum* in Mexico that have long-lanceolate or almost cirrhate leaves may have isodiametric distal laminal cells or they may be short-rectangular. Many Australian collections of *L. viticulosoides* (e.g., Australia, New South Wales, D. H. Vitt & H. Ramsay 27611, MO, spores heterogeneous) that I have seen have high distal laminal papillae quite like those of var. *exasperatum* (Card.) R. H. Zander of Mexico. The distal laminal papillae of Australian plants are 5–7 μm in height or half the thickness of the lamina, within the range of those of var. *exasperatum,* which are 5–7(–15) μm in height. The Mexican populations vary to a greater extent in length, occasionally as long as is the thickness of the lamina, to 15 μm, but there is complete intergradation. In Mexico, var. *exasperatum* may have spores of either one or two size classes. Apparently the two isolated populations are polytopic in origin, with the same genes for this one trait probably separately involved. Interspecific polytopic origin signals the possibility of pre-adaptation (exaption) in *L. viticulosoides*, and a name at the varietal level is appropriate. The key trait is that, in var. *exasperatum,* the laminal papillae (including those on the abaxial surface of the costa) are columnar in side view and about 1/2 the width of the lamina or more, not hemispherical as in the typical variety, which only occasionally has distinctly columnar papillae (in Chinese collections of the var. *viticulosoides*) but these are no more than 1/4 the width of the lamina. One specimen (Australia, Queensland, D. H. Norris 37092, MO) has hemispheric distal laminal papillae on most plants but some plants have the typical columnar papillae, particularly close to the costa, and this is here considered a minor variation exhibited in the variety as an intermediate.

A theory of the adaptive value of tall papillae is that such may be self-cleaning through the "lotus effect" (Barthlott & Neinhuis 1997; Niklas & Spatz 2012: 346). Leaves (e.g., those of the lotus) become hydrophobic by water perching on epidermal papillae 10–20 μm in height, these, however, with epicuticular waxes. Dust particles are mostly too large to land between the papillae and instead adhere to water droplets, which, when they drain, clean the leaves. The var. *exasperatum,* indeed, occurs in dusty environments. This should be examined because the establishment of functions for most elements of moss morphology would be a major advance in understanding.

Distribution: North America (Mexico); Central America (Guatemala, Costa Rica); South America (Ecuador); Australia; soil, rock, decaying logs, branches and trunks of trees, at moderate elevations.

Representative specimens examined (see also Zander 1972)
Australia: New South Wales, Moppy Lookout, 1060 m, D. H. Vitt & H. Ramsay 17611, October 12, 1981, MO; New South Wales, Gloucester Tops, 37 km WSW of Gloucester, tree trunk, 1300

m, H. Streimann 1568, January 25, 1975, MO; New South Wales, Weeping Rocks Track, New England National Park, branches of *Nothofagus*, 1500 m, H. Streimann 65181, September 11, 1999, MO; Queensland, S Bald Rocks, Girraween National Park, moist boulder, granite, 1200 m, D. H. Norris 37014, March 22, 1974, MO; Queensland, Wilson's Peak, boulder, 800 m, D. H. Norris 37092, March 23, 1974, MO.

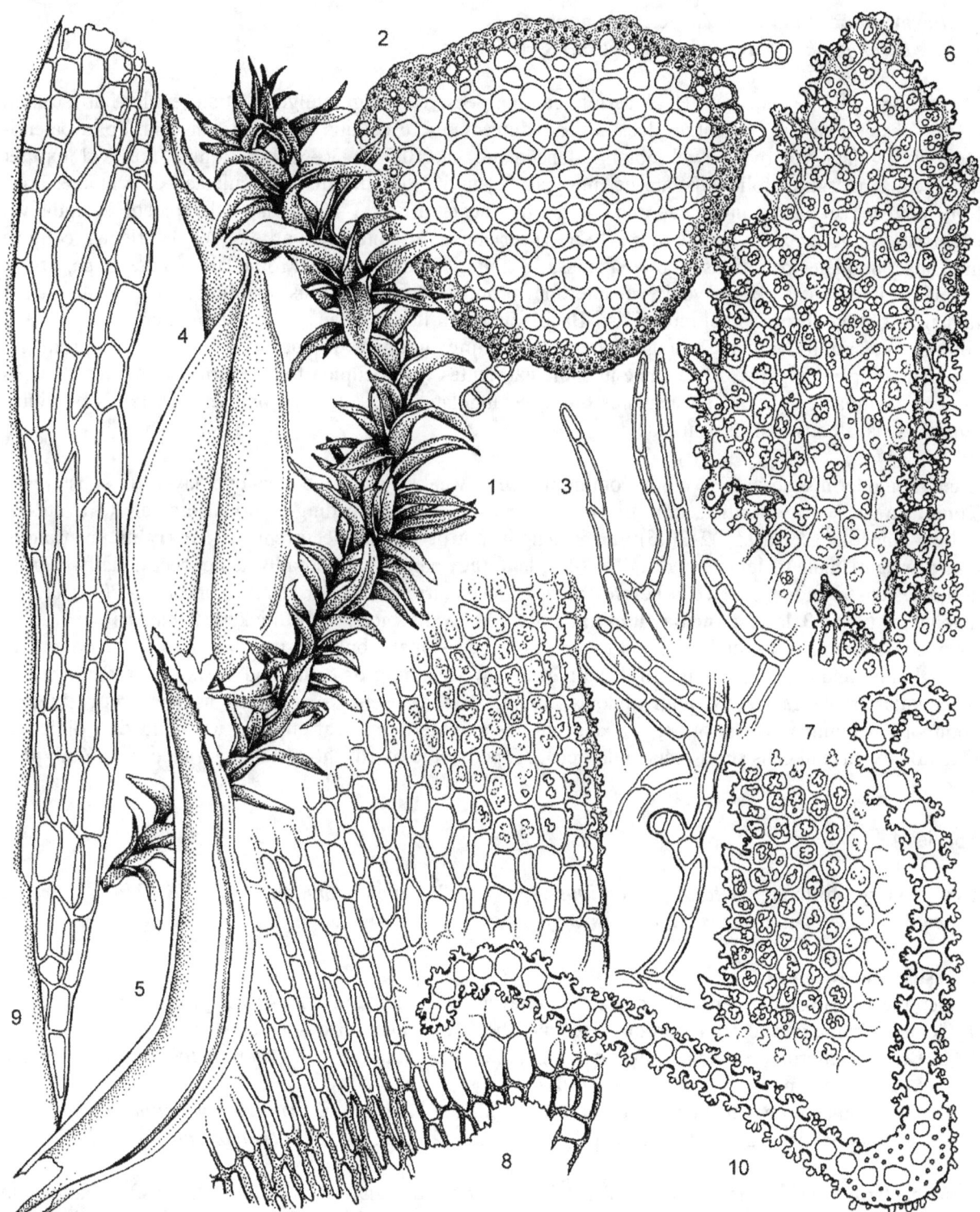

Figure 13-4. *Leptodontium scaberrimum.* 1. Habit, wet. 2. Stem section with two leaf decurrencies. 3. Tomentum, elongate and short-cylindric terminal cells. 4–5. Cauline leaves. 6. Ligule at leaf apex. 7. Distal laminal margin. 8. Basal laminal cells, decurrency torn off when leaf stripped from stem. 9. Leaf decurrency remaining attached to stem on its left. 10. Leaf section.

5. Microleptodontium R.H. Zander, **gen. nov.**
Type: *Microleptodontium flexifolium* (Dicks.) R.H. Zander

Microleptodontium gen. nov. Hyalodermis caulina deest vel magis vulgo exillime evoluta uti stratum unum substereidarum. Tomentum deest vel albidum, tenue, e cellulis elongatis constans. Folia patenti-recurva, ligulata vel ovato- vel oblongo-lanceolata, ca. (0.5–)2–3(–6) mm longa, non vel exiliter decurrentia; margines folii unistratosae, integrae vel in 1/2–1/4 distalis denticulatae vel dentatae; apex folii anguste vel late acutus, interdum gemmifer; costa usque 3–4 cellulas sub apicem evanida vel excurrens, strato adaxiali stereidarum e 1–2(–4) seriebus constante, abaxiali e (1–)2–3 seriebus; cellulae medianae laminales parietibus modice incrassatis, luminibus quadratis vel rotundato-quadratis, parietibus superficialibus planis vel convexis vel interdum protuberantibus, papillis simplicibus vel 2-fidis et per lumen 2–4 atque dispersis vel multifidis; Basis folii humiliter vaginans vel ovata exiliterque distincta, e cellulis rectangulis translucentibus, fenestrellis (miniwindows) desunt, vitiis desunt. Reproductio asexualis propria vulgaris, uti gemmae in caule exorientes vel in cupula foliosa terminali. Status sexualis dioicus. Cellulae exotheciales capsulares brevi-rectangulatae, ca. 2:1; peristomium e 16 ramis brevibus, tenuiter striatis constans. Calyptra laevis.

Stem hyalodermis absent or more commonly very weakly developed as one layer of substereids. **Tomentum** absent or whitish, thin, of elongate cells. **Leaves** spreading-recurved, ligulate to ovate- or oblong-lanceolate, ca. (0.5–)2–3(–6) mm in length, not or weakly decurrent; **leaf margins** unistratose, entire or denticulate to dentate in distal 1/2–1/4; **leaf apex** narrowly to broadly acute, occasionally bearing gemmae; **costa** ending 3–4 cells before the apex to excurrent, adaxial stereid band of 1–2(–4) layers, abaxial of (1–)2–3 layers; **medial laminal cells** with walls moderately thickened, lumens squared or rounded-quadrate, superficial walls flat to convex or occasionally bulging, papillae simple to 2-fid and 2–4 per lumen and scattered or multifid; **leaf base** low sheathing or ovate and weakly differentiated, of rectangular translucent cells, miniwindows absent, stripes absent. **Specialized asexual reproduction** common, of gemmae borne on stem, excurrent costa, or terminal leaf cup. **Sexual condition** dioicous. **Capsule** exothecial cells short-rectangular, ca. 2:1; peristome of 16, short, lightly striated rami. **Calyptra** smooth.

Species 4.

Evolutionary formula: ***Microleptodontium* sp.** > (*Rubroleptodontium*, (***M. flexifolium*** > (*M. gemmascens, M. stellaticuspis, M. umbrosum*)))

Classical key to species
1. Gemmae borne on the stem only, costa not excurrent .... 1. *Microleptodontium flexifolium*
1. Gemmae borne on excurrent portion of costa of the distal leaves, also often on stem.
   2. Leaves dimorphic, gemmae claviform .... 4. *Microleptodontium umbrosum*
   2. Leaves monomorphic, gemmae obovoid.
      3. Gemmae borne on excurrent costa .... 2. *Microleptodontium gemmascens*
      4. Gemmae in an apical toothed cup .... 3. *Microleptodontium stellaticuspis*

1. Microleptodontium flexifolium (Dicks.) R. H. Zander, **comb nov.**
Bryum flexifolium Dicks., Fasc. Pl. Crypt. Brit. 4: [29]. 1801, basionym. See Karttunen, Taxon 37: 156–157. 1988 for discussion of Dickson as authority. Neotype chosen by Frahm and Schu-macker (1986) as "nr. 39 *Bryum flexifolium*" in Dickson Herbarium (BM).

Leptodontium flexifolium (Dicks.) Hampe in Lindb., Öfvers. Förh. Kongl. Svenska Vetensk.-Akad. 21: 227. 1864.

Leptodontium erectifolium Dixon, Farlowia 1: 30. 1943 fide Norris & Koponen, Acta Bot. Fenn. 137: 105. 1987.

Leptodontium filicaule Dixon, J. Bot. 76: 249. 1938 fide Sloover, Bull. Jard. Bot. Natl. Belgique 57(3/4): 428. 1987.

Leptodontium nakaii S. Okamura, J. Coll. Sci. Imp. Univ. Tokyo 36(7): 9. 1915 fide K. Saito, J. Hattori Bot. Lab. 39: 463. 1975.

Leptodontium subfilescens Thér. & Naveau, Bull. Soc. Roy. Bot. Belgique 60: 27. 9. 1927 fide Sloover, Bull. Jard. Bot. Natl. Belgique 57(3/4): 428. 1987.

Leptodontium styriacum (Jur.) Limpr., Laubm. Deutschl. 1: 565. 1888 fide Frahm, Arch. Bryol. 144: 1–4. 2012. Lectotype chosen by Frahm and Schumacker (1987): Austria, auf Gimmer-schiefer-Felsen, 7000′, Steiermark, 25, August (1) 869, Breidler," W.

Leptodontium tenerascens var. *planifolium* P. Varde & Thér., Bull. Soc. Bot. France 87: 354. 1940. Syntypes: volcan Karisimbi, 4,200 m; Volcan Muhavura, 3,500 m; Massif du Ruwen-zori, 1500–1800 m, M. H. Humbert, 1929 (lectotype selected here—volcan Karisimbi, Humbert 7743, P, herb. I. Thériot).

Leptodontium tenerascens var. *subfilescens* (Thér. & Nav.) Thér., Bull. Soc. Bot. France 87: 355. 1940 fide Sloover, Bull. Jard. Bot. Natl. Belgique 57(3/4): 428. 1987.

Bryoerythophyllum yichunense C. H. Gao, Fl. Musc. Chinae Bor.-Oc. 379. 1977 fide Sollman, Trop. Bryol. 21: 65–66. 2002.

Bryoerythrophyllum dentatum (Mitt.) P. C. Chen, Hedwigia 80: 5. 253. 1941, see Zander, Bryologist 75: 231. 1972, and Zander, Bryo-logist 84: 546. 1972.

For additional synonymy, see Zander (1972).

Illustrations: Allen (2002: 105); Aziz & Vohra (2008: 247); Chen (1941: 316); Gangulee (1972: 784); Norris & Koponen (1989: 104); Saito (1975: 265); Zander (1972: 232; 1993: 134; 1994: 262).

Plants turf-forming, yellow-brown distally, brown proximally. **Stem** length 1–3(–5) cm, section rounded-pentagonal, central strand absent, sclerodermis present in 1(–2) layers, hyalodermis weakly differentiated, thin-walled on surface in larger plants, not collapsed. **Axillary hairs** 8–10 cells long, basal 2 brownish. **Tomentum** whitish or brownish, thin, of elongate cells. **Leaves** erect-appressed when dry, spreading-recurved when wet,

lingulate, 1–2(–3.5) mm in length, distal lamina carinate to rather flat; **leaf margins** recurved in proximal 1/2–2/3, dentate to denticulate in distal 1/3; **leaf apex** broadly acute, often mucronate, apex keeled to somewhat flattened; **leaf base** scarcely sheathing, rectangular; **costa** concolorous, ending 3–8 cells before apex, adaxial costal cells elongate, rows of cells across costa at midleaf viewed in section 2–4, width at base 50–65 μm, section reniform to semicircular in shape, adaxial stereid band in 2–4 layers, abaxial band in 1–2 layers, fully included guide cells 4; **medial laminal cells** same across leaf or leaf border sometimes differentiated as 3–5 rows of thick-walled, less papillose cells, distal lamina occasionally spotted throughout with single, isolated brownish cells, marginal cell thickness in section same compared to medial cells, medial cell width 11–15 μm, 1:1, cell walls evenly and moderately thickened, superficial cells walls flat; medial laminal papillae low, simple and hollow (cup- to c-shaped in optical section) or flat, 2- to 3-fid, 4–7 per lumen; **leaf base** differentiated medially in p portion of leaf, miniwindows absent, stripes absent, basal cells differentiated evenly across base or in inverted U, mostly 2–3:1, not porose. **Specialized asexual reproduction** by gemmae borne on short stalks on stem in leaf axils, obovoid, 50–75 μm long, with (2–)3 transverse walls, 1 longitudinal wall. **Sexual condition** dioicous. **Perigonia** terminal. **Perichaetial leaf length** 2–3(–4) mm. **Seta** 1–1.5 cm in length. **Capsule** cylindric, 1.5–2 mm, exothecial cells short-rectangular, 30–60 μm wide, 1(–2):1, stomates phaneroporous, at base of urn, annulus of 2–3 rows of differentiated cells; **peristome** linear or lanceolate, 200–240 μm, lightly striate, basal membrane absent, **operculum** conic-rostrate, cells straight. **Calyptra** cucullate, smooth, calyptra length ca. 3 mm. **Spores** yellowish brown, rounded in diameter, 13–17 μm, lightly papillose. **KOH laminal color reaction** yellow.

Syntypes of *Leptodontium tenerascens* var. *planifolium* at (P) were examined. The "Volcan Muhavura" specimen is apparently a *Zygodon* sp. with a virulent red-purple coloration that is foreign to Pottiaceae but common in Bryaceae. The specimen was similar to *L. flexifolium* but has no marginal denticulation and the laminal papillae are simple and scattered, never bifid; the stem has neither sclerodermis nor central strand. Given that Thériot segregated the Volcan Muhavura specimen as a form, and given that the specimen from Massif du Ruwenzori is missing, the lectotype is chosen here as volan Karisimbi, Humbert 7743 (P, herb. I. Thériot) although it is scrappy. Taxonomically the lectotype is a flagellate form of *L. flexifolium*.

Leptodontium pergemmascens Broth. of China was synonymized by Zander (1972) with *L. flexifolium* but Norris and Koponen (1989) and also Saito (1975) were of the opinion that this is a good species, distinguished by obovoid gemmae, broadly ligulate leaves, strongly dentate leaf margins, and bark habitat (see illustration of Saito, 1975: 465). I have seen the type of the former name (Zander 1972), and still feel it is correctly synonymized. That type and many other specimens of large *L. flexifolium* from China (e.g., Yunnan, Redfearn 33358, MO, Redfearn & Su 2143, MO) are impressive in size (to 2 mm in length) and elaboration of the leaf dentition (up to 25 teeth on each side of a leaf). Smaller plants in China (e.g., Yunnan, Y. Jia J07930, PE) with broad-apex and ca. 1.5 mm in length, have less and smaller teeth (ca. 6 on each side of the leaf) on the laminal margins. There are no combinations of traits distinguishing these and other specimens at the species level if the range of collections worldwide is considered. The lower stem leaves are often much narrower above and weakly dentate in broad-leaved plants; specimens with narrower leaves usually have obovoid gemmae, while the broad-leaved form may have elliptic, conic-pointed gemmae. There are intermediates, such as China (Yunnan, Jiang 83211, KUN) with ca. 6 small teeth on each side of the leaf. while the shape of the leaf and degree of dentition worldwide is correlated within the narrow-leaved expression, including flagellate plants with entire or nearly entire leaves. Interestingly, *Leptodontium viticulosoides* also has a robust form in central China, with less strongly thickened laminal cells walls. I consider both broad-leaved forms as mesic habitat ecotypes or otherwise robust variants common in China. Perhaps both species retain in China ancient genetic material specialized for highly mesic forests now rare elsewhere.

Anagenetic (phenetically complex) species may speciate through isolation of large and small variants in a reduction series, such as in the Pottiaceae with the reduction series *Molendoa hornschuchiana* > *M. sendtneriana* > *M. boliviana,* although specialized functions soon develop in the smaller variants, e.g. *Gymnostomum aeruginosum* > (*G. calcareum* with bulging capsule walls and *G. viridulum* with gemmae). Neither scenario applies to *Microleptodontium flexifolium.* A similar cline of large, well-developed specimens concentrated in China may be observed in *Leptodontium viticulosoides.*

Specimens with very thin laminal cell walls, such as those growing on bark in shade, may exhibit no marginal border (e.g. Yunnan, Ma 12–4086, KUN).

Bryoerythophyllum hostile (Herz.) P.-C. Chen of Yunnan, China (e.g. Ma & Yin 10–1199, KUN) has most of the superficial traits of *L. flexifolium,* e.g., ligulate dentate leaves with thickened margins in 3–4 rows, and differentiated leaf base, but is immediately distinguished by its red color (enhanced in KOH), stem with central strand, and costa with epidermis.

Flagellate branches of *Williamsiella aggregata* are very similar to mature plants of *Microleptodontium flexifolium.* The leaves of the latter are, however, less strongly keeled, blunt and usually have a broad marginal band of thicker walled cells, but the telling distinction is the only weakly inflated hyalodermis of *M. flexifolium.*

Zygodon reinwartii (Hornsch.) A. Braun in B.S.G. may be confused with *Leptodontium flexifolium* because of its similar leaf shape and elongate gemmae, but it is distinguished by its red tomentum, lack of a marginal border of thicker-walled, smoother cells, and the marginal teeth emerging radially (orthogonally) from the margin rather than tilted forward.

Not seen:

Leptodontium flexifolium fo. *compacta* Hessel. ex Rosevinge, Bot. Iceland 2: 452. 1918.

Leptodontium flexifolium fo. *gemmipara* Frahm, Nov. Hedw. 24: 418, 1973 [1975].

Distribution: North America (U.S.A., Mexico), Central America (Guatemala, Costa Rica), South America (Venezuela, Colombia, Ecuador, Peru, Bolivia), Europe (U.K., France, Belgium, Germany, Switzerland, Austria, Russia); Africa (Cameroon, Congo, Ruwanda, Kenya, Lesotho, Tanzania, Malagasy Republic); Asia (India, Nepal, China, Formosa, Japan); Indian Ocean Islands

(Réunion); Southeast Asia (Indonesia, Papua-New Guinea); lava, limestone, rock, soil, litter, rotten log, straw, bark, roofs, mostly moderate to high elevations.

Representative specimens examined (see also Zander 1972):
Austria: Styria, Prope Irdning, Breidler, s.n., MO. **Borneo:** N. Borneo, West Coast Res., Mt. Kinabalu, summit zone, W. Meijer B 10384, July 21, 1960, MO. **Cameroon:** Mt. Cameroun, S slope, lava rock, 3000 m, D. Balazs 80/r, Nov. 27, 1967, EGR. **China:** Fujian Prov., Mt. Wuyi, Huang Guang Hill, rock, 1900 m, Deng-ke Li, Cai-hua Gao 13065, Sept. 27, 1981, PE; Taiwan, Pingtung Co., Kwai-ku to Mt. Pei-ta-wu-shan, soil, dry slope, 2190–3090 m, Ching-chang Chuang 1529, July 19, 1968, MO; Setschwan, SW, Muli, tree trunk, 2850 m, H. Handel-Mazzetti 3181b, s.d., MO; Sichuan Prov., Dao Cheng County, Ju Long District, soil, 3450–3500 m, Si He 31703, MO; Sichuan Prov., Omei Shan, trail to Golden Summit, subalpine coniferous forest, soil over limestone, 2430–3000 m, P. L. Redfearn 34612, August 22, 1988, MO; Sichuan Prov., Kangdingxian, 3200 m, Y. Jia 2281, August 25, 1997, PE; Sichuan Prov., Dujiangyan, 900 m, M. Z. Wang 58347, Aug. 20, 2002, PE; Sichuan Prov., Dujiangyan, 980 m, Y. Jia J06384, Aug. 12, 2002, PE; Sichuan Prov., Wenchuan, 2100–2400 m, M. Z. Wang 58398, Aug. 22, 2002, PE; Sichuan Prov., Maerkangxian, 3000–3200 m, Y. Jia 03242, Sept. 8, 1997, PE; Shaanxi, 3350 m, Z. P. Wei 6248, Sept. 8, 1963, PE; Taiwan, Hua-Lien, Xiu-lin Village, Shi-Men Shan Mt., soil, 3237 m, Si He 36012, August 9, 2002, MO, Si He 36043, August 9, 2002, MO; Taiwan, Pingtung Co., Kwai-ku to Mt. Pei-ta-wu-shan, cliff, 2190–3090 m, Ching-chuang Chuang 1505, 1529, July 19, 1968, PE; Taiwan, Ilan Co., Chi-li-tieng to Mt. Nan-hu-ta-shan, humus, meso slope, 2400–3120 m, C. C. Chuang 1706, Aug. 23, 1968, PE; Taiwan, Taichung Co., Trail to Snow Mt., litter, 2000–2500 m, J. R. Shevock 17998, April 23, 1999, MO; Yunnan Prov., Kunming Municipality, Qiongzhusi Temple, trunk of tree, 2190 m, P. L. Redfearn & Y.-G. Su 2143, July 30, 1984, MO; Kunming Municipality, vicinity of Daxiao, shaded rotten log, 2180 m, P. L. Redfearn et al. 2008, July 28, 1984; Yunnan Prov., montis Piepun, trunk of tree, 3800–3850 m, H. Handel-Mazzetti 3181a, August, MO; Yunnan Prov., Kunming, Panong Distr., trunk of Cupressaceae, 1955 m, Ma W. Z. 12–4086, Aug. 29, 2012, KUN; Yunnan Prov., Kunming, vicinity of Institute of Botany, trunk of *Thuja*, P. L. Redfearn 33358, October 25–26, 1983, MO; Yunnan Prov., between Mekong and Salween, margins of bamboo grove, 3600–3950 m, H. F. Handel-Mazzetti 8364, 9983, H; Yunnan Prov., Shangri-La Co., Xiao-Zhong-Dian, 3200 m, X.-J. Li 81–946, June 14, 1981, KUN; Yunnan Prov., Wei-Xi Co., Li-Ki-Ping, dead wood, 3173 m, W. Z. Ma 10–1312, July 27, 2010, KUN; Yunnan Prov., Yunlong County, Tianchi Forest Nature Reserve, J. Shevock 32147, MO; Yunnan Prov., vicinity of Institute of Botany, Kunming, P. L. Redfearn, 33358, October 25–26, 1983, MO; Yunnan Prov., Lanping Co., dead wood, 2679 m, M. A. Wen-Zhang 10–1489, July 29, 2010, KUN; Xizang, Bomixian, 2600 m, Y. Jia 05667a, July 31, 2000, PE; Xizang, Leiwuqi, 3907 m, Y. Jia J07930, Aug. 10, 2004, PE. **India:** Mekteshwar, Lord Shiva's Temple, wall, 2660 m, J. P. Srivastava 4214, October 1972, MO; Uttaranchal, Nainital Distr., Nainital, Kumaon Himalaya, U.P., epiphytic on *Cedrus deodara* bark, 2000 m, S. D. Tewari 163, May 12, 1989, MO; Madari, on way to Pindari glacier of Kumaun Himalaya, over animal dung, 3600 m, S. D. Tewari 169, June 15, 1983, MO. **Kenya:** North Nyeri Distr., Mount Kenya, rotting log, 3400 m, C. C. Townsend 85/258, January 25, 1985, MO; Mt. Kenya, Sirimon Track, ground, 3300 m, S. Rojkowski 103, Dec. 22, 1975, EGR; Mt. Kenya, W slope, rock, 3000 m, D. Balazs 103/c, Jan. 20, 1968, EGR; Mt. Kenya, lower Rupingazi Valley, moraine, 3050 m, J. Spence 2754/C, July 1, 1984, EGR. **Lesotho:** Pass of Guns, on field mark, 3100–3200 m, B. O. Van Zanten 76.09.930, September 17, 1976, MO. **Madagascar:** SE Madagascar, Andringitra Mts, nature reserve near Antanifotsi, rocks, 1900–2000 m, T. Pocs 9463/AM, September 24, 1994, MO. **Malawi:** Sapitwa, rock, 2740 m, Z. Magombo M4292a, June 25, 1991, MO. **Nepal:** Lirung Glacier, terminal moraine, 4060 m, G. & S. Miehe 11380, September 9, 1986, MO. **Papua-New Guinea:** Eastern Highlands, Bismark Ranges, Mount Wilhelm, tussock grassland, 3600 m, W. A. Weber & D. McVean B–32245, July 5, 1968, MO; Eastern Highlands, Bismark Ranges, Mount Wilhelm, steep slope, 3600 m, W. Weber & D.

McVean 32245, July 5, 1968, EGR; Morobe Prov., Mt. Sarawaket Southern Range, 3 km E of L. Gwam, slope, 3500 m, T. Koponen 32470, July 8, 1981, CA; Morobe Prov., Kewieng No. 1, 4 km S or Teptep airstrip, on roof, 2070–2200 m, T. Koponen 34456, July 27, 1981, CA. **Réunion:** St. Benoit, Piton des Neiges, soil banks, 2370 m, T. A. Hedderson 16634, March 26, 2008, MO. **Russia:** Southern Siberia, Transbaikalia, Agin-Buryat Autonomous Area, vertical surface rock, 772 m, O. M. Afonina 228, July 12, 2006, MO. **Rwanda:** Ruhengeri, Mt. Karisimbi, on Senicio, 3600–3900 m, T. Pocs 8102, September 14, 1991, EGR, MO. **Switzerland:** Zinal, Valais, chalet roofs, P. G. M. Rhodes 290, June 25, 1914, MO; Gipfel des Hexstein, auf verwitterten Grasstocken, 2550 m, J. Baumgartner 2210, MO; Berner Oberland, Haslital, Guttannen, auf einen Schindeldach im Ort, 1300 m, J.-P. Frahm 762266, August 22, 1979, MO. **Tanzania:** Shira Plateau, Kilimanjaro, rocks, 4000 m, K. Rasmussen 109, April 7, 1970, MO; Kilimanjaro Mts., below Machame Hut, stone, 2950–3040 m, T. Pocs 86131/S, MO; Kilimanjaro Mts., Machame route, peaty soil, 3150 m, B. van Zanten 86.08.481, Aug. 8, 1986, EGR; Kilimanjaro Mts., Marangu Route, rocks, 3760 m, S. Rojkowski 313, Jan. 17, 1976, EGR; Mt. Meru, SW slope, soil, 3300–3400 m, T. Pocs 8687/CI, June 15, 1986, EGR; Mt. Meru, SW slope, ground, 3400–3500 m, T. Pocs et al. 88152/Z, June 24–25, 1988, EGR.

2. Microleptodontium gemmascens (Mitt.) R. H. Zander, comb. nov.

Leptodontium gemmascens (Mitt.) Braithw., Brit. Moss Fl. 1: 256. 1887.

Didymodon gemmascens Mitt., Mem. Lit. Soc. Manchester 3: 235. 1868, basionym. Neotype chosen by Frahm and Schumacker (1986): England, on old thatch, Amberley, Sussex, April 1857, Davies s.n., NY.

Didymodon flexifolius var. *gemmascens* (Mitt) Hobk., Syn. Brit. Mosses 60. 1873.

Leptodontium flexifolium var. *gemmascens* (Mitt.) Braithw., J. Bot. 8: 394. 1870.

Didymodon flexifolius var. *gemmiferus* Schimp., Syn. ed. 2: 164. 1876.

Illustrtation: **Figure13-5.**

Plants caespitose or loose mats, yellow-green distally, reddish brown proximally. **Stem** length 1–1.5 cm, section rounded-pentagonal, central strand absent, sclerodermis absent, hyalodermis absent. **Axillary hairs** apparently are the hyaline stalks of the gemmae. **Tomentum** absent, rhizoids absent or of elongate cells. **Leaves** incurved, once-twisted when dry, spreading, recurved at top of leaf base when wet, ovate-lanceolate, 4–5(–6) mm in length, distal lamina broadly channeled and narrowly channeled at costa; **leaf margins** recurved in proximal 1/3, dentate in distal 1/3; **leaf apex** acute to short-acuminate, apex keeled; **leaf base** oblong, weakly differentiated medially as a rounded M, cells thin-walled, 12–20 μm, 2–4:1; **costa** concolorous, costa excurrent in a short gemmiferous awn, adaxial costal cells adaxial costal cells elongate, rows of cells across costa at midleaf viewed in section 2(–4), width at base 70–80 μm, section semicircular to oblate in shape, adaxial stereid band in 1 layer, abaxial band in 1–2 layers, fully included guide cells 2(–3); **medial laminal cells** rounded-quadrate, leaf border of 1–2 rows of less papillose cells or non-papillose and walls somewhat thickened, marginal cell thickness in section same compared to medial cells, medial cell width 9–12 μm, 1(–2):1, cell walls thin-walled, superficial cells walls flat to weakly convex; medial laminal papillae simple, hollow, 4–6 per lumen. **Specialized asexual reproduction** by gemmae borne on costa apex and also in leaf axils, obovoid, 60–75 μm long, 5–7 celled, mostly with 2 transverse septa. **Sexual condition** unknown. **Sporophyte** not seen. **KOH laminal color reaction** yellow.

Leptodontium gemmascens (Mitt.) Braithw. is similar to *L. umbrosum* in the leaves which bear propagula on an excurrent costa, but the former is considered a distinct species as the leaves are not dimorphic—all have an excurrent costa—and the propagula are obovoid, never claviform. A summary of what is known of this "Thatch Moss" was given by Porley (2008), see also Driver (1982) and Werner and Sauer (1994).

From Zander (1972): "Though not directly the concern of this treatment, an important range extension for *L. gemmascens* (Mitt.) Braithw. should be noted. Previously known only from Great Britain, the species was collected on Marion Island in the southern Indian Ocean by Huntley, and distributed as *L. proliferum* (Huntley 795, 2015, 2016, 2047, NY). A number of mosses are known to have bipolar distributions, among them *Sphagnum magellanicum* Brid., *Distichium capillaceum* (Sw.) B.S.G., and *Tortula papillosa* Wils. ex Spruce. ... It is possible that [*Leptodontium*] sect. *Verecunda*, to which *L. gemmascens* belongs, was once very widely distributed, and the older segregates of the group now consist of disjunct or endemic populations. Colonization of at least some disjunctive stations by long distance dispersal cannot be ruled out, especially in view of the variety of structural modifications for asexual reproduction in sect. *Verecunda*," a synonym of *Microleptodontium*.

Distribution: Europe (Belgium, Denmark, France, Germany, Luxemburg, The Netherlands, U.K.; Indian Ocean Islands (Marion and Prince Edward islands, Kerguelen), growing on thatch (wheat straw and reed), grasses, decaying vegetation and other organic detritus.

Specimens examined: U.K.: England, Sussex District, Amberley, pr. Brighton, Davies, s.n., MO (isoneotype from Schimper Collection at BUF); Sussex District, Amberley, G. Davies, April 1858, MO, ex BM; Sussex District, near Blackdown, rotten wood on roof of a shed, W. E. Nicholson, s.n., April 1911, MO; Sussex, Amberley, thatched roof, G. E. Wallace, November 22, 1975, MO; Hurstpierpoint, in tectis stramineis, W. Mitten, s.n., 1865, MO, ex BM. **Kerguelen:** Ile Australia, Golfe du Morbihan, overlooking Lac Alicia, detritus in grassland, 110 m, R. Ochyra 2786/06, December 19, 2006, MO. **Prince Edward Islands:** Marion Island, B. Huntley 795, 2015, 2016, 2047, NY.

3. Microleptodontium stellaticuspis (E. B. Bartram) R. H. Zander, **comb. nov.**
Leptodontium stellaticuspis E. B. Bartram, Bull. Brit. Mus. (Nat. Hist.), Bot. 2: 55. 1955, basionym. Type: Ecuador, Pichincha, *Bell 728* (BM!—holotype; FH!—isotype).

Illustrations: Churchill & Linares (1995: 687); Zander (1972: 237).

Plants in a low turf, green to yellowish brown. **Stem** length 1–2 cm, section rounded-pentagonal, central strand absent, sclerodermis present, 1 layer, hyalodermis present but not collapsed and cells similar to substereids but in larger plants with superficial walls thin. **Axillary hairs** 3–5 cells long, with 2–3 darker basal cells. **Tomentum** absent, rhizoids of elongate cells, few, all of elongate cells. **Leaves** erect-appressed, flexuose when dry, spreading-recurved when wet, oblong-lanceolate, 0.5–1.2 mm in length, distal lamina carinate, leaf base continuous with outer layer of stem, often with an alar tab when stripped off; **leaf margins** recurved in lower 1/2–2/3, distantly dentate in distal 1/3; **leaf apex** acute, apex carinate; **leaf base** ovate, weakly sheathing, inner basal cells differentiated, basal cells 11–13 μm long, 1–2:1; **costa** concolorous, percurrent, ending in a 4–5-toothed cup, adaxial costal cells elongate, rows of cells across costa at midleaf viewed in section 2, width at base 60–80 /μm, section semicircular to circular in shape, adaxial stereid band in 1 layers, abaxial band in 2 layers, fully included guide cells 2; **medial laminal cells** rounded-quadrate, leaf border not differentiated or less papillose or smooth and cells rectangular and thicker-walled in one row in distal 1/4 of leaf, marginal cell thickness in section about same compared to medial cells, medial cell width 9–11 μm, 1–2:1, cell walls evenly thickened, lumens rounded, superficial cells walls bulging on both sides; medial laminal papillae 2–3-fid, crowded towards center of lumen. **Specialized asexual reproduction** by gemmae, obovoid, 25–35 × 45–65 μm, mostly 2 transverse septa, borne in a dentate gemmae cup terminal on leaf and less commonly also in leaf axils. **Sexual condition** apparently dioicous. **Perigonia** unknown. **Perichaetial leaf length** 1.2–1.8 mm. **KOH laminal color reaction** yellow.

Microleptodontium stellaticuspis has been collected on Kerguelen by R. Ochyra. This is another of a set of essentially Andean Streptotrichaceae species that also occur in southern Atlantic or Indian ocean islands, possibly inadvertently distributed by humans. Although one sometimes finds gemmae in the terminal cups, they are not oriented as though they were generated there. It is possible that only swollen hyphae come from rhizoid initials in the cups. The gemmae are often abundantly borne laterally on the stem on short much-branched multicelled stalks. Several mosses, such as *Tortula pagorum* (Milde) De Not., produce propagula which are held in cup-like rosettes formed by enclosing leaves, and, in *Tetraphis pellucida* Hedw., this leafy "gemma cup" is elevated above the plant by a leafless stalk formed by the stem. But, apparently, *M. stellaticuspis* is the only moss with such a propagula-bearing cup formed terminally on a leaf.

Distribution: South America (Colombia; Ecuador; Venezuela); Indian Ocean Islands (Kerguelen), on thatch, grasses, low to high elevations.

Specimens examined (see also Zander 1972): **Ecuador:** Napo Prov., Papallacta, thatched roof, 3200 m, T. Arts 16/054, July 16, 1991, MO. **Kerguelen:** Terre Grande, Presq'Ile Bouquet de la Grye, Port Couvreux, moist grassland, 20 m, R. Ochyra 378/06, November 19, 2006, MO.

4. Microleptodontium umbrosum (Dúsen) R. H. Zander, **comb. nov.**
Tortula umbrosa Dúsen, Ark. Bot. 6(10): 9, 3 f. 4–12. 1907, basionym.

Leptodontium proliferum Herz., Biblioth. Bot. 87: 33, fig. 8. 1916. Type: Bolivia, Cochabamba, Tunariseen, Herzog 3429, JE-lectotype; BM, CANM, H, NY, M, s-PA-isolectotypes.
Streptopogon rzedowskii M. Á. Cárdenas, Phytologia 61: 297. 1986

Illustrations: Herzog (1916: 33); Zander (1972: 237; 1994: 263).

Plants turf-forming, green to yellowish brown. **Stem** length 1–2, section rounded-pentagonal, central strand absent, sclerodermis absent or 1 layer substereid present, hyalodermis present, weakly developed, not collapsed. **Axillary hairs** 3–6 cells in length, basal 1–2 thicker walled. **Tomentum** whitish or brownish, thin, of all elongate cells. **Leaves** erect-appressed when dry, spreading-recurved when wet, dimorphic, distal leaves oblong to ovate-lanceolate, proximal leaves oblong to ovate-lanceolate, 2–3.5 mm in length, lamina carinate distally; **leaf margins** recurved in proximal 1/2–3/4, distal leaves mostly entire, proximal leaves denticulate in distal 1/4–1/3; **leaf apex** acuminate to a narrowly blunt apex, apex flat; **leaf base** scarcely sheathing, inner basal cells

differentiated, basal cells rectangular, 2–4:1, filling leaf base, narrower in 3–4 rows along margins; **costa** concolorous, distal leaves excurrent in a short awn, proximal leaves with costa percurrent to subpercurrent, adaxial costal cells elongate, rows of cells across costa at midleaf viewed in section 4, width at base 70–90 µm, section reniform to semicircular in shape, adaxial stereid band in 1 layer of substereids, abaxial band in 2 layers of substereids, fully included guide cells 4; **medial laminal cells** rectangular, mostly 15–20 × 25–35 µm, leaf border often differentiated as 2–4 rows of thick-walled, less papillose cells, marginal cell thickness in section about same compared to medial cells, medial cell width 15–20 µm, 1–2:1, cell walls thin to evenly thickened, superficial cells walls weakly convex; medial laminal papillae low, 2-fid, 2–4 per lumen. **Specialized asexual reproduction** by gemmae claviform, 30–40 × 85–110 µm, with mostly 3 transverse septa, borne in spheric to cylindrical clusters on sterigmata-like projections of the excurrent costa of the youngest of the upper leaves. **Sexual condition** dioicous. **Perigonia** terminal. **Perichaetial leaf length** to 4 mm. **Seta** 1–1.1 cm in length. **Capsule** cylindric, 1–1.5 mm, exothecial cells short-rectangular. ca. 12 × 26 µm, thick-walled, stomates phaneropore, at base of capsule, annulus of 2–3 rows of differentiated cells; **peristome** teeth linear, 140–160 µm, yellowish brown, indistinctly striate, ca. 6 articulations, basal membrane absent; **operculum** conic-rostrate, ca. 0.4 mm, cells straight. **Calyptra** not seen. **Spores** rounded, 13–15 µm in diameter, light brown, weakly papillose. **KOH laminal color reaction** yellow.

Microleptodontium umbrosum is much like *M. gemmascens,* but the former differs in the dimorphic leaves, having weak, thin whitish tomentum, leaf apex less strongly keeled, medial cell width 15–20 µm (versus 9–12 µm), gemmae longer and claviform, 85–110 µm (versus obovoid, 60–75 µm), and is known to produce sporophytes.

A proposal was made (Cano & Gallego 2008) to conserve the name *Leptodontium proliferum* Herzog against the earlier *Tortula umbrosa* Dusén, which was taxonomically the same species. This proposal failed in committee, thus a new combination here in *Microleptodontium* uses the earlier epithet. The illustration provided with the original description of *Tortula umbrosa* is taxonomically that of *L. proliferum*. It has been reported for South Lancashire, Britain, on decaying grass (Porley & Edwards 2010) as almost certainly an introduced species. Cárdenas (1987) reported this species for Mexico.

Distribution: Mexico; South America (Columbia, Bolivia); Europe (Britain); Africa (Lesotho); organic soil, high elevations.

Specimens examined (see also Zander 1972). **Lesotho:** Mahlasela Pass, 12 km SE of New Oxbow Lodge, peaty soil, 3000 m, J. G. Duckett & H. W. Matcham s.n., April 14, 1994, MO.

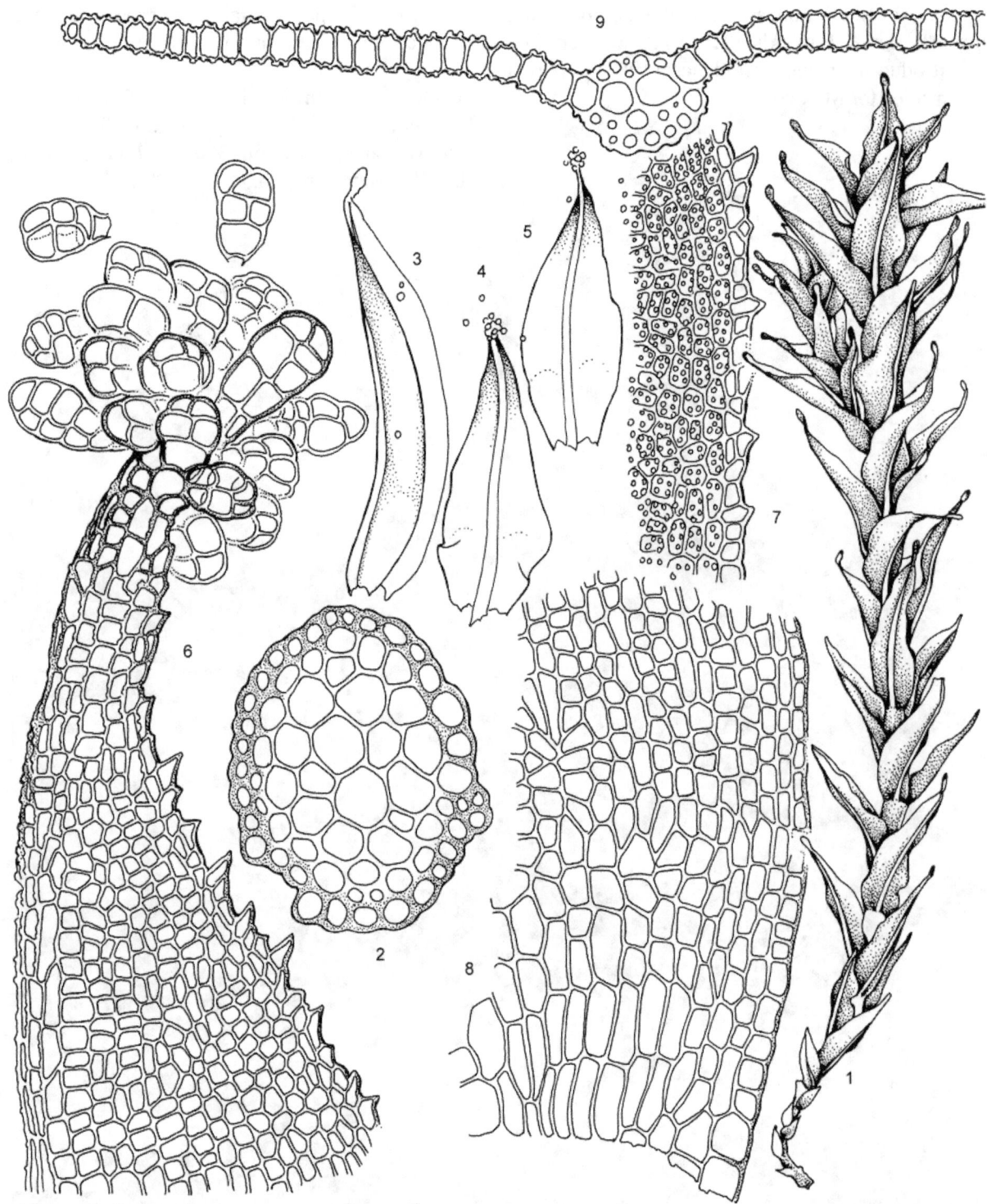

Figure 13-5. *Microleptodontium gemmascens.* 1. Habit, moist. 2. Stem section. 3–5. Leaves. 6. Leaf apex with gemmae. 7. Detail of distal leaf margin. 8. Leaf base. 9. Leaf section.

6. WILLIAMSIELLA E. Britt., Bryologist 12: 62. 1909.
Williamsia Broth., Nat. Pflanzenfam. 1(3): 1190. 1909; non *Williamsia* Merrill, Philippine J. Sci. 3: 165. 1908. Type: *Williamsia tricolor* (Williams) Broth.

Stem hyalodermis present. **Tomentum** absent or arbusculate, brown to whitish, thin or dense, of elongate cells or ending in short-cylindric cells, sometimes arising in lines along stem. **Leaves** squarrose to spreading-recurved, ovate-lanceolate to lanceolate, ca. (2.5–)3–5(–8) mm in length, not or weakly decurrent; **leaf margins** unistratose, dentate in distal 1/2–1/3, occasionally to near base; **leaf apex** narrowly acute to narrowly obtuse; **costa** ending 2–4(–8) cells before the apex, adaxial stereid band of usually 1 layer, abaxial of 1–2 layers; **medial laminal cells** with walls moderately thickened, lumens quadrate to short-rectangular, superficial walls convex or weakly bulging, papillae simple to multifid and 4–6 per lumen and scattered or scablike; **leaf base** sheathing or ovate and broad, of rectangular translucent cells or hyaline as large windows, differentiated miniwindows absent, stripes absent or present. **Specialized asexual reproduction** in most species, of gemmae borne on stem. **Sexual condition** dioicous. **Capsule** exothecial cells short-rectangular, ca. 2–3:1; peristome of 16, smooth or lightly striated rami, occasionally low-spiculose. **Calyptra** smooth.

Species 4.

Evolutionary formula: *W. araucarieti* > (*W. tricolor*, (*W. aggregata* > (*Leptodontium, W. lutea, Stephanoleptodontium*)))

Classical key to species
1. Leaves strongly keeled distally.
  2. Leaves weakly recurved, dentate in distal 1/2 to 1/3; plants occasionally with gemmae ....1. *Williamsiella aggregata*
  2. Leaves strongly recurved to circinnate, not to weakly dentate; plants lacking gemmae .... *Williamsiella interrupta*
1. Leaves somewhat merely channeled distally, often twisted so distal portion lies flat, recurved below.
  3. Leaves 4–8 mm long, leaf base inserted orthogonally to the stem resulting in a "baggy" appearance in the leaf base; dentate in distal 2/3 or nearly to insertion, sometimes erose on the lower margins, laminal cells 10–13 μm in width .... 3. *Williamsiella lutea*
  3. Leaves 3–4 mm long, leaf base inserted at an angle to stem and base not baggy; dentate in distal 1/2–1/3, laminal cells 9–12 μm in width.
    4. Leaf base rectangular, medial basal cells similar to distal cells in width and color, but longer .... 2. *Williamsiella araucarieti*
    4. Leaf base broadly ovate, medial basal cells hyaline, inflated, and iridescent ....4. *Williamsiella tricolor*

The genus *Williamsiella* is here considered central and basal to the *Leptodontium* and *Stephanoleptodontium* lineages constituting a large portion of Streptotrichaceae. The gametophyte of *W. araucarieti* approaches closely that of *Streptotrichum ramicola* but the peristome is reduced to that of most advanced species of the family. The traits of the genus are generalist and intermediate (excepting the inflated basal laminal cells of *W. tricolor*), and are given in its description. The main feature are the low, scattered distal laminal papillae. *Austroleptodontium*

interruptum is similar to *W. aggregata* in many respects but differs strongly in the presence of antrorse prorulae (forward-pointing slanting papillae) terminal of the calyptra, which is primitive.

1. **Williamsiella aggregata** (Müll. Hal.) R. H. Zander, **comb. nov.**
Trichostomum aggregatum Müll. Hal., Syn. Musc. Frond. 1: 580. 1849, basionym. Type: Java, Zollinger 2130 (NY–isotype).

Didymodon aggregatus (Müll. Hal.) A. Jaeger, Ber. Thatigk. St. Gallischen Naturwiss. Ges. 1871–72: 362 (Gen. Sp. Musc. 1: 210).

Leptodontium aggregatum (Müll. Hal.) Kindb., Enum. Bryin. Exot. 63. 1888.

Leptodontium aggregatum var. *hyalinum* (Fleisch.) Broth., Nat. Pfl. 1(3): 1190, 1909 fide Norris & Koponen, Acta Bot. Fenn. 137: 106. 1987.

Leptodontium hyalinum Fleisch., Musci Fl. Buitenz. 1: 369. 1904 fide Norris & Koponen, Acta Bot. Fenn. 137: 106. 1987.

Leptodontium taiwanense Nog., J. Jap. Bot. 20: 144. 1944. Type: Formosa, Ozaki 8498 (US).

Illustrations: Chen (1941: 321); Norris & Koponen (1989: 107). **Figure 13-6.**

Plants caespitose or loose mats, yellow-brown distally, brown proximally. **Stem** length 7–10 cm, section rounded-pentagonal, central strand absent, sclerodermis present, of 2–3 cell layers, hyalodermis present. **Axillary hairs** with all hyaline or basal cells becoming more brownish with age, 7–11 cells in length. **Tomentum** absent or occasional as groups of short, reddish brown rhizoids, absent or short-rectangular to definitely short-cylindric cells. **Leaves** weakly spreading when dry, squarrose when wet, lanceolate from an oblong sheathing base, (2.7–)3–4 mm in length, distal lamina broadly channeled and narrowly channeled at costa; **leaf margins** plane to weakly recurved distally, broadly recurved at mid leaf (at shoulder), revolute in proximal 1/2 of leaf base, dentate in distal 1/3 with evenly spaced sharp weakly papillose teeth of 3–5 cells each; **leaf apex** long-acuminate to a narrowly blunt or sharp apex, apex flat to broadly channeled; **leaf base** oblong, high-sheathing, little decurrent, cells differentiated across leaf base, gradually intergrading into quadrate distal cells at about mid leaf; stripes absent, basal cells ca. 7 µm wide, 4–6:1, walls about as thick as lumens grading to thin-walled, porose or not in same collection; **costa** concolorous, ending 3–4 cells before apex, adaxial costal cells elongate, rows of cells across costa at midleaf viewed in section 4–5(–6), width at base (50–)65–80 µm, section flattened, reniform in shape, adaxial stereid band in 1(–2) layers, abaxial band in 1–2 layers, fully included guide cells 4; **medial laminal cells** irregularly subquadrate,

occasionally transversely or longitudinally rectangular, leaf border not differentiated, marginal cell thickness in section about same compared to medial cells, medial cell width (7.5–)9–13(–15) µm, 1(–2):1, cell walls thickened at corners or evenly thickened or little thickened, lumens rounded-quadrate to rounded triangular or rectangular, angular, superficial cells walls nearly flat to weakly bulging; medial laminal papillae 4–6 per cell, simple to low and scablike. **Specialized asexual reproduction** of axillary obovoid gemmae of 5–6 cells, two transverse walls, ca. 60 µm long; fragile elongate flagellae occasion, with distant leaves like those of *L. flexifolium.* **Sexual condition** dioicous. **Perigonia** lateral as small buds. **Perichaetial leaf length** 5–6 mm. **Seta** 0.9–1.1(–1.7) mm in length. **Capsule** cylindric, 2.5–2.7 mm, exothecial cells evenly thickened, stomates rudimentary, at base of capsule, annulus of 4–6 rows of reddish brown cells, not revoluble; **peristome** straight, of 16 teeth divided to base, 550–600 µm, spirally striate or nearly smooth, 9–11 articulations, basal membrane absent, **operculum** long-conic, 1–1.2 mm, cells straight. **Calyptra** cucullate, smooth, calyptra length 3.8–4 mm. **Spores** rounded to elliptical, 15–20 µm in diameter, brown, papillose. **KOH laminal color reaction** yellow, cells at leaf insertion red.

Williamsiella aggregata and *Crassileptodontium pungens* have similar laminal papillae and may be confused. *Crassileptodontium pungens* never has gemmae, the costa has usually two full layers of stereid cells adaxially (not one or one and a half), and the juxtacostal basal cells are generally hyaline at the leaf insertion (as miniwindows). *Stephannoleptodontium capituligerum* is easily counfounded but has much larger laminal cells and dense, short tomentum of many short-cylindric cells.

The specimen from Papua New Guinea: Eastern Highlands, Bismark Ranges, Mount Wilhelm, tussock grassland, subalpine, 11300 feet, W. A. Weber and D. McVean B-32244, 5 July 1968, MO, has elongate flagellate branches with distant leaves like those of *L. flexifolium* (short ovate, costa ending 4–6 cells below apex) but these lack that species' distal laminal border of thick-walled cells. These branchlets come from much larger plants clearly of *L. aggregatum*, and the branchlets may end in enlarged leaves much like those of *L.*

aggregatum. Some plants of *L. aggregatum* have an obtuse apex with costa ending 4–6 cells before apex. Plants of genuine *L. flexifolium* from exactly the same area Papua New Guinea: Eastern Highlands, Bismark Range, Mount Wilhelm, tussock grassland, 12,000 ft., W. A. Weber & D. McVean B-32245, July 5, 1968 (MO), as *L. erectifolium* Dix., have flagellate branchlets with much the same appearance, but all but the smallest leaves have the characteristic thickened laminal border, while fertile plants have the typical non-flagellate leaves of the species. Unusually robust plants of *L. flexifolium,* Papua New Guinea: Morobe Prov., Mt. Sarawaket Southern Range, alpine grassland, 3400 m, T. Koponen 32533, July 8, 1981 (MO), as *L. aggregatum,* look much like *W. aggregata* but are distinguishable by the green coloration in nature, broadly blunt mature leaves with large, almost erose multi-cell teeth, and smooth celled margin of 4–6 slightly thick-walled cells. The margins of *L. aggregatum* are of few celled, more distant teeth, and are bordered by 1–3 rows of less or non-papillose cells, these occasionally with thickened walls as seen from above or in section. *Austroleptodontium interruptum* is quite similar but lacks the strongly dentate upper leaf margins.

Distribution: Asia (China, Taiwan); Southeast Asia (Indonesia: Java, Sulawesi), New Guinea, Papua New Guinea)

Representative specimens examined:
China: Taiwan, Hualien Co., trail to Hohuanshan North, soil, 3400 m, J. R. Shevock 14424, September 17 1996, MO;Taiwan, Nantou Co., W of Taroko National Park, roadbank of metamorphic rock, 2500 m, J. R. Shevock 17891, April 22, 1999, MO; Taiwan, Miaoli Co., W of Green Lake, litter, humus, 3600 m, J. R. Shevock 18096, April 26, 1999, MO; China, Taiwan, Miaoli Co., NW summit of Snow Mountain, litter and humus, 3600 m, J. R. Shevock 18096, April 26, 1999, MO, Reniform costa, 1 layer; Taiwan, Nantou Co., Ren-ai Village, rocks, 2925 m, Si He 36219, August 9, 2002, MO; Taiwan, Nantou Co., Ren-ai Village, soil, 2000–2200 m, Si He 36463a, August 8, 2002, MO; Taiwan, Pingtung Co., Kwai-ku, 2190 m, C. C. Chuang 1393, July 18, 1968, PE. **Indonesia:** Java Occ., Res. Priangan, G. Gede, in decl. G. Pangrango, 2700-3060 m, F. Verdoorn 11731,

1910, EGR, MO; Java Occ., Res. Prangan, G. Gede, in decl. G. Pangrango, cacumine, 3060 m, F. Verdoorn 24, Aug. 1930, CA; Java Occ., Res. Prangan, G. Gede, in decl. G. Pangrango, terricola, 3020 m, F. verdoorn 72, Sept. 1930, CA; Indonesia, Java, Kandang Badak to Pangrango top, ground under brush, 3000 m, H. S. Yates 2744, November 22, 1927, CA, MO, c.fr.; Java, Pangrango, Moeller, April 26, 1897, MO, c.fr.; Java, Pangrango, E. Nyman 142, July 16, 1898, CA; Java, West Java, Pangrango, 2000–3000 m, J. Motley, 1854, July & October 1854, CA. **Papua New Guinea:** Morobe Prov., Mt. Sarawaket Southern Range, 3 km SE of L. Gwam, open alpine grassland with limestone sinkholes, 3400 m, T. Koponen 32533, July 8, 1981, CA, MO; Morobe Prov., Mt. Sarawaket Southernb Range, L. Gwam, humus, 3400 m, T. Koponen 32250, July 7, 1981, CA; Eastern Highlands, Bismark Ranges, Mount Wilhelm, tussock grassland, short Lake Aunde, subalpine, 3350 m, W. A. Weber & D. McVean B-32244, July 5, 1968, EGR, MO.

2. Williamsiella araucarieti (Müll. Hal.) R. H. Zander, **comb. nov.**
Trichostomum araucarieti C. Mull., Bull. Herb. Boissier 6: 93. 1898, basionym. Lectotype: Brazil, Santa Catarina, Serra Geral, Ule, Bryoth. Brasil. 57) (FH-lectotype; JE, M, NY, S-PA, US—isolectotypes).
Leptodontium araucarieti (Müll. Hal.) Par., Ind. Bryol. Suppl. 224. 1900.
For more synonymy see Zander (1972).

Illustrations: Zander (1972: 267).

Plants in loose mats, yellow-brown to greenish brown. **Stem** length 6–9 cm, section rounded-pentagonal or occasionally grading to rounded-triangular, central strand absent, sclerodermis present, of 2–3 cell layers, hyalodermis present. **Axillary hairs** to 10–11 cells in length, basal 2–3 cells brownish. **Tomentum** arbusculate, very pale brown to white, ending in short-cylindric cells. **Leaves** spreading-recurved when dry, spreading-to squarrose-recurved when wet, lanceolate to ovate-lanceolate with acute apex, 3–4 mm in length, distal leaf straight, distal lamina broadly channeled; **leaf margins** recurved in proximal 1/2–3/4, dentate in distal 1/4–1/2; **leaf apex** acute, apex flat to broadly channeled; **leaf base** oblong, high-sheathing, little decurrent, cells differentiated straight across at leaf shoulders, marginal cells enlarged in recurvature, stripes present, basal cells 9–11 μm, 2–6:1; stalked, rudimentary leaves, lanceolate, usually costate, mostly 0.5–1 mm, occasionally present on the stem; **costa** concolorous, ending 4–8 cells before apex, adaxial costal cells elongate, rows of cells across costa at midleaf viewed in section 6–8, width at base 60–70 μm, section flattened, reniform in shape, adaxial stereid band in 1 layer, abaxial band in 1–2 layers, fully included guide cells 4; **medial laminal cells** subquadrate to occasionally longitudinally elongated, leaf border not differentiated, marginal cell thickness in section about same compared to medial cells, medial cell width (7–)9–11 μm, 1(–2):1, cell walls little and evenly thickened, somewhat bulging, superficial cell walls weakly bulging; medial laminal papillae simple to multifid, scattered or weakly grouped over the center of each lumen. **Specialized asexual reproduction** by gemmae, obovoid, 30–50 × 55–65 μm, with mostly 2 transverse septa, borne on short stalks on the stem. **Sexual condition** dioicous. **Perigonia** lateral. **Perichaetial leaf length** (4.5–)5–6 mm. **Seta** (0.8–)1–1.3(–2) mm in length. **Capsule** cylindric, 0.5–0.6 × 2–2.5 mm, urn ca. 1.8–2.3 mm, exothecial cells evenly thickened, 2–3:1, 25–30 μm wide, stomates phaneroporous at base of urn, annulus of 4–6 rows of reddish brown cells; **peristome** straight, 16, 300–650 μm, spirally striate and low-spiculose, of 9–11 articulations; preperistome occasionally present, short, basal membrane absent, **operculum** conic-rostrate, ca. 1 mm, cells straight. **Calyptra** cucullate, length ca. 3 mm. **Spores** rounded, (7–)11–15 μm in diameter, light brown, lightly papillose. **KOH laminal color reaction** yellow.

Although *Williamsiella araucarieti* is quite similar to *Streptotrichum ramicola* in gametophyte morphology, a detailed comparison shows many differences. The former species has axillary hairs 10–11 cells in length; tomentum arbusculate, leaves broadly channeled distally; costa ending 4–8 cells before apex; laminal border of epapillose cells absent; distal laminal cells 9–11 μm wide; gemmae often present. *Streptotrichum ramicola*, on the other hand, has axillary hairs 4–9 cells in length; thin tomentum; leaves keeled distally; costa percurrent (or appearing so because of the keeled

leaves); distal laminal cells 10–15 μm wide; and gemmae absent. The sporophyte morphology is quite different between the two taxa.

Williamsiella araucarieti is much like *W. aggregata* of Australasia. The type of *W. araucarieti* has longer leaves (3–4 mm vs. 2.7–3.2 mm in *W. aggregata*) and thinner, slightly smaller laminal cells, but other specimens from Brazil (Reitz 2537–US) have the large, thick-walled distal laminal cells with angular lumens characteristic of Javan *W. aggregata*. *Williamsiella aggregata* differs in the strongly keeled distal portion of the leaves, as against the channeled or flattened distal portion of the leaves of *W. araucarieti*, which often lie flat. Bud scales are large in *W. aggregata*, smooth celled, strongly dentate. In South America, *Stephanoleptodontium longicaule* var. *microruncinatum* (Dus.) R.H. Zander may be mistaken for *W. araucarieti* because of the low distal laminal papillae but the former is distinguished easily by the strongly protuberant branching papillae along the basal laminal margins.

Distribution: South America (Peru, Bolivia, Brazil).

Representative specimens examined (see also Zander 1972):
Bolivia: Depto. La Poz, Prov. Nor Yungas, Kolini, road between Cumbre de la Poz and Unduavi, between Laguna Kolini and Cuadrilla 35, humid high Andean grasslands, grassy areas., 4350 m, M. Lewis 88-749 d-4, June 6, 1907, MO. **Brazil:** Santa Catarina, N of Curitibanos, sandy soil, 1010 m, D. H. Vitt 21036, Sept. 4, 1977, MO.

3. Williamsiella lutea (Taylor) R. H. Zander, **comb. nov.**
Didymodon luteus Taylor, London J. Bot. 5: 48. 1846, basionym.
Leptodontium luteum (Taylor) Mitt., J. Linn. Soc., Bot. 12: 50. 1869.
Trichostomum luteum (Taylor) Hampe, Flora 45: 450. 1862.
For additional synonymy see Zander (1972).

Illustrations: Allen (2002: 109); Zander (1972: 263). **Figure 13-7.**

Plants in loose mats, greenish to yellowish brown distally, brown proximally. **Stem** length 3–20 cm, section rounded-pentagonal or occasionally triangular, central strand absent, sclerodermis present, hyalodermis present. **Axillary hairs** of ca. 18 all hyaline cells, basal 3–4 thicker-walled. **Tomentum** absent or very thin and lacking short-cylindric cells. **Leaves** erect to spreading when dry, spreading-recurved when wet, lanceolate with a narrowly obtuse apex, with an elliptical sheathing base, 4–8 mm in length, distal lamina slightly carinate; **leaf margins** revolute in proximal 1/2–2/3, dentate from apex to near the insertion, occasionally only in distal 1/2 of leaf, often becoming erose near leaf insertion; **leaf apex** narrowly obtuse, apex carinate to broadly channeled; **leaf base** elliptical-sheathing, baggy like that of *Stephanoleptodonium longicaule*, base rectangular, 4–5:1, papillose to base or nearly so, cell walls thick and porose to straight and only moderately thickened, no stripes, basal cells 7–11 μm wide, 3–6:1; **costa** concolorous, ending 2–4 cells before apex, adaxial costal cells elongate, rows of cells across costa at midleaf viewed in section 8–10, width at base 70–90 μm, section flattened-reniform in shape, adaxial stereid band in 1 layer, abaxial band in 1–2 layers, fully included guide cells 4; **medial laminal cells** subquadrate to short-rectangular, leaf border often differentiated as 1–5 rows of short-rectangular, thick-walled cells, occasionally enlarged near leaf base, marginal cell thickness in section about same compared to medial cells, medial cell width 9–11(–13) μm, 1:1(–2), cell with evenly and slightly to moderately thickened walls, superficial cells walls flat to weakly convex; medial laminal papillae crowded, low, delicate, simple to multifid or often flattened, usually 4–6 over each lumen. **Specialized asexual reproduction** occasional, by gemmae, obovate, 30–40 × 45–65 μm, with mostly 2 transverse septa, borne on short stalks on the stem distally. **Sexual condition** dioicous. **Perigonia** lateral. **Perichaetial leaf length** 7–8 mm. **Seta** 1.2–1.7(–3) mm in length. **Capsule** cylindric, exothecial cells evenly thick-walled, short-rectangular, 2:1, stomates phaneropore, at base of capsule, annulus of about 6 rows of reddish brown cells; **peristome** linear, 550–650 μm, smooth or weakly obliquely striated, with mostly 10–15 articulations, basal membrane absent, **operculum** conic, 1–1.5 mm, cells straight. **Calyptra** cucullate, smooth, calyptra length ca. 4 mm. **Spores** rounded, 12–14(–20) μm in diameter, brown, lightly papillose. **KOH laminal color reaction** yellow.

The major traits of *Williamsiella lutea* are the low laminal papillae scattered over the lumens and the leaf margins weakly dentate to near the base. This may be one of the most primitive species, having rather long peristomes, basal cells longer but otherwise not particularly differentiated and are papillose to near the insertion, and the medial laminal papillae are not specialized. The leaves are large for the genus, narrowly obtuse, often distinctively wide-spreading when dry (though twisted). The leaf base just at the insertion is somewhat convex, with a "baggy" appearance, which is due to the extreme leaf base being inserted at right angles to the stem, an unusual trait, which helps distinguish this species from *W. araucarieti,* when the marginal dentition does not extent all the way to the leaf base. The leaves commonly have a distinct elongate marginal decurrency. The axillary hairs may or may not have 1–2 brownish basal cells on the otherwise hyaline uniseriate-celled hairs. The marginal cells may be longitudinally elongate, more thick-walled, and less papillose in 2–4 rows but this is not always the case. This species is distinctive in the deep yellow stain commonly, but not always, exuding from material mounted in 2% KOH solution.

This species is one of the few in *Streptotrichaceae* to produce sporophytes in Africa (*Microleptodontium. flexifolium* and *Leptodontium viticulosoides* are the others). Sporophytes are known from only one collection, Tanzania, Kilimanjaro, above Mandara, 2800 m, T. Pócs 6245/A, Sept. 18, 1970, EGR. The sporophytes are produced 1-5 per perichaetium. No perigonia were located.

Although *Williamsiella lutea* has tomentum absent or thin and lacking short-cylindric cells, It does have a characteristic cluster of small papillae on the abaxial base of the costa, as does *Williamsiella tricolor* and *Stephanoleptodontium capituligerum*. The distinctions between these species are clear, however. This papilla formation is in much the same area of the caulogram, but parsimony of nesting in the analytic key indicates no direct evolutionary descent.

Distribution: South America (Colombia, Bolivia, Ecuador, Peru); Africa (Kenya, Tanzania, Uganda).

Specimens examined (see also Zander 1972):
Kenya: Mt. Kenya Natl. Park, Percival's Bridge, W slopes of Mt. Kenya, sunny rock, 2750 m, M Crosby & C. Crosby 13332a, December 30, 1972, MO; Mt. Kenya National Park, near summit of Naro Moru National Park Road, W slope of Mt. Kenya, roadbank, M. Crosby & C. Crosby 13319, Dec. 30, 1972, EGR, MO; Mt. Kenya National Park, vicinity of Percival's Bridge, sunny rock, M. Crosby & C. Crosby 13332A, Dec. 30, 1972, EGR; **Tanzania:** Kilimanjaro, trail to Kibo Peak, soil, 3000 m, A. J. Sharp 7376, July 27, 1968, UC, MO; Kilimanjaro Mts., above Mandara Hut, mossy ground, 2800 m, S. Pocs & T. Pocs 6245/A, Sept. 18, 1970, EGR; in monte Meru, 3300 m, C. Troll 276, 1934, CA, MO; Mount Meru, SW slope below summit, terricolous, 3400–1500 m, T. Pocs et al. 88152/M, June 24-25, 1988, EGR; Arusha Natl. Park, Mt. Meru E slope, 2560 m, O. Mørtensson, s.n., Jan. 7, 1977, EGR; Tanzania, Arusha District., Mt. Meru, Engare Narok gorge, soil, dry situation, 3400–3500 m, T. Pocs & R. Ochyra 88152/M, June 24-25, 1988, EGR, MO; above Mandara, Kilimanjaro, forest floor, 2800 m, T. Pocs 6245, September 18, 1970, MO; Kilimanjaro Mts, Marangu Route, 2800 m, T. Pocs (Bryophyta Exsic. Fasc. 4, 171), July 8, 1976, CA, EGR. **Uganda:** NE part of caldera of Mt. Elgon, Suam Valley, around Hot Springs, humid gorge, ground, 3600 m, Bence Pocs 9220/AC, January 16, 1992, MO; District Toro U2, Mt. Ruwenzori, near Nyamileju Hut, boulders in open forest, 3300 m, K. A. Lye B-36419, December 30, 1968, MO; Suam Valley, Hot Springs, Mt. Elgon, ground, stream bank, 3600 m, M. Chuah et al. 9220/AO, Jan. 16, 1992, EGR.

4. **Williamsiella tricolor** (Williams) E. Britt., Bryologist 12: 62. 1909.
Syrrhopodon tricolor Williams, Bull. New York Bot. Gard. 3: 114. 1903, basionym. Type: Bolivia, La Paz, Williams 2846, NY, holotype.
Leptodontium tricolor (Williams) R.H. Zander in R.H. Zander & E. Hegewald, Bryologist 79: 20. 1976.
Williamsia tricolor (Williams) Broth., Nat. Pflanzenfam. 1(3): 1191. 1909.

Leptodontium calymperoides Thér., Rev. Bryol. Lichenol. 9: 16. 1936. Type: Ecuador, Pichincha, Benoist, 1931 (S-PA), Benoist 4348, BM, isosyntypes.

Illustrations: Zander (1972: 276).

Plants in loose mats, green distally, tan basally. **Stem** length 4–10 cm, section rounded-pentagonal, central strand absent, sclerodermis of 1–2 layers, hyalodermis present. **Axillary hairs** all hyaline, of 10–20 cells, the basal 2–3 thicker-walled, occasionally brownish. **Tomentum** dense, arbusucate or arising directly from stem in lines, deep red, of both elongate and short-cylindric cells. **Leaves** incurved, slightly twisted when dry, spreading-recurved from a sheathing base when wet, lanceolate, 2.5–4 mm in length, distal lamina carinate, weakly cucullate; **leaf margins** narrowly recurved in proximal 1/2–3/4, dentate in distal 1/3; **leaf apex** bluntly acute to narrowly acute, apex broadly channeled; **leaf base** widely elliptic to obovate, with large hyaline windows in lower 1/3 of leaf, rising higher medially in a double-arch, stripes absent, basal cells hyaline, porose, thin-walled, ca. 17–20 μm wide, 3–5:1; **costa** red at base, ending 2–4 cells before apex, adaxial costal cells elongate, rows of cells across costa at midleaf viewed in section 2–4(–6), width at base 60–90 μm, section reniform in shape, adaxial stereid band in 1 layers, abaxial band in 2 layers, fully included guide cells 4; **medial laminal cells** rounded-quadrate, leaf border 1 row of smooth cells, marginal cell thickness in section about same compared to medial cells, medial cell width 9–11 μm, 1:1, cell walls evenly and moderately thickened, superficial cells walls weakly convex; medial laminal papillae scablike, low 2-fid, crowded, 3–4 per lumen. **Specialized asexual reproduction** absent. **Sexual condition** apparently dioicous. **Perigonia** not seen. **Perichaetial leaf length** 4–7 mm. **Seta** 1.1–1.3 mm in length. **Capsule** cylindric, ca. 2.5 mm, exothecial cells thin-walled, 2–3:1, stomates phaneropore, at base of capsule, annulus of 2–4 rows of reddish cells; **peristome** of 16 teeth, bifid or 3–4-fid, smooth, straight, ca. 950 μm, smooth, whitish, of ca. 17 articulations, basal membrane absent, **operculum** 0.9–1 mm, cells straight. **Calyptra** cucullate (orig. descr.), smooth (orig. descr.), calyptra length 4 mm long (orig. descr.). **Spores** spheric, 15–17 μm in diameter, colorless, papillose. **KOH laminal color reaction** yellow.

Although synonymized with *Leptodontium captiluligerum* by Zander (1972). *L. tricolor* (Williams) R.H. Zander was recognized as a good species by Zander and Hegewald (1976). Recognized here are *Williamsiella tricolor,* it is distinguished from *S. capituligerum* by lanceolate leaves, with a narrowly acute or obtuse apex, iridescent (viewed under the dissecting microscope) basal cells forming a hyaline fenestration of thin-walled but porose cells taking up most of the leaf base, elongate weakly papillose cells bordering the basal fenestration in a narrow cartilaginous band, and distal laminal cells smaller, essentially flat-surfaced, with spiculate or scab-like simple to bi- or tri-fid papillae scattered over the lumens. The papillae are like those of *W. lutea,* and *W. tricolor* is related assuming convergent evolution of species with large hyaline windows in the leaf base.

Distribution: South America (Bolivia, Ecuador, Peru); humus, bark, rock, tree trunk.

Additional specimens examined:
Bolivia: Toncoli, Herzog 4380, FH, Cochabamba, Herzog, 1908 (M) and Herzog, 1911 (FH). **Peru.** Depto. Ancash, Prov. Huari: Quebrada Pucavado, E. & P. Hegewald 7715, MO; Prov. Yungay, Laguna Llanguanuco, E. & P. Hegewald 7542, MO; Dpto. La Libertad, Prov. Otuzco, Usquil, E. & P. Hegewald 5287, MO.

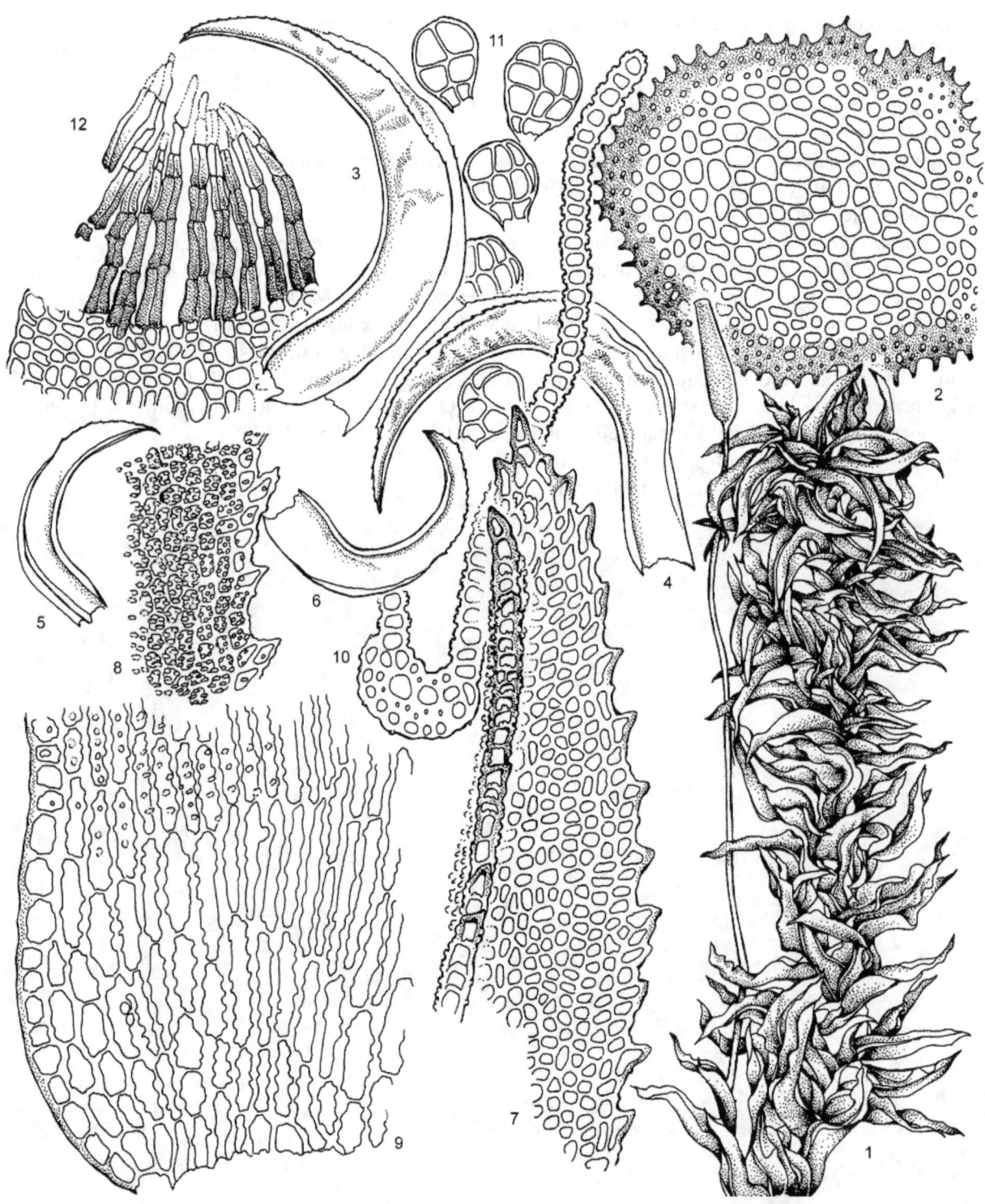

Figure 13-6. *Williamsiella aggregata.* 1. Habit, moist. 2. Stem section. 3–6. Leaves. 7. Leaf apex. 8. Detail of leaf distal margin. 9. Leaf base. 10. Leaf section. 11. Gemmae. 12. Peristome, portion.

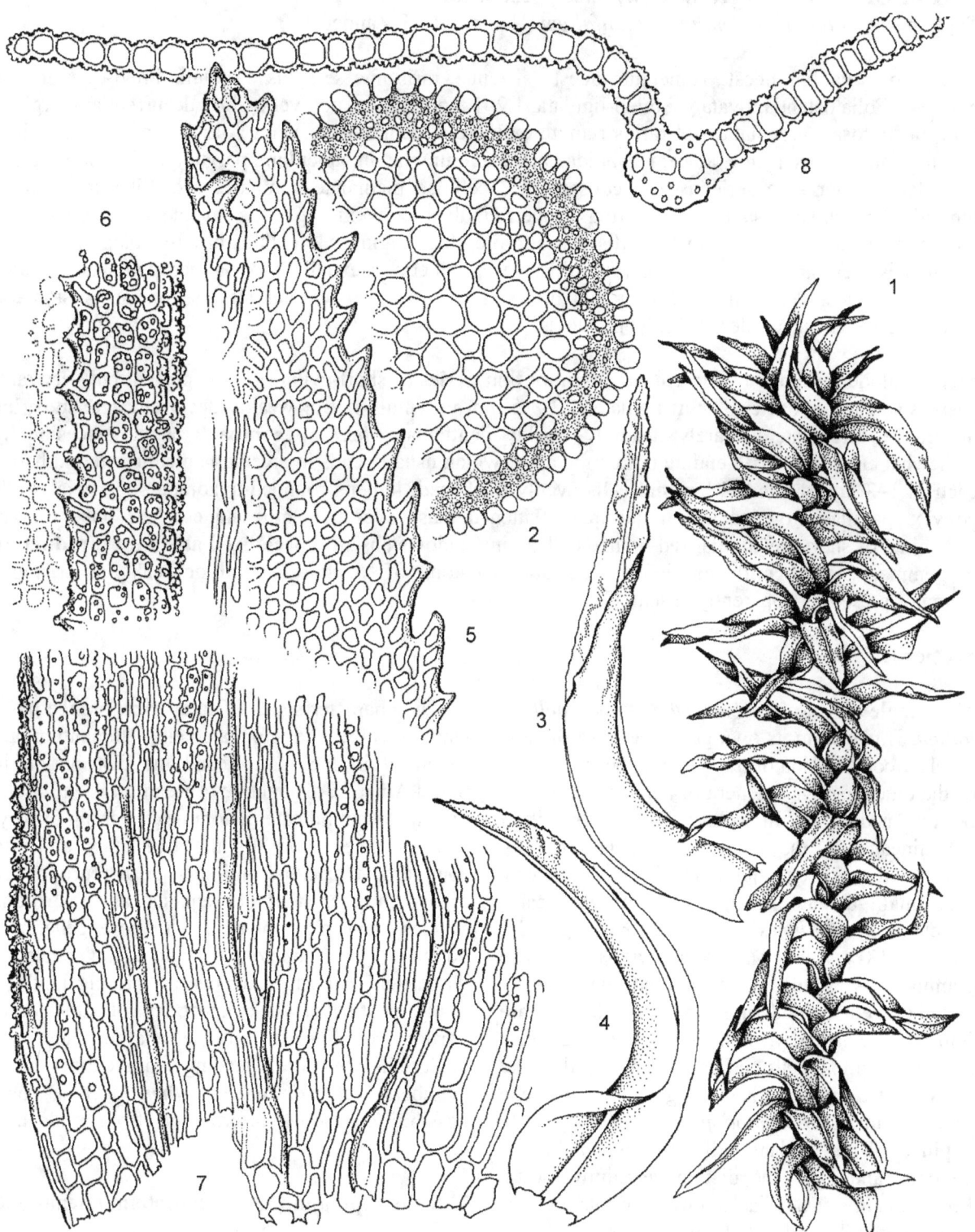

Figure 13-7. *Williamsiella lutea.* 1. Habit, wet. 2. Stem section. 3–4. Cauline leaves. 5. Leaf apex. 6. Marginal laminal cells near leaf base. 7. Marginal portion of leaf base. 8. Leaf section.

7. RUBROLEPTODONTIUM R. H. Zander, **gen. nov.**
Type species: *Rubroleptodontium stellatifolium* (Hampe) R. H. Zander.

Hyalodermis caulina deest. Tomentum deest vel tenue, rarissime seriebus e cellulis brevi-cylindraceis praesens. Folia patentia, ovata vel brevi-ligulata, 1.2–1.5 mm longa, non vel exiliter decurrentia; margines folii unistratosae, crenulatae vel rarius remote exiliterque denticulatae; apex folii late acutus, interdum subcucullatus; costa ante apicem evanida, apice plerumque in apiculum parvum terminante, stratis stereidarum inter se e seriebus 1–2 constantibus; cellulae medianae laminales aequaliter incrassatae, luminibus rotundato-quadratis, parietibus superficialibus convexis, papillis in massam applanatam multiplicem supra lumen totum confertis; basis folii exiliter vaginans, e cellulis brevi-rectangulis e rubro aurantiacis, fenestrellis (miniwindows) nullis. Reproductio asexualis propria per gemmas. Status sexualis dioicus. Peristomium e 16 ramis brevibus, laevibus vel tenuiter striatis constans; stomatia praesentia vel ut videtur desunt. Calyptra laevis.

Stem hyalodermis absent. **Tomentum** absent or thin, series of short-cylindric cells present but very rare. **Leaves** spreading, ovate to short-ligulate, 1.2–1.5 mm in length, not or weakly decurrent; **leaf margins** unistratose, crenulate to rarely distantly weakly denticulate; **leaf apex** broadly acute, occasionally somewhat cucullate; **costa** ending before the apex, which usually ends in an small apiculus, stereid bands each in 1–2 layers; **medial laminal cells** evenly thickened, lumens rounded-quadrate, superficial walls convex, papillae crowded into a multiplex flattened mass over the entire lumen; **leaf base** weakly sheathing, of short-rectangular red-orange cells, miniwindows absent. **Specialized asexual reproduction** by gemmae. **Sexual condition** dioicous. **Capsule** peristome of 16, short, smooth or weakly striate rami; stomates present or apparently absent. **Calyptra** smooth.

Species 1.

The single species, *Rubroleptodontium stellatifolium*, is placed in its own genus because there is no clear combination of traits that embeds it in one of the other segregate genera of *Leptodontium*, and because it does not fit well at the ends or within any lineage in the analytic key. The closely crowded leaves are unique in the Streptotrichaceae. The small, dentate leaf-like enations are nearly matched in *Williamsiella araucarieti* (Zander 1972: 218). Like *Microleptodontium*, it has gemmae, is small in size, and has poorly differentiated hyalodermis, but the leaves are imbricate when dry, and have small, densely papillose laminal cells and orange basal cells. Like *Crassileptodontium*, it has orange basal cells (one species in *Crassileptodontium*) and densely papillae distal laminal cells, but is small, has gemmae and lacks the distinctive miniwindows. Like *Leptodontium*, it lacks a hyalodermis, but has densely papillae distal laminal cells, and has gemmae. Like *Stephanoleptodontium*, it has gemmae, but differs in orange basal cells, plant size and distal laminal cell papillae.

The character states of small size, orange basal laminal cells, and elongate-celled tomentum are advanced in the family, and the distribution in South America disjunctive to Réunion Island in the Indian Ocean indicates a complex vicariant history. The variable nature of the leaf apex and full complement (asexual and sexual) of reproductive organs implies that it remains an anagenetic species (multiple biotypes) with core generative potential. There apparently are no extant species in its particular lineage, nor can a "missing link" be easily postulated to connect it with the basal portions of lineages in the analytic key. An evolutionary position leading from extinct basal members of *Microleptodontium* is the best possibility pending information from molecular data.

1. Rubroleptodontium stellatifolium (Hampe) R. H. Zander, **comb. nov.**
Anacalypta stellatifolia Hampe, Vidensk. Meddel. Dansk. Naturhist. Foren. Kjøbenhavn 34: 37. 1872. Type: Brazil, Rio de Janeiro, Glaziou,

s.n., sub num. 5205 (BM-holotype; S-PA-isotype), basionym.

Leptodontium stellatifolium (Hampe) Broth., Nat. Pflanzenfam. 1(3): 400. 1902.
For additional synonymy see Zander (1972).

Illustrations: Allen (2002: 114); Zander (1972: 232). **Figure 13-8.**

Plants forming a low, dense turf, brown above, reddish to yellowish below. **Stem** length 2–3 cm, section rounded-pentagonal, central strand absent, sclerodermis present in 1 layer, hyalodermis absent; often with chaffy epapillose entire or dentate pseudoparaphyllia at base of branches and stem. **Axillary hairs** with basal 1–2 cells brown, ca. 8 cells in length. **Tomentum** thin or absent, all terminal cells elongate or occasionally with series of 3–4 short cells ending the thicker rhizoids. **Leaves** usually very crowded, appressed-incurved and imbricate when dry, spreading-recurved when wet, ligulate to ovate, 1.2–1.5 mm in length, distal lamina weakly carinate but deeply channeled along costa, leaves at base of stem ovate-lanceolate, with ca. 2 bifid papillae per lumen, acuminate to a long sharp single-celled apiculus; **leaf margins** reflexed in proximal 1/2, edentate but minutely crenulate in distal 1/2 by projecting cell walls and papillae; **leaf apex** acute to obtuse and apiculate, apiculus short, broad and flat or occasionally absent, occasionally 1–2 small marginal teeth near apiculus, apex keeled or occasionally weakly cucullate; **leaf base** ovate, differentiated across leaf, no stripes, 2–4 rows marginal cells shorter and less colored, basal cells filling only 1/4–1/3 of leaf, not porose, no miniwindows, basal cells 9–11 µm, 4–5:1; **costa** when mature green with orange to reddish basal cells, costa ending (2–)4–6 cells before apex, much wider at insertion than at midleaf, occasionally spurred, adaxial costal cells elongate, rows of cells across costa at midleaf viewed in section 3–4, width at base 50–80 µm, section flattened-elliptic in shape, adaxial stereid band in 1–2 layers, abaxial band in 1 layer, 2–4 fully included guide cells; **medial laminal cells** subquadrate, leaf border not differentiated, marginal cell thickness in section about same compared to medial cells, medial cell width 9–11 µm, 1:1, cell walls moderately thickened, superficial cells walls bulging; medial laminal papillae flattened, simple to 2- or 3-fid, 2–

3 per lumen, or multifid and somewhat fused. **Specialized asexual reproduction** by gemmae, obovoid, 35–45 × 75–90 µm, with 2–4 transverse and 1 longitudinal septa, borne on stem. **Sexual condition** dioicous. **Perigonia** terminal or subterminal by innovation. **Perichaetial leaf length** 2–2.5 mm. **Seta** 8–10 mm in length. **Capsule** cylindric, 1.25–1.5 mm, exothecial cells short-rectan-gular, 1–1.5:1, stomates present or apparently absent, annulus 2–4 rows of reddish brown cells; **peristome** linear, 400–450 µm, indistinctly striate, 5–7 articulations, preperistome occasionally present, short, basal membrane absent, **operculum** short-conic, ca. 0.4 mm, cells straight. **Calyptra** cucullate, smooth, calyptra length 2–2.5 mm, clasping the seta. **Spores** round, 15–18 µm in diameter, colorless, weakly papillose. **KOH laminal color reaction** distally yellow to red-orange, basal cells and costa strongly colored red-orange.

Rubroleptodontium stellatifolium occurs in Costa Rica, as reported by Allen (2002: 111), as well as in Colombia and southeastern Brazil. The Costa Rican material entirely lacks the clear, one-celled apiculus, whilst plants from Brazil only occasionally lack this feature. The Costa Rican plants also have more flattened distal superficial laminal cell walls, giving the appearance of that of *Williamsiella lutea*. A herbarium specimen from Colombia (Churchill et al. 14095, MO) is also *R. stellatifolium,* differing as an extreme form with strongly rounded apex, no apiculus and costa ending ca. 6 cells before the apex. A specimen from Réunion (Hedderson 16643, BOL, MO) is intermediate, with rounded, cucullate apex, costa ending 2–3 cells before the apex, but with the apiculus present though reduced to 2–4 cells (Zander & Hedderson 2017).

Rubroleptodontium stellatifolium is easily distinguished by the crowded leaves, bright green distal laminal cells contrasted with the red-orange costa and basal cells, and, except for the clear conical apical cell (which may be lacking), entire but minutely crenulate leaves. The stem sclerodermis has no trace of a thin outer cell walls while more robust plants of the somewhat related *Microleptodontium flexifolium* (Dicks.) Hampe have a weakly distinguishable hyalodermis. *Micro-/lepdotonium flexifolium,* doubtfully present on Réunion, is quickly distinguishable by more distantly inserted leaves, presence of marginal teeth in larger plants, usual presence of a marginal border of 2–4 rows of cells with thickened and less papillose walls, lack of differentially colored basal leaf cells, and larger distal laminal cells 11–15 µm in width. Flagellate forms of *R. stellatifolium* with smaller, distant leaves (e.g. Hedderson 15805) may morphologically phenocopy *Microleptodontium flexifolium* in lacking the red-orange basal cells and distal laminal cells enlarged and bifid-papillose, but are distinguishable by completely lacking any marginal laminal teeth or enlarged pseudoparaphyllia. The abaxial surface of the costa is also more strongly simply papillose in *R. stellatifolium* than in *M. flexifolium.*

Hyophila nymaniana (Fleisch.) Menzel (= *H. rosea* Williams) of tropical and subtropical regions may be confused with this species, especially by its similar leaf shape and red basal cells, but that species has stellate gemmae and the costa in section shows a clearly differentiated adaxial epidermal layer in addition to the adaxial stereid band. The genus *Zygodon* Hook. & Taylor (Orthotrichaceae) may be similar in aspect but is distinguished by the ribbed capsule, perichaetial leaves similar to the cauline leaves and usually homogeneous costal section.

Dentate epapillose pseudoparaphyllia occur at bases of stems or buds as, but are also occasionally found isolated as phylloids on the stem or below the perichaetium. They are nearly matched by those of *Williamsiella araucarieti* (Müll. Hal.) Paris (Zander, 1972: 218). Although similar though smaller pseudoparaphyllia may occur at the base of branches of other Streptotrichaceae species, e.g. *Leptodontium excelsum* (Sull.) E. Britton, and *M. flexifolium,* the sometimes scattered position of those of *Rubroleptodontium stellatifolium* and the common narrowing of the base to a single stalk cell is unique. The sharp contrast between the dentate epapillose pseudo-paraphyllia and the entire, papillose cauline leaves is distinctive, and compared to the size of the plants the former are rather large. The chaffy phylloids are similar to but not as simple as the elongate bi(tri)seriate axillary hairs of *Tortella humilis* (Pottiaceae).

The closely crowded leaves of *Rubroleptodontium stellatifolium* are unique in the family, and show up commonly as two attached sectioned leaf bases when sectioning the stem. This, plus the character states of small size and orange basal laminal cells, indicate a derived status for this species. The widely disjunctive distribution and isolated combination of morphological traits imply that this is a fairly successful remnant of a larger, ancient assemblage of similar species.

Distribution: Central America (Costa Rica); South America (Colombia, southeastern Brazil); Indian Ocean Islands (Réunion); soil, at high elevations.

Representative specimens examined (see also Zander 1972):
Costa Rica: San José, M. R. Crosby 3900B, MO.
Colombia: Antioquia, Santa Rosa de Osos, soil, 2500 m, S. P. Churchill 14095, July 6, 1986, MO.
Brazil: Rio de Janeiro, Parque Nacional Itatiaia, from Agulhas Negras to Brejo da Lapa, 2000–2300 m, J. P. Frahm 1182, July 25, 1977, MO. **La Réunion:** N sloping plateau of la Roche Ecrite, S of St. Denis, terricolous, subalpine ericaceous bush, 2100 m, G. Kis 9420/CB, Aug. 23, 1994, EGR; St. Benoit, St. Benoit, Piton des Neiges, soil banks, 2370 m, T. A. Hedderson 16634, March 26, 2008, MO; Commune Sainte Rose, Piton de la Fournaise, Cratere Commerson, trait to Caverne des Lataniers, Erica-dominated vegetation, on basalt, 2370 m, T. A. Hedderson 15805, 1 Dec. 2004, MO, PRE; Comune Le Tampon, Plaine de Caffres, walk from Mare Piton de la Fournaise, Cratere Commerson, trait to Caverne des Lataniers á Boue to Piton des Neiges, *Erica*-dominated vegetation, on volcanics, 1730 m, T. A. Hedderson 18941, 26 June 2016, MO, PRE; Comune St. Denis, Sentier du Roche Ecrit, from les Haute du Brûlé, rain forest with *Cyathea,* over volcanic rock, 1400 m, T. A. Hedderson 16579, 11 Dec. 2007, MO, PRE; Comune Saint Joseph, Route to Piton de la Fournaise, Puy la Pas de Sables, Ericaceous vegetation in dried out lake bed, 2370 m, T. A. Hedderson 18897, 27 May 2015, MO, PRE.

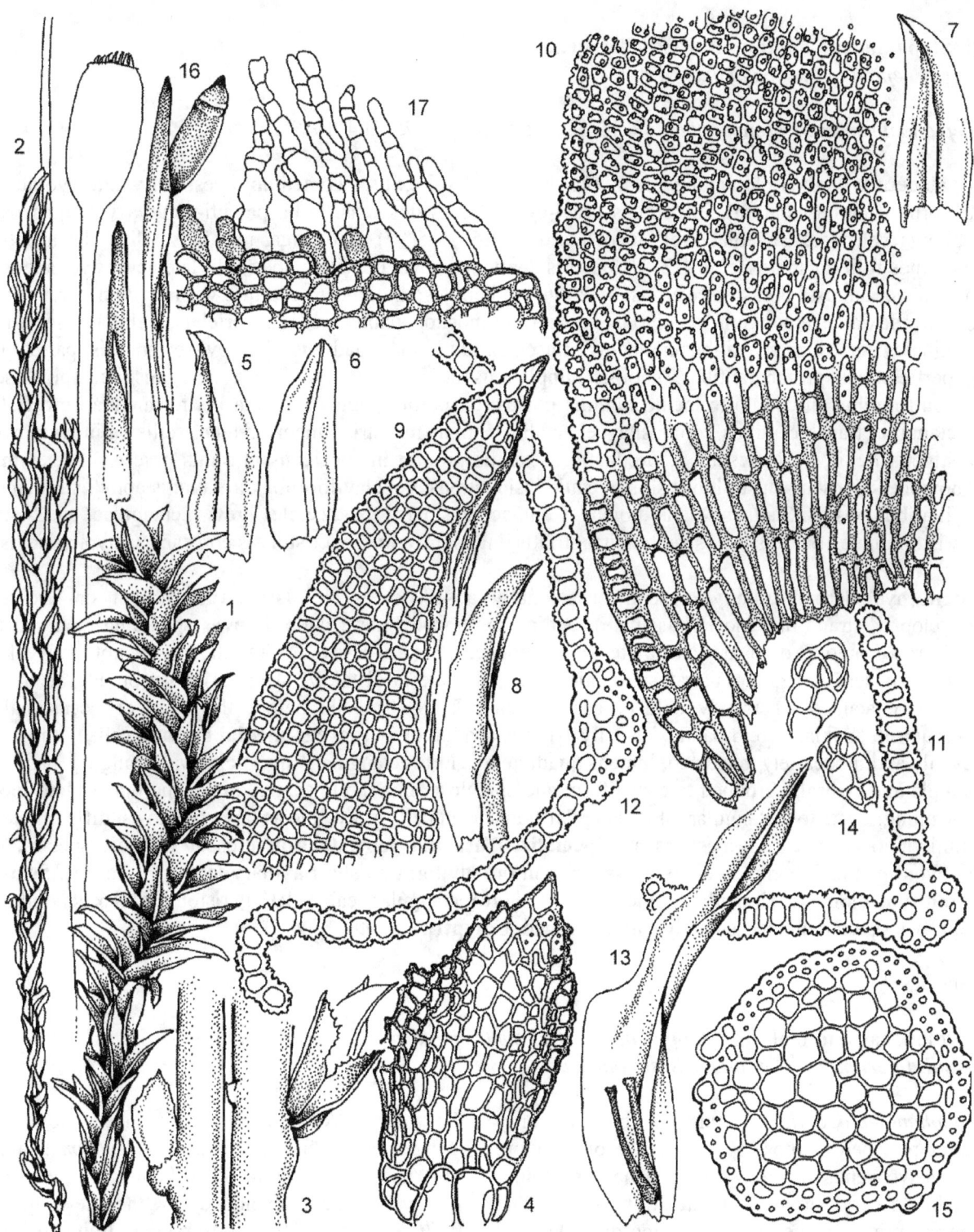

Figure 13-8. *Rubroleptodontium stellatifolium.* 1. Habit, wet. 2. Habit, dry. 3. Pseudoparaphyllia on stem. 4. Pseudoparaphyllium. 5–8. Cauline leaves. 9. Leaf apex cells. 10. Basal laminal cells. 11–12. Leaf sections. 13. Perichaetial leaves and archegonia. 14. Gemmae. 15. Stem section. 16. Calyptra, and calyptra adherent to operculate capsule. 17. Peristome, with prostome.

8. Stephanoleptodontium R. H. Zander, **gen. nov.**
Type: *Leptodontium longicaule* Mitt.

Leptodontium sect. *Coronopapillata* R. H. Zander, Bryologist 75: 264. 1972.

Hyalodermis caulina plerumque adest, interdum deest vel exilis. Tomentum deest vel valide evolutum, filiforme (inarbusculatum), in cellulas brevi-cylindraceas desinens. Folia patenti-recurva vel squarrosa, oblonga vel ovato-lanceolata vel lanceolata, ca. (1.5–)3–5(–7) mm longa, non vel exiliter decurrentia; margines folii unistratosae, in 1/2–1/4 distalis denticulatae vel dentatae; apex folii late acutus vel anguste acutus vel anguste obtusus; costa usque 2–5(–10) cellulas sub apicem evanida vel interdum percurrens, strato adaxiali stereidarum e vulgo serie una composito, abaxiali e 1–3 seriebus; cellulae medianae laminales parietibus aequalibus ac incrassatiusculis, luminibus quadratis vel brevi-rectangulis, parietibus superficialibus protuberantibus, papillis simplicibus ac in centro luminis congregatis vel spiculoso-coroniformibus in circulo in medio luminis positis; basis folii vaginans, ovata vel rectangula, e cellulis rectangulis translucentibus composita vel hyalinis uti in fenestrise magnis, fenestrellis (miniwindows) distinctis nullis, vitiis desunt. Reproductio asexualis propria in speciebus plurimis praesens, e gemmis obovoideis constans in stipitibus sitis in caule insidentibus, vel claviformibus in ramulis aphyllis brevibus axillaribus positis. Status sexualis dioicus. Cellulae capsulares exotheciales brevi-rectangulae, ca. 2–3:1; peristomium e 16 ramis laevis vel tenuiter striatis, intersum spiraliter cristulatis constans. Calyptra laevis.

Stem hyalodermis usually present, occasionally absent or weak. **Tomentum** absent or strongly developed, strand-like (not arbusculate), ending in short-cylindric cells. **Leaves** spreading-recurved to squarrose, oblong or ovate-lanceolate to lanceolate, ca. (1.5–)3–5(–7) mm in length, not or weakly decurrent; **leaf margins** unistratose, denticulate to dentate in distal 1/2–1/4; **leaf apex** broadly acute to narrowly acute or narrowly obtuse; **costa** ending 2–5(–10) cells before the apex or occasionally percurrent, adaxial stereid band of usually 1 layer, abaxial of 1–3 layers; **medial laminal cells** with walls evenly and moderately thickened, lumens quadrate to short-rectangular, superficial walls bulging, papillae simple and centrally grouped to coroniform and spiculose, in a circle centered over the lumen; **leaf base** sheathing, ovate to rectangular, of rectangular translucent cells or hyaline as large windows, differentiated miniwindows absent, stripes absent. **Specialized asexual reproduction** present in most species, of obovoid gemmae borne on short stalks on stem or claviform on short axillary leafless branches. **Sexual condition** dioicous. **Capsule** exothecial cells short-rectangular, ca. 2–3:1; peristome of 16, smooth or lightly striated rami, occasionally spirally ridged. **Calyptra** smooth.

Species 7.

Evolutionary formula: ***S. longicaule*** > (*S. syntrichioides,* (***S. brachyphyllum*** > *S. filicola*), (***S. capiluligerum*** > (*S. latifolium, S. stoloniferum*)))

Stephanoleptodontium may be shortly circumscribed as those species in the older, more inclusive genus *Leptodontium* that have spiculose or bi- to trifid laminal papillae arranged in a ring on the protuberance of each bulging distal laminal cell, and which are also do not have arbusculate tomentum (which excludes the anomalous *Leptodontium scaberrimum*), and which have no longitudinal stripes on the leaf base.

Three species that had been included by Zander (1972) in *Leptodontium* sect. *Coronopapillata* in the New World are not included here: *L. araucarieti* (Müll. Hal.) Paris, *L. luteum* (Tayl.) Mitt. and *L. tricolor* (R. S. Williams) R. H. Zander. These are here excluded to the new genus *Williamsiella* by the conservative trait of low-crustose papillae crowding the superficies of the distal laminal cells. The Old World species *L. aggregatum* (Müll. Hal.) Kindb. is also referred to *Williamsiella*.

Classical key to species

1. Leaves 2–3 mm long, ovate to short-lanceolate, distal laminal cells 9–11 μm wide; stems 1–3 cm long; gemmae common.
   2. Leaves distally keeled and often strongly recurved .... 1. *Stephanoleptodontium brachyphyllum*
   2. Leaves distally plane to weakly keeled or broadly channeled.
       3. Leaves monomorphic, basal cells inflated-hyaline, gemmae borne on short, leafless branches .... 6.*Stephanoleptodontium stoloniferum*
       3. Leaves dimorphic, either ovate and bluntly acute or lanceolate and narrowly acute; basal cells weakly differentiated, gemmae borne on stem............................ 3. *Stephanoleptodontium filicola*
1. Leaves (2–)3–5(–6) mm long, lanceolate, distal laminal cells 11–15 μm wide; stems 3–10 cm long; gemmae occasional.
   4. Stems without hyalodermis, or hyalodermis of cells about the size of the sclerodermis cells and little collapsed when old.
       5. Hyalodermis absent; tomentum absent; leaves not decurrent, costa concolorous, not thickened at the base ....7. *Stephanoleptodontium syntrichioides*
       5. Hyalodermis reduced in size but distinct, of superficially thin-walled cells, little collapsed when old; tomentum short and furry, in longitudinal dense strips; costa blackened and thickened near leaf base .... 4. *Stephanoleptodontium latifolium*
   4. Leaves with hyalodermis of enlarged thin-walled cells that are collapsed in mature parts of stem.
       6. Leaves with hyaline fenestrations formed by sharply demarcated inner basal cells, these thin-walled .... 2. *Stephanoleptodontium capituligerum*
       6. Leaves without sharply demarcated hyaline fenestrations, inner basal cells with moderately thickened or porose-thickened walls .... 5. *Stephanoleptodontium longicaule*

1.   Stephanoleptodontium   brachyphyllum (Broth. & Thér.) R. H. Zander, **comb. nov.**

Leptodontium brachyphyllum Broth. & Thér., Bull. Acad. Int. Geogr. Bot. 16(196): 40. 1906, basionym. Type: Colombia, Cundinamarca, Bogotá, *Apollinaire-Marie s.n., 1904* (FH, H, NY, S-PA—isotypes).

Leptodontium pusillum M. T. Colotti & Schiavone, Lindbergia 33: 47. 2008 [2009]. Type: Argentina, Tucumán, Carapunco, M. Schiavone & A. Biasuso 1797 (isotype–MO), syn. nov.

Zygodon simii Dix., Trans. Roy. Soc. S. Afr. 8: 198, pl. 11, fig. 8. 1920. Holotype: Sim 8698 (BM) = *L. longicaule* var. *longicaule* fide Sloover 1987; seen by RZ, labeled as 8698 as "type" in BM but cited as 8690 in original publication, and is taxonomically *L. brachyphyllum* Broth. & Thér.

Illustrations: Allen (2002: 95); Magill (1981: 190); Zander (1972: 267).

Plants in loose mats, green to yellowish brown. **Stem** length to 8 cm, section rounded-pentagonal, central strand absent, sclerodermis of 1–2(–3) layers, hyalodermis present. **Axillary hairs** 5–15 cells long, basal 1–2 cells thicker-walled or brown. **Tomentum** thin, red-brown, of both elongate and short-cylindric cells. **Leaves** erect, twisted when dry, squarrose-recurved when wet, ovate-lanceolate, apex acute, 2.5–3.5 mm in length, distal lamina carinate; **leaf margins** recurved in proximal 1/2–2.3, dentate in distal 1/3, teeth smooth, large; **leaf apex** acute, apex carinate; **leaf base** short-sheathing, inner basal cells differentiated in proximal 1/3–1/2 of sheathing base, stripes lacking, basal cells 10–15 μm in length, 3:1; **costa** concolorous, percurrent or ending 1–3 cells before apex, adaxial costal cells elongate, 6–7 rows of cells across costa at midleaf viewed in section, width at base 60–75 μm, section flattened in shape, adaxial stereid band in 1 layer, abaxial band in 2(–3) layers, fully included guide cells 4(–5); **medial laminal cells** subquadrate, leaf border undifferentiated or differentiated in 1 marginal row of less papillose to smooth cells, marginal cell thickness in section same compared to medial cells, medial cell width 9–11 μm, 1:1, cell walls evenly and moderately thickened, superficial cell walls bulging; medial laminal papillae simple to 2-fid, usually centrally grouped in a crown-like ring over each lumen. **Specialized asexual reproduction** of obovoid gemmae with 2 transverse and one longitudinal septa. **Sexual condition** almost surely dioicous. **Perigonia** not seen. **Perichaetial leaf length** 4–5.5 mm. **Seta** 1.2–1.5 cm in length. **Capsule** cylindric, weakly curved, 2.2–2.5 mm in length, exothecial cells 20–26 μm wide, 2–3:1, thin-walled, stomates phaneropore, at base of capsule, annulus differentiated as 4–6 rows of reddish brown cells; **peristome** teeth linear, straight, of 32 paired rami, 280–350 μm, smooth, 8–13 articulations, basal membrane absent, **operculum** conic-rostrate to long-conic, 0.5–1.2 mm in length, cells straight. **Calyptra** cucullate, smooth, calyptra length 2–3.4 mm. **Spores** spheric, 11–13(–15) μm in diameter, colorless to light brown, lightly papillose. **KOH laminal color reaction** yellow.

Streptoleptodontium brachyphyllum is common in South Africa. It fruits rarely. Contrary to the artificial key by Zander (1972) but as noted in the description, the distal laminal cells have spiculose papillae on a central salient. It differs from *S. longicaule* by the leaves smaller, shortly ovate-lanceolate, strongly recurved throughout above the leaf base, distal laminal cells 9–11 μm in width, leaf base little differentiated, merely ovate, weakly sheathing, basal cells occasionally differentiated only near insertion, stems commonly shorter. *Streptoleptodontium longicaule* has larger leaves, 3–5 mm in length, weakly recurved or straight above the sheathing leaf base, and distal laminal cells 11–13 μm in width. *Leptodontium filicola* has similar leaves lower in the stem but these are not keeled or only weakly so, and the costa ends about 6 cells before the leaf apex rather than 2–4 cells before the apex, and distally differentiated, triangular leaves associated with enhanced gemmae production.

The differentiated basal cells of *S. brachyphyllum* are thin-walled and reminiscent of those of *S. capituligerum* but are usually in a much smaller group, and *S. capituligerum* has longer leaves and larger distal laminal cells. It is possible that *S. brachyphyllum* is derived from such an ancestor by reduction, although the reverse is also possible.

Streptoleptodontium brachyphyllum is similar to *Crassileptodontium pungens,* which may be difficult to distinguish when the ventral costal stereid band is of only one layer of cells but the papillae are more distant and centered over each lumen; *C. pungens* has laminal cells that are in leaf section vertically elongate and crowded, the papillae are crowded, the basal cells are firm and a small "miniwindow" of hyaline cells is usually evident next to the costa at the leaf insertion.

In *S. brachyphyllum* the leaf base is often "baggy," meaning there are ca. 4 loose pleats across the leaf base. The leaf basal cells are red (in KOH) in 2–3 rows across the insertion.

Distribution: North America (Mexico); Central America (Guatemala); South America (Colombia; Peru; Bolivia; Argentina); Africa (South Africa; Lesotho); Asia (China); tree trunks, soil, rocky slopes, at moderately high elevations.

Representative specimens examined (see also Zander 1972):
Argentina: Tucumán, Carapunco, soil, 2600 m, M. Schiavone & A. Biasuso 1797, May 16, 1995, MO. **China:** Guizhou Prov., Yinjiang Co., CN, Fanjianshan Range, dense hardwood forest, rock wall, 2245 m, J. R. Shevock 35182, May 17, 2010, MO; Taiwan, Xiu-lin Village, Shi-Men Shan Mt., soil, 3237 m, Si He 36020, August 9, 2002, MO. **Congo:** Prov. Kivu, Kahuzi-Biega Nat. Park, Mt. Biega, ground, 2700 m, J.-P. Frahm 6931, Aug. 28, 1991, EGR. **Lesotho:** near Pass of Guns, shaded peaty soil, 3100–3200 m, B. O. Van Zanten 7609925, September 17, 1976, MO; Maloti Mts., Moteng Pass, alpine heath-grassland, soil over rock, 2850 m, J. Van Rooy 2912, February 1987, MO; Sani Pass, marsh flats, rock, 2800 m, R. E. Magill 4417, December 6, 1977, MO. **South Africa:** KwaZulu-Natal, top of Swaartskop, nearly bare rock, 1520 m, J. Sim 8698, June 1917, BM; Transvaal, Graskop, scrubby forest area, s.n. 2430 DD, April 1979, CA, MO; Transvaal, Woodbush, H. Wager 11654, s.d., EGR; Transvaal, Woodbush Forest Reserve, R. Magill 6519, May 23, 1981, CA, EGR; Transvaal, Mariepskop, quartzite slab, 1900 m, P. Vorster 794, June, 11, 1969, US; Transvaal, Mariepskop, fynbos, sandy soil, 1970 m, P. Vorster 731, June 3, 1969, UC; Transvaal, Hebron Mountain plateau, soil, 1700 m, P. Vorster 1647, Nov. 23, 1969, UC. **Tanzania:** near Meru Crater, Arusha National Park, decaying log, 2400 m, A. Sharp et al. 665/A, July 13, 1968, EGR; Mt. Meru, Laikinoi approach, SW slope, soil, 3000-

3300 m, T. Pocs et al. 88150/A, June 24, 1988, EGR; Ngorongoro Crater, 2300 m, T. Pocs 89039/C, January 24, 1989, MO; Ngorongoro Crater, wood, 2000–2100 m, T. Pocs & S. Chuwa 89027/AC, Jan. 28, 1989, EGR.

2. Stephanoleptodontium capituligerum (Müll. Hal.) R. H. Zander, comb. nov.

Leptodontium capituligerum Müll. Hal., Linnaea 42: 323. 1879, basionym. Type: Argentina, Siambon, Lorentz s.n., NY, lectotype.

For additional synonymy, see Zander (1972), who incorrectly included *L. tricolor* in *L. capituligerum* (see Zander & Hegewald 1976).

Illustration: Allen (2002: 97); Zander (1972: 274; 1994: 268).

Plants in mats, green to yellow-brown. **Stem** length 2–9 cm, section rounded-pentagonal, central strand absent, sclerodermis of 2(–3) layers, hyalodermis present. **Axillary hairs** ca. 12 cells in length, basal 3–5 cells brownish. **Tomentum** light to reddish brown, arising from same point of attachment as gemmae stalk-cells at leaf insertion, with short-cylindric cells. **Leaves** erect when dry, squarrose-recurved when wet, ovate-lanceolate to lanceolate, 3.5–4 mm in length, lamina carinate distally; **leaf margins** recurved in proximal 1/2–2/3, dentate in distal 1/3; **leaf apex** acute to broadly acute, apex carinate; **leaf base** sheathing, decurrent, inner basal cells rectangular, basal cells 11–14 μm, 5:1, not porose; **costa** concolorous, ending 2–5 cells before apex, adaxial costal cells elongate, 4 rows of cells across costa at midleaf viewed in section, width at base 90–100 μm, section reniform to semicircular in shape, adaxial stereid band in 1 layer, abaxial band in 2(–3) layers, fully included guide cells (3–)4; **medial laminal cells** subquadrate, leaf border of 1 row of cells less papillose distally, elongate proximally, marginal cell thickness in section about half as thick through in 1(–3) cell border compared to medial cells, medial cell width 11–15 μm, 1:1, cell walls lumens angular, walls little thickened, superficial cells walls bulging; medial laminal papillae sometimes scattered, crowded, low, granular, but usually spiculose in a crown-like ring over cell lumen, raised on the salient of the bulging cell wall, often with a high columnar base. **Specialized asexual reproduction** by obovoid

gemmae, mostly 30–35 × 50–75 μm, with mostly 2 transverse septa, borne on short stalks on the stem distally. **Sexual condition** apparently dioicous. **Perichaetial leaf length** 6–7 mm. **KOH laminal color reaction** yellow. **Sporophytes** unknown.

Stephanoleptodontium capituligerum differs from *S. latifolium* by the costa not or little thickened or darker at the base of the leaf, stem with a hyalodermis of enlarged cells, and tomentum tufted, arising from abaxial sides of leaf buttresses (not in linear furry patches), mainly from base of costa and margins of leaf, essentially the same places that in other leaves generate gemmae on stalks. The tomentum of *S. captiuligerum* is possibly homologous with gemmiferous stalks. It differs from *L. longicaule* by the shorter, less acuminate leaves and basal cells somewhat enlarged, thin-walled (not porose-thickened) and papillose only in upper 1/3 to 1/2 (seldom nearly to the base). Under the dissecting microscope the hyaline leaf bases are easily discerned as glistening white areas. *Leptodontium brachyphyllum* may occasionally have enlarged, whitish basal cells, but the leaves are short, 2–3 mm in length, and the distal laminal cells are smaller, 9–11 μm wide.

Stephanoleptodontium latifolium (Congo and Rwanda) is apparently a descendant of *S. capituligerum*, possibly with an intermediate unknown shared ancestor with the unusual trait of small hyalodermal cells little larger than the stereid cells, or these apparently undifferentiated. Both *S. latifolium* and *S. capituligerum* have dense, red, but not arbusculate tomentum with some ends differentiated as a series of 2–6 short-cylindric or short-rectangular cells. This distinctive trait is matched in *Williamsiella tricolor* (R. S. Williams) R.H. Zander. It occurs elsewhere in the family, however, commonly in *Crassileptodontium subintegrifolium* and often in *Leptodontium excelsum*.

Distribution: North America (Mexico); Central America (Guatemala, Costa Rica); South America (Colombia, Ecuador, Bolivia, Peru, Argentina, Uruguay, Brazil); Africa (Ethiopia, Tanzania); Indian Ocean Islands (Réunion); soil, bolder, branches, wood, at moderately high elevations.

Representative specimens examined (see also Zander 1972):

Ethiopia: Shewa, Mount Entotto, n of Addis Ababa, 3100 m, C. Puff 820911, November 9, 1982, MO; Shewa, watershed of basins of Nile River, border of crater of volcano Wonchi, shaded soil, banks of deep narrow gully, 3150 m, D. A. Petelin 28-15, February 17, 1991, MO; w of Kara Deema, Garamba valley, Philippia woodlands, boulder field, 3740 m, G. & S. Miehe 1743, March 2, 1990, MO; below Gorba Guracha, Togona Valley, boulder field, 3600 m, G. & S. Miehe 2929, February 27, 1990, MO; Harenna escarpment, exposed branches, 2850 m, G. & S. Miehe 92, December 23, 1989, MO; Addis Ababa, British embassy, 2500 m, L. Boulos 11269b, September 24, 1977, MO. **Réunion:** Ile Bourbon, J. Rodriquez, June 4, 1898, MO. **Tanzania:** Arusha Distr., W slope Mt. Meru, subalpine, ground, 3100 m, B. Pocs 8666/Y, May 26, 1986, MO; T. Pocs 38, May 25, 1986, CA, MO.

3. Stephanoleptodontium filicola (Herzog) R. H. Zander, **comb. nov.**

Leptodontium filicola Herzog, Biblioth. Bot. 87: 34, pl. 9, fig. 2. 1916, basionym. Type: Bolivia, Santa Cruz, Herzog 4512, JE, holotype.

Leptodontium planifolium Herzog, Biblioth. Bot. 87: 37. 1916. Type: Bolivia, Río Saujana, Herzog 3225, JE, holotype; S-PA, isotype.

For additional synonymy, see Zander (1972).

Illustrations: Allen (2002: 103); Zander (1972: 242, 276). **Figure 13-9.**

Plants in procumbent mats, greenish to yellowish brown. **Stem** length 2–4, section rounded-pentagonal, central strand present, sclerodermis present, hyalodermis present. **Axillary hairs** ca. 12 cells in length, the basal 2–3 thicker-walled and brownish. **Tomentum** stringy, with short-cylindric cells. **Leaves** catenulate-incurved when dry, spreading when wet, ovate to ovate-lanceolate, dimorphic, 1.5–3.5 mm in length, distal lamina broadly channeled, specialized leaves associated with strongly gemmiferous regions of the stem smaller, 1–2 mm long, catenulate-incurved to make an opening at leaf when dry, appressed against the gemmae when wet, acuminate-lanceolate from an elliptic base, apex acute to narrowly obtuse, costa ending 2–3 cells before the apex; **leaf margins** narrowly recurved in proximal 1/2, denticulate;

leaf apex broadly acute to narrowly obtuse, apex flat; **leaf base** ovate, weakly sheathing, differentiated mainly juxtacostally, stripes absent, miniwindows lacking, basal cells 10–12 µm wide, 3–5:1; **costa** of proximal leaves ending 4–6(–15) cells before the apex, adaxial costal cells elongate, rows of cells across costa at midleaf viewed in section 4–6, width at base ca. 70 µm, section flattened, adaxial stereid band in 1–2 layers, abaxial band in 1–2 layers, fully included guide cells (2–)4; **medial laminal cells** subquadrate, leaf border not differentiated, marginal cell thickness in section 1/2 thick in one row compared to medial cells, medial cell width 9–11 µm, 1:1, cell walls moderately and evenly thickened, superficial cell walls bulging; medial laminal papillae simple to 2-fid, centrally grouped. **Specialized asexual reproduction** by gemmae, usually numerous, obovoid, 30–45 × 35–55(–175) µm, with usually 2(–6) transverse septa, borne on short stalks in dense clusters distally on the stem. **Sexual condition** dioicous. **Perigonia** terminal. **Perichaetial leaf length** 4–5 mm. **Seta** 1.4–1.7 cm in length. **Capsule** cylindric, 1.5–2 mm long, exothecial cells rectangular, 2:1, stomates phaneropore at base of capsule, annulus differentiated in 2 rows of yellowish brown cells; **peristome** of 16 of paired rami, 400–500 µm, striate, basal membrane absent, **operculum** conic-rostrate, ca. 1 mm, cells. **Calyptra** cucullate, calyptra length 2.5–3 mm, smooth. **Spores** spheric, 11–13 µm in diameter, light brown, lightly papillose. **KOH laminal color reaction** yellow.

Stephanoleptodontium filicola is considered a species distinguished from its closest relatives by small habit size; relatively small (2–3 mm long), broadly channeled leaves, which are dimorphic, being either ovate and flexuose-appressed when dry, spreading when wet, or short-acuminate and much smaller, incurved catenulate over densely gemmiferous stem portions when dry, appressed over the gemmae when wet; and distal laminal cells mostly 9–11 µm in diameter.

Although *Leptodontium planifolium* Herzog was recognized as distinct from *S. filicola* (as *L. filicola*), with reservations, by Zander (1972), additional collections available since then demonstrate complete intergradation, with some collections having the facies ascribed to *L. planifolium*, namely largely ovate leaves flexuose-appressed with dry, some almost entirely of acuminate lanceolate leaves that are incurved-catenulate when dry as with *S. filicola*, and some with sections of the stem with one type of leaf or the other. Here, *L. planifolium* is reduced to synonymy. Both species were described in the same publication and any one may be chosen as correct; *S. filicola* is here selected, being best known to the original author given his illustrations and discussion.

Stephanoleptodontium filicola occurs in the New World from Costa Rica (Allen 2002: 102) to Argentina (Colotti & Schiavone 2011). It was reported from Ecuador by Arts and Sollman (1998). In the Old World, it is uncommon, found only in southeastern South Africa and Tanzania, where it has not been previously reported, although two of the specimens cited by Magill (1981: 192) as the similar *Stephanoleptodontium brachyphyllum* (which is not uncommon in South Africa) are actually *S. filicola* (i.e., Crosby & Crosby 7770, and Vorster 1745B).

The leaves of strongly gemmiferous parts of stems of *Stephanoleptodontium filicola* are open-catenulate when dry. This condition is matched in forms of *Microleptodontium flexifolium* and some few other species of Pottiaceae, e.g., *Barbula indica* (Hook.) Spreng. and *Geheebia mascha-logena* (Ren. & Card.) R. H. Zander, that have stem-borne gemmae and most of which also generate sporophytes.

Stephanoleptodontium filicola fruits only in South America, and stations in Africa may represent remnants of a larger distribution, or may represent the results of long-distance dispersal of the gemmae from South America. The leaves when wet press the gemmae against the stem (see illustration), but open up like a sieve as the leaf apex flexes inward when drying. If the gemmae are more resistant in dry conditions than spores, then long-distance dispersal may be the reason for the distribution of asexual forms in Africa. Because this condition is cognate with other gemmiferous species, a species concept combining the aspects of both *S. filicola* and *Leptodontium planifolum* provides a similar process-based function. Because it is a Popperian bold hypothesis, more interesting to science, this evolutionary scenario is preferred over a static view of two intergrading species with separate and isolated evolutionary means.

The earlier revision of *Leptodontium* (Zander 1972) cited *L. filicola* only from New World locations, and did not mention that leaves of *L. filicola* are dimorphic, ovate and similar to those of *L. brachyphyllum* (but are more plain and concave than keeled) in non-gemmiferous areas of the stems, but distinctively smaller, lanceolate or elliptical and bottle-nosed, marginal teeth largely at the very apex, the leaves densely crowded in gemmiferous lengths of the stem usually at the apex.

The larger leaves differ from those of *L. brachyphyllum* in being less keeled or recurved, like those of *L. planifolium* but more ovate-lanceolate than ovate, and having a costa ending 2-4 cells below the leaf apex rather than 4–6(–8) cells as in *L. planifolium*. The smaller leaves are either straight or incurved when dry. Specimens of *L. filicola* from Africa are identical with those of the New World. The stem may be gemmiferous in areas of the stem away from the smaller, lanceolate leaves, and the gemmiferously modified areas of the stem are usually terminal but may be intercalary. The modified leaves on portions of the stems of *L. filicola* are catelulate when dry and appear to release the gemmae at that time, while they are appressed when wet, and seem to retain the gemmae then. This salt-shaker-when-dry adaptation is the opposite of Pottiaceae capsules, in which the peristome largely opens when wet. The unrelated gemmiferous species *Didymodon mas-chalogena* (Ren. & Card.) Broth. has much the same adaptation.

Distribution: Central America (Costa Rica); South America (Venezuela, Colombia, Ecuador, Peru, Bolivia, Argentina, Brazil, Chile); Africa (South Africa, Tanzania); soil, humus, rock, high elevations.

Representative specimens examined (see also Zander 1972):

Bolivia: La Paz, Aylulaya, Siete Vueltas, soil and humus, 2800 m, A. Fuentes et al. 10753, May 17, 2006, MO; Chuquisaca, Chirimolle, 30 km SE of Azurduy, soil, shade, 2627 m, R. Lozano et al. 2503, June 24, 2007, MO; Santa Cruz, Monte Paulo, rotten trunk, 2363 m, S. Carreño & M. Huanca 1051, Aug. 20, 2011, MO; Santa Cruz, Florida, Municipio Samaipata, soil, shady, 2120 m, S. Churchill et al. 21817, Aug. 29, 2002, MO;

Santa Cruz, Manuel M. Caballero, Serraniá de Siberia, ravine, 2100 m, S. Churchill et al. 22450-A, April 11, 2003, MO; Cochabamba, Carrasco, Serranía Siberia, rock, 2100 m, M. Decker 567, January 19, 2005, MO. **Brazil:** Minas Gerais, Mun. Caparaó Nova, Parque Nac. do Caparaó, 1970–2350 m, D. Vital & W. Buck 11714, Sept. 16, 1984, MO; Santa Catarina, São Juaquim, old Auracaria, 1420 m, Schafer-Verwimp & Verwimp 10574, Dec. 23, 1988, MO. **Colombia:** Dpto. del Putumayo, Mpio. de Colón, Reserva Natural La Rejoya, shade, 2750 m, B. Ramirez 10.215, Nov. 10, 1996, MO; Boyaca, Carretera Ráquira-La Candelaria, soil, 2450 m, A. Cleef et al. 3477, May 6, 1972, MO. **Peru:** Pasco, Oxapampa, Parque Nacional Yanachaga-Chemillén, wall, moist, shady, 2420-2540 m, J. Opisso et al. 16119, March 15, 2003, MO. **South Africa:** KwaZulu-Natal, 71 km N of Kranskop, along dirt road to Vryheid, wattle grove, on humus, 1400 m, R. E. Magill 5100, August 24, 1978, MO; Natal, Wattle grove, 71 km N of Kranskop, humus, 1400 m, R. Magill 2830 DB, Aug. 24, 1978, MO; Northern Transvaal, W of entrance to Morgenzon Forest Reserve, rock in forest, 1750 m, M. R. & C. A. Crosby 7770, January 14, 1973, MO; Transvaal, 30 km SE of Carolina, humus, R. Magill 2630 AB, March 10, 1977, MO; Transvaal, Mt. Maripskop, summit, 1900 m, P. Vorster 1745B, Feb. 12, 1969, MO; Swaziland, Hhohho Distr., dirt road to Piggs Peak, forested area, rock, R. E. Magill 3542, December 3, 1977, MO, R. E. Magill 2531AA, Mar. 12, 1977, MO; Natal, 71 km N of Kranskop, along dirt road to Vryheid, wattle grove, on humus, 1400 m, R. E. Magill 2830DB, Aug. 14, 1978, MO, R. E. Magill 2630AB, Mar. 10, 1977, MO. **Tanzania:** Ngorongoro Crater, SE outer slope, lignicolous, 2300 m, T. Pocs & S. Chuwa 89039/C, Jan. 24, 1989, EGR.

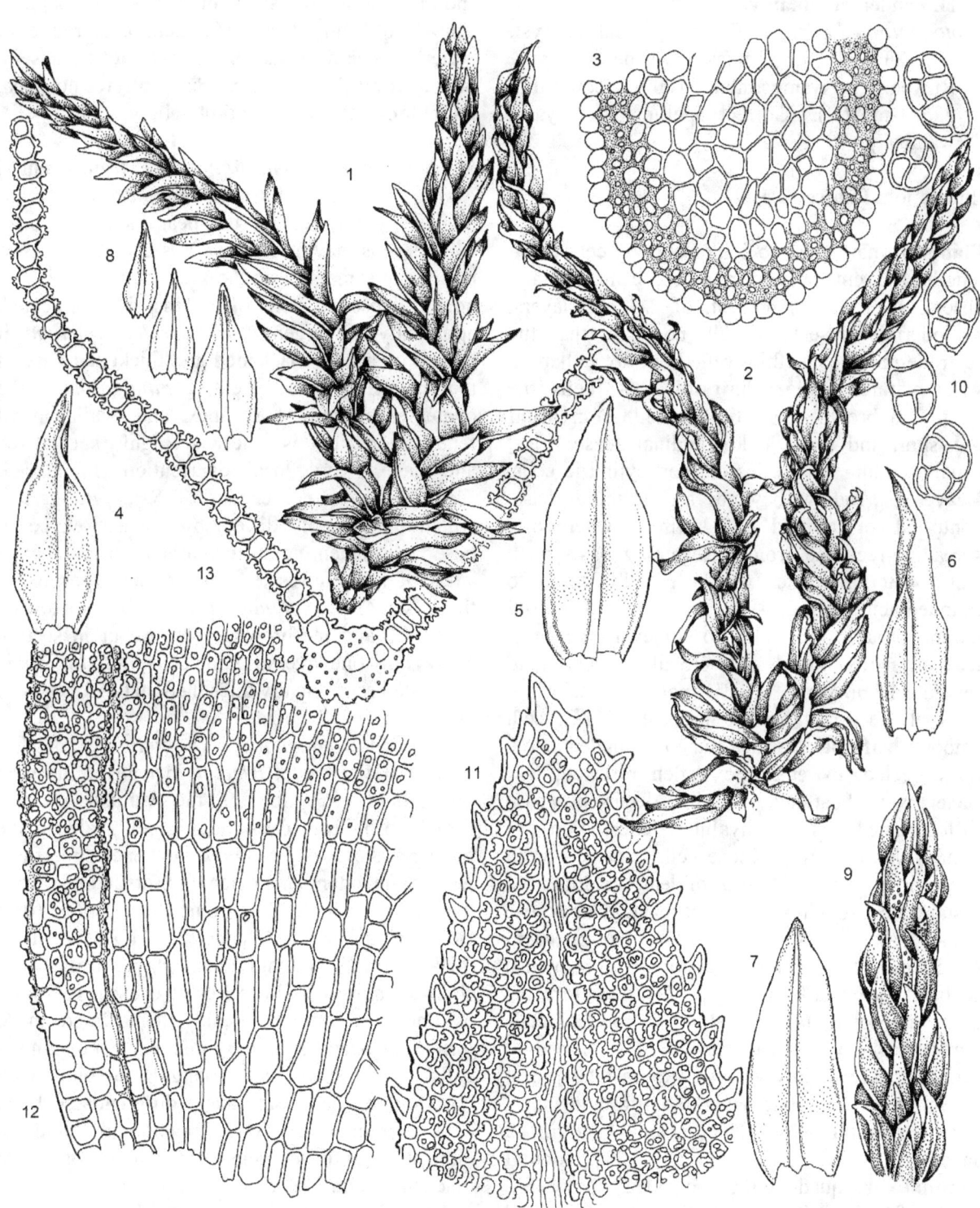

Figure 13-9. *Stephanoleptodontium filicola.* 1. Habit, wet. 2. Habit, dry. 3. Stem section. 4–7. Stem leaaves. 8. Three leaves of gemmiferous branch. 9. End of gemmiferous branch. 10 Gemmae. 11 Leaf apex. 12. Leaf base. 13. Leaf section.

4. Stephanoleptodontium latifolium (Broth.) R. H. Zander, **comb. nov.**

Leptodontium latifolium Broth. Bot. Jahrb. Syst. 24: 239. 1897, basionym. Type: Uganda, Ruwenzori, Nyamwambu, 11000′, Scott Elliot 275, s.d. (H). [Correct spelling is "Nyamwamba"]

Illustration: **Figure 13-10.**

Plants in mats, dark to light yellow-green. **Stem** length 4–6 cm, section rounded, central strand absent, sclerodermis present, of 2–3 cell layers, hyalodermis present but cells no larger than the sclerodermous cells, thin-walled but not collapsed when mature. **Axillary hairs** of ca. 13 cells, the basal 2–6 brown, more firm-walled. **Tomentum** red, short and furry, in longitudinal dense strips, usually ending in a series of short-cylindric cells. **Leaves** spreading beyond a sheathing base, contorted or twisted to 2 times when dry, spreading-recurved from a sheathing base when wet, ovate-lanceolate with a broadly acute to rounded-acute apex, 4–5 mm in length, distal lamina broadly channeled to somewhat carinate; **leaf margins** weakly to broadly recurved to revolute in proximal 1/3–2/3, not or occasionally narrowly decurrent, dentate in distal 1/4–1/3, teeth smooth; **leaf apex** broadly acute to rounded-acute, apex keeled, extreme apex often reflexed; **leaf base** ovate-sheathing, margins decurrent, base differentiated as a hyaline fenestration of rectangular weakly papillose cells, rounded M in shape in proximal 1/4–1/3 of leaf, grading into distal cells through 4–6 rows of rectangular porose cells, basal cells 10–15 μm, 4–6:1; **costa** concolorous to glistening green distally, brownish to black proximally, ending 5–7 cells before the apex, adaxial costal cells elongate, rows of cells across costa at midleaf viewed in section 3–4, width at base rapidly widening to ca. 250 μm at insertion, section semicircular to reniform in shape, adaxial stereid band in 1 layer, abaxial band in 1–2 layers, fully included guide cells 3–4; **medial laminal cells** quadrate-rhomboid, leaf border 1–2 rows of non-papillose cells including marginal teeth, walls only slightly thickened, marginal cell thickness in section about 1/2 as thick in 1–4 rows at margins compared to medial cells, medial cell width 11.5–12.5 μm, 1:1, cell walls distally thin-walled medially, thick-walled towards apex and

margins, superficial cells walls strongly convex and bulging on both sides; medial laminal papillae simple to bifid, 2–6 per lumen, centered over lumen. **Specialized asexual reproduction** absent. **Sexual condition** not seen. **Sporophytes** not seen. **KOH laminal color reaction** yellow.

Stephanoleptodontium latifolium, from central Africa, is similar in appearance to *S. capituligerum,* but differs in the near lack of a stem hyalodermis, having hyalodermis cells thin-walled but of the same size as the sclerodermal cells and not collapsed when old. This is duplicated, apparently convergently, in *Microleptodontium.* It has an unusually blackened and thickened base of the costa. It is quite like *S. capituligerum* in the ovate leaves with sharp apex, and hyaline basal fenestrations, but is otherwise highly specialized and has a very local distribution. It is well supported as a descendant of *S. capituligerum,* which immediately differs by the narrow costal base and tomentum tufted, arising from abaxial sides of leaf buttresses. The proximal portion of the costa in *S. capituligerum* is often somewhat thickened and red-brown in color, but in section shows only one layer of guide cells; in *L. latifolium* the costa is sufficiently thickened that 2–3 layers of guide cells may be evident.

Stephanoleptodontium latifolium matches closely *Williamsiella tricolor* in many traits with the salient exception of the laminal papillae form—coroniform over bulging cell walls in the former and crowded bifid over flattened superficial wall in the latter. The matching traits include weak hyalodermis in at least some *W. tricolor* (e.g., Bolivia, La Paz, C. Aldana 429, MO); tomentum thick, formed in longitudinal strips, furry, with short-cylindric cells in series at ends; leaves with hyaline fenestrations (weakly demarcated in *S. latifolium* because of the enlarged distal laminal cells); leaf apex broadly triangular to blunt with costa ending before the apex; differentiated basal cells bordered by an intramarginal band of elongate papillose cells; and base of costa blackened and enlarged (greatly so in *S. latifolium*). Given that *W. tricolor* and *S. latifolium* are highly specialized, rarely encountered, and restricted to the New World and to central Africa respectively, it is felt better to derive these separately in the caulogram.

Stephanoleptodontium latifolium differs from *S. longicaule* in its unusually large medial laminal cells, large hyaline fenestrations of thin-walled basal cells bordered by a few rows of narrow marginal cells, and dark brown to black base of the costa, which is thickened near the insertion on the stem. It is similar to *S. syntrichioides,* a rare species of Latin America, in the large size of the plants and lack of a hyalodermis, but *S. syntrichioides* differs in lack of tomentum; costa not much thickened (110–125 µm wide at base versus160–165 µm), base not blackened or dark brown except at extreme base; leaves long-lanceolate, not ovate-lanceolate; leaf apex narrowly blunt, not broadly and sharply acute, distantly dentate in distal 1/3–1/4 (not closely dentate-erose in distal 1/2); laminal basal cell walls glistening in surface view but not forming fenestrations, not sharply delimited from medial cells, thick-walled. *Stephanoleptodontium longicaule* is superficially similar but has basal lamina cells papillose to near the leaf insertion.

Distribution: Africa (Republic of Congo; Rwanda; Uganda); lava, rock, high elevations.

Representative specimens examined in addition to type:
Congo: massif des Birunga, N of Karisimbi, humus, 3700 m, J. L. De Sloover 13.165, January 26, 1972, EGR, MO. **Rwanda:** massif des Birunga, sommet du Gahinga, versant interne du cratere, lava, 3450 m, J. L. De Sloover 13.536, February 14, 1972, EGR, MO. **Uganda:** Muteinda rock shelter, Kansonge Valley, Ruwenzori Mts, 3845 m, J. P. Loverridge 315, January 16, 1962, MO.

5. Stephanoleptodontium longicaule (Mitt.) R. H. Zander, **comb. nov.**
Leptodontium longicaule Mitt., J. Linn. Soc., Bot. 12: 51. 1869, basionym. Type: Ecuador, Pichincha, Spruce 30b, NY, holotype; BM, isotype.
Didymodon longicaulis (Mitt.) A. Jaeger, Ber. Thätigk. St. Gallischen Naturwiss. Ges. 1: 209. 1873.

For additional synonymy, see Zander (1972).

Illustrations: Zander (1972: 267). Allen (2002: 108); Magill (1981: 190); Zander (1972: 267).

Plants in loose mats, green to yellow-brown distally, brown proximally. **Stem** length to 12 cm, section rounded-pentagonal, central strand absent, sclerodermis 1–2 layers, hyalodermis present. **Axillary hairs** of 9–10 cells, the basal 1–2 cells thicker-walled. **Tomentum** thin, red-brown, many with series of short-cylindric cells. **Leaves** spreading-recurved, flexuose when dry, spreading-recurved when wet, ovate-lanceolate to lanceolate with an acute apex, 3–5 mm in length, distal lamina carinate; **leaf margins** recurved to revolute in proximal 1/2–3/4, dentate in distal 1/4; **leaf apex** acute, apex keeled; **leaf base** broad, sheathing, scarcely decurrent, baggy, inner basal cells short- to long-rectangular, no stripes, basal cells 8–13 µm, 3–5:1; **costa** concolorous, ending 4–6 cells before apex, adaxial costal cells elongate, rows of cells across costa at midleaf viewed in section 6–9, width at base 70–90 µm, section flattened-reniform in shape, adaxial stereid band in 1 layer, abaxial band in (1–)2 layers, fully included guide cells 4; **medial laminal cells** subquadrate, leaf border not differentiated, marginal cell thickness in section 1/2 as thick compared to medial cells, medial cell width 11–15 µm, 1:1, cell wall lumens rounded-hexagonal, walls evenly and moderately thickened, superficial cells walls bulging; medial laminal papillae spiculose, branching, arranged in a crown-like ring over the cell lumen, raised on the salient of the bulging cell wall. **Specialized asexual reproduction** by gemmae, obovoid, mostly 25–33 × 45–55 µm, with usually 2 transverse septa, borne on short stalks on the stem. **Sexual condition** dioicous. **Perigonia** not seen. **Perichaetial leaf length** 10–15 mm. **Seta** 1–1.5 cm in length. **Capsule** cylindric, 2–2.5 mm, exothecial cells short-rectangular, 20–25 × 30–55 µm, stomates phaneropore, at base of capsule, annulus of 4–5 rows of reddish brown cells; **peristome** linear, 15–20 µm wide, 440–460 µm in length, reddish brown, spirally lightly striate, with 8–10 articulations, basal membrane absent, **operculum** conic to conic-rostrate, 0.65–1 mm, cells straight. **Calyptra** cucullate, smooth, calyptra length ca. 3 mm. **Spores** rounded, 13–17 µm in diameter, brown, papillose. **KOH laminal color reaction** yellow.

The leaf base of *Stephanoleptodontium longicaule* is inserted nearly at right angles with the stem, making the leaf base "baggy" as is the case with *Williamsiella lutea,* which differs in the laminal papillae scattered over a flat or weakly convex cell lumen and the marginal dentition extending nearly to the leaf base. *Stephanoleptodontium longicaule* var. *longicaule* occasionally has gemmae, as in Malawi, Cent. Prov., Dedza Distr., Dedza Mt., J. Pawek 11567C, MO.

Distribution: Central America (Guatemala, Costa Rica); South America (Colombia, Ecuador, Peru, Bolivia); Africa (Ethiopia, Congo, Kenya, Malawi, Tanzania, Zimbabwe, South Africa, Malagasy Republic); Réunion; Comoro Archipelago; slopes, meadows, rock, soil, organic debris, moderate elevations.

Representative specimens examined (see also Zander 1972):
Comoro Archipelago: Grand Comore Island, W slope of Kartala, summit, ground and on branches, 2150 m, T. Pocs 9159/BY, March 19, 1991, EGR, MO. **Congo:** Prov. Kivu, Mt. Kahuzi, ground, 2900 m, J.-P. Frahm 7895, Feb. 9, 1991, EGR. **Ethiopia:** Shewa, border of crater of volcano Wonchi, 20 km SSE of Ambo, shaded soil, afro-alpine meadow, 3150 m, D. A. Petelin 11-9, October 14, 1990, MO. **Kenya:** South Nyeri District, Aberdare Mts., above Karura Falls, soil, 3500 m, C. C. Townsend 85/492, February 13, 1985, MO; Aberdare Natl. Park, near entrance, steep slope, grass and forbs, 3175 m, P. Kuchar B8487, January 3, 1979, MO. **Malagasy Republic;** Fianarantsoa, Andringitra, NE slopes, 37 km S of Ambalavao, 1700–2000 m, M. R. Crosby 6993G, November 2, 1972, MO. **Malawi:** Cent. Prov., Dedza Dist., Dedza Mt., rocks, wet, shade, 2190 m, J. Pawek 11566 E, August 16 1976, MO,

J. Pawek 11567 C, August 16 1976, MO; Southern Prov, Mulanje Distr., Mulanje Mts., Lichenya Plateau, 1800–2000 m, L. Ryvarden 11778, Mar. 9–10, 1973, EGR, soil, 1750–1840 m, T. Pocs 9186/R, Apr. 17, 1991, EGR. **Réunion:** Piton de la Fournaise, resting place W of the mountain, soil, E. & P. Hegewald 11554, January 3, 1991, MO; Cirque de Cilaos, Piton des Neiges, soil, 1850 m, J. De Sloover 17.600, Dec. 22, 1973, EGR, MO; Foret de Bebour, valley of Marsoins River, soil, 1300–1550 m, G. Kis & A. Szabo 9437/CF, Aug. 8-Sept. 1, 1994, EGR; Cirque de Salazie, E slope of Roche Ecrite, soil, 1150–1250 m, A. Szabo 9414/CC, Aug. 21, 1994, EGR, soil, 1900–1990 m, A. Szabo 9419/CD, Aug. 23, 1994, EGR, MO. **Rwanda:** Prefecture de Ruhengeri, volcan Gahinga, base of crater, soil, 3400 m, J. L. De Sloover 19.408, June 1988, EGR, MO. **South Africa:** Transvaal, Wolkberg Wilderness Area, Serala Peak, ground, 1980 m, F. Venter 11390, January 24, 1986, MO; Mpumalanga, Mount Sheba Nature Reserve, Golagola Trail, soil in forest, 1500 m, J. Van Rooy 4235, January 25, 2007, MO; Transvaal, God's Window, 1800 m, Lübenau SA 241, September 16, 1990, MO. **Tanzania:** Kilimanjaro, above Nkweseko, closed *Myrica salicifolia* forest, ground and bark, 1800 m, E. W. Jones & T. Pocs 6359/AC, 6358/C, January 1, 1971, MO; S Uluguru Mts., Lukwangule Plateau, Cyperaceae bogs, 2400 m, T. Pocs 6826, 2002, EGR, MO; S Uluguru Mts., N of Luk-wangule Peak, soil, 2600 m, T. Pocs et al. 86142/AM, Aug. 17, 1986, EGR; Arusha District., W slope of Mt. Meru, ground, 3100 m, T. Pocs et al. 8666/Y, May 25, 1986, EGR; Mt. Meru, Laikinoi approach, SW slope, soil, 2800–3000 m, T. Pocs et al. 88149/M, June 24-25, 1988, EGR. **Zimbabwe:** Eastern districts (north), between Penhalonga and Inyangani, Mt. Sheba Forest Estate, W. S. Lacey 16, 1958, MO.

Key to varieties
1. Distal laminal and basal marginal papillae not clearly dimorphic, distal papillae arranged in a central crown, basal marginal cells less strongly papillose … *Stephanoleptodontium longicaule* var. *longicaule*
1. Distal laminal and basal marginal cell papillae dimorphic, distal papillae low and weakly centered or reduced to a low lens, basal marginal papillae strong … *Stephanoleptodontium longicaule* var. *microruncinatum*

5a. Stephanoleptodontium longicaule var. microruncinatum (Dus.) R. H. Zander, **comb. nov.**

Leptodontium microruncinatum Dus., Ark. Bot. 6(8): 10, pl. 5, fig. 1–3. 1906, basionym. Type: Chile, Chiloé, Guaitecas, Dusén s.n., 1897, S-PA, lectotype; FH, isolectotype).

Dicranum stellatum Brid., Muscol. Recent. Suppl. 4: 63. 1819 [1818]. Type: Île de Bourbon, Bory St. Vincent, 1803 (B).

Leptodontium longicaule subsp. *stellatum* (Brid.) Sloover, Bull. Jard. Bot. Natl. Belgique 57(3/4): 444. 1987.

Leptodontium stellatum (Brid.) Ren., Rev. Bot. Bull. Mens. 9: 213. 1881.

For additional synonymy, see Zander (1972).

Illustrations: Zander (1972: 270, 274).

Leaves long-lanceolate, to 6 mm, dentate in the distal 1/2; **distal laminal cells** with thickened walls, papillae often obscure or fused into an irregular lens-shaped cap over each lumen; **capsules** somewhat larger, to 3.5 mm long; **spores** somewhat larger, 18–22 µm in diameter.

Stephanoleptodontium longicaule var. *microruncinatum* has a more austral distribution than the typical variety. Additional evidence that it is only weakly distinguished from var. *longicaule* is the fact that the odd teeth on the distal medial lamina occurs in both varieties.

Leptodontium stellatum is identical to *S. longicaule* var. *microruncinatum* in the long leaves, densely papillose basal leaf margin, and low distal laminal papillae. It has been characterized (Sloover 1987) by scattered 1(–3)-celled teeth on both adaxial and abaxial sides of the lamina in distal 1/8 of leaf, these probably developmentally related to distal lamina ripples with seemingly overgrown, concave and convex areas. Leaves are recurved in lower 7/8 of leaf. Stem-borne gemmae are often produced, clavate, 6–8-celled. The type of *L. stellatum* and many specimens from Réunion are clearly this laminally dentate variant of var. *microruncinatum*. The lectotype at B has gemmae, dentate distal laminae, but is otherwise var. *microruncinatum* with spiculose basal laminal margins and distal cells with central blunt salient. In addition, specimens from Tanzania (e.g., S-

Uluguru Mts., Lukwangule Plateau, T. Pócs 6826/B, MO, same locale, T. Pócs et al. 86142/AM, EGR; Kilimanjaro, above Nkweseko, E. W. Jones & T. Pócs 6359/AC, MO) have the rippled leaves and distal medial laminal teeth but are clearly var. *longicaule* (shorter leaves, blunt apex, large laminal cells with central branching papillae, no rasp-like basal margins). A specimen of *S. longicaule* (Papua New Guinea, Koponen 32533, MO) rather far outside its range has fleshy leaves and occasional laminal teeth near the margin, largely on the adaxial surface of the leaves, and differs from *S. longicaule* in Africa only by the poorly developed laminal papillae (but in this work a single trait is insufficient to distinguish a species).

The medial laminal teeth may be viewed as a minor genetic variant or susceptibility found in some African and Indian Ocean island specimens of both var. *longicaule* and var. *microruncinatum*. Thus, *L. stellatum* is a synonym of var. *microruncinatum*. A specimen of *S. syntrichioides* from Colombia (Tolima, PNN de Los Nevados, S del volcán, en bosque andino, nublado, 3350 m, T. Hammen & R. Jaramillo 3285, August 12, 1975, MO) also has medial teeth near the leaf apex and in addition enlarged thick-walled smoother cells in 4–5 rows as in *L. flexifolium*; these traits are considered neutral or nearly neutral mutational features associated with the genus.

Distribution: South America (Venezuela, Ecuador, Peru, Bolivia, Chile, Juan Fernandez Islands); South Atlantic Islands (Tristan da Cunha, Gough Island); Africa (Kenya, Tanzania, South Africa); Indian Ocean Islands (Réunion, Crozet Islands); soil, bark, moderate elevations.

Representative specimens examined (see also Zander 1972):

Réunion: Cirque de Salazie, 1900-1990 m, A. Szabo 9419/CD, August 23, 1994, MO; forest above St. Denis, 1400–1800 m, K. Een 375, October 11, 1962, MO; above Denis, zone de Tamarin, 1400-1800 m, K. Een 3311, Oct. 11, 1962, EGR; Foret de Bebour, 1300–1550 m, G. Kis 9437/CF, August 31, 1994, MO; Arrt. du Vent, summit of La Roche Ecrite, ground, slope, 2270 m, M. R. Crosby 8264, November 25, 1972, EGR, MO, trail to La Roche Ecrite, soil, M. Croby & C. Crosby 8260, Nov. 26, 1972, EGR; E edge of

Cirque de Mafate, soil, 1820–1878 m, S. Orban 9424/CF, Aug. 26, 1994, EGR; Treu de fer trail, soil, 1500–1540 m, T. Pocs 08061/Q, Sept. 12, 2008, EGR. **Tanzania:** Kilimanjaro, forest above Nkweseko, ground and on bark, 1800 m, E. Jones & T. Pocs 6359/AC, Jan. 1, 1971, EGR.

6. Stephanoleptodontium stoloniferum (R. H. Zander) R. H. Zander, **comb. nov.**

Leptodontium stoloniferum R. H. Zander, Bryologist 75: 239. 1972, basionym. Type: Colombia, Cundinamarca, Paramo Choachi, Lindig 2127, M, holotype, separated from isotype of *Leptodontium filescens*).

Illustrations: Allen (2002: 115); Zander (1972: 242; 1993: 134).

Plants mat-forming. green to yellowish brown. **Stem** length 1–3 cm, section rounded to elliptic, central strand absent, sclerodermis present, hyalodermis present. **Axillary hairs** of ca. 8 cells, the basal thicker-walled, brownish. **Tomentum** thin, brownish, ending in elongate or short-cylindric cells. **Leaves** recurved at sheathing base, incurved distally when dry, spreading-recurved when wet, oblong to ovate-lanceolate, 3–3.5 mm in length, distal lamina broadly channeled, lying flat on a microscope slide; **leaf margins** recurved in proximal 1/2, denticulate to dentate in distal 1/4–1/3; **leaf apex** broadly acute, apex broadly channeled; **leaf base** sheathing, with inner cells differentiated as double arches of hyaline windows, no stripes, basal cells 15–18 μm wide, 4–6:1; **costa** in proximal half of leaf base reddish brown, ending 2–10 cells before apex, adaxial costal cells elongate, 4 rows of cells across costa at midleaf viewed in section, width at base 70–80 μm, section reniform to semicircular in shape, adaxial stereid band in 1 layer, abaxial band in 2–3 layers, fully included guide cells 4; **medial laminal cells** subquadrate, leaf border of ca. 1 row of less papillose or smooth cells, marginal cell thickness in section about half that of medial cells, medial cell width 9–11 μm, 1:1, cell walls evenly thickened and moderately thickened at corners, superficial cell walls convex to bulging; medial laminal papillae simple to 2–3-fid, 4–6 per lumen. **Specialized asexual reproduction** by gemmae, claviform or long-elliptic, 30–35 × 110–175 μm, mostly with 4–6 transverse septa, borne on leafless branchlets to 2 mm long emerging from between distal leaves. **Sexual condition** probably dioicous. **Sporophyte** not seen. **KOH laminal color reaction** yellow.

Stephanoleptodontium stoloniferum has some-what enlarged distal laminal cells, (9–)11–13 μm, sharply defined hyaline basal cells, and leafless gemmiferous branches, with long-ellipsoidal gemmae. The unspecialized leaves have a costa ending 6–8 cells before the leaf apex. The hyaline fenestrations are much like those of *L. capituligerum* but that species has much more acute apices, costa subpercurrent, and leaves larger, (2.5–)3–4 mm in length, and lacks specialized gemmiferous branches. The long-elliptic gemmae are rather different than those of most Streptotrichaceae, which are short-obovoid to clavate. The gemmae are rather clearly modifications of the short tomentum of short-cylindric to short-rectangular cells arising from the stem in linear series, and do not have the hyaline stalks of the gemmae of other species.

Distribution: Central America (Costa Rica); South America (Colombia; Ecuador); thatched roofs, high elevations.

Representative specimens examined (see also Zander 1972):
Costa Rica: Cartago, Cartago Canton, Cordiller de Talamanca, Paramo Buenavista, epiphyte, 3300 m, G. Dauphin et al. 2228, Feb. 26, 1996, MO, Dauphin & V. Ramirez 2263, Feb. 27, 1996, MO.

7. Stephanoleptodontium syntrichioides (Müll. Hal.) R. H. Zander, **comb. nov.**

Trichostomum syntrichioides Müll. Hal., Linnaea 38: 602. 1874, basionymn. Type: Colombia, Antioquia, Sonson, G. Wallis s.n., 1872, NY, isotype.

Leptodontium syntrichioides (Miill. Hal.) Kindb., Enum. Bryin. Exot. 63. 1888.

Trichostomum syntrichioides C. Miill., Linnaea 38: 602. 1874. Type: Colombia, Antioquia, Sonson, Wallis, 1872, NY, isotype.

For additional synonymy, see Zander (1972).

Illustrations: Allen (2002: 116); Zander (1972: 270).

Plants in loose mats, green to yellowish brown. **Stem** length to 10 cm, section rounded-pentagonal, central strand absent, sclerodermis of 1–2 layers, hyalodermis absent. **Axillary hairs** all hyaline, stout, 20–25 µm in width at 3 or 4th cell from base, of 13–16 cells of which basal 2–3 are slightly thicker-walled. **Tomentum** none. **Leaves** flexuose to contorted when dry, spreading-recurved when wet, ovate- to oblong-lanceolate, 4–7 mm in length, distal lamina broadly channeled; **leaf margins** broadly recurved to revolute in proximal 2/3–4/5, dentate in distal 1/4; **leaf apex** narrowly obtuse, apex broadly channeled to weakly cucullate; **leaf base** rectangular-sheathing, slightly baggy, differentiated and filling sheathing base, stripes absent, basal cells 11–15 µm wide, 4–7:1; **costa** concolorous, ending 4–8 cells before apex, adaxial costal cells elongate, 4–6 rows of cells across costa at midleaf viewed in section, width at base 90–110 µm, section reniform in shape, adaxial stereid band in 1 layer, abaxial band in 1–2 layers, fully included guide cells 4; **medial laminal cells** subquadrate, leaf border cells of margin in 1 row less papillose, marginal cell thickness in section about same compared to medial cells, medial cell width 13–17 µm, 1:1, cell walls moderately and evenly thickened, superficial cell walls bulging; medial laminal papillae spiculose, branching, in a crown-like ring over center of each cell lumen. **Specialized asexual reproduction** not seen. **Sexual condition** probably dioicous. **Perigonia** not seen. **Perichaetial leaf length** 7–8 mm. **Seta** ca. 2 cm in length. **Capsule** cylindric, 3.4–3.6 mm, exothecial cells short-rectangular, mostly 2–3:1, stomates not seen, annulus differentiated in 4–6 rows, reddish brown; **peristome** teeth linear, 420–470 µm, reddish orange, spirally ridged, of 7 articulations, basal membrane absent, **operculum** not seen. **Calyptra** not seen. **Spores** spheric, 22–24 µm in diameter, papillose. **KOH laminal color reaction** yellow.

Stephanoleptodontium syntrichioides differs from *S. longicaule* var. *longicaule* by its fleshy leaves to 1.8 mm wide at midleaf, compared to up to 1.2 mm for *L. longicaule*. Both species have leaves of the same length. *Stephanoleptodontium syntrichioides* lacks a well-differentiated hyalodermis, and the cortex has little thickening as a sclerodermis. The hyalodermis of *S. longicaule* is well differentiated, being of larger cells compared to the inner cells, and a sclerodermis is obvious.

Distribution: Central America (Costa Rica), South America (Ecuador, Colombia, Peru); on soil, high elevations.

Representative specimens examined (see also Zander 1972):
Colombia: Tolima, PNN de Los Nevados, vertiente S del volcán, camino El Rancho a La Cueva, cloud forest, 3550 m, T. Hammen & R. Jaramillo 3285, 12 August 1975, MO. **Peru:** Dept. Junin, Prov. Huancayo, 14 km to Chilifruta, 3000m, humus, P. and E. Hegewald 9220, 6 July 1977 (Bryophyta Neotropica Exsiccata 165), MO.

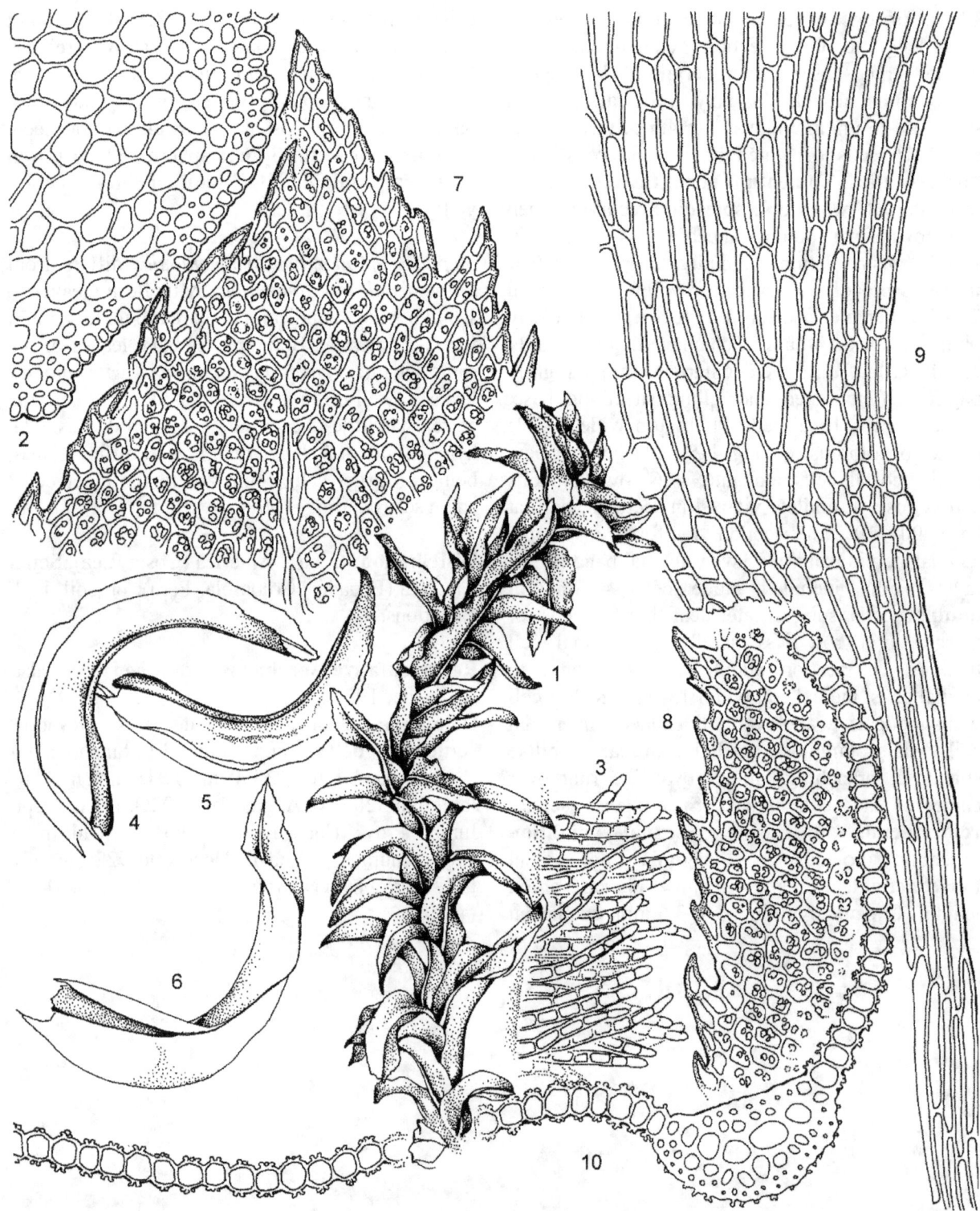

Figure 13-10. *Stephanoleptodontium latifolium.* 1. Habit. wet. 2. Partial stem section. 3. Tomentum. 4–6. Cauline leaves. 7. Leaf apex. 8. Distal leaf margin. 9. Leaf marginal base and decurrency.

9. STREPTOTRICHUM Herzog, Biblioth. Bot. 87: 37. 1916.
Type: *Streptotrichum ramicola* Herzog.

Stem hyalodermis present. **Tomentum** thin or absent, when present ending in short-cylindric cells. **Leaves** lanceolate from oblong-sheathing base, ca. 3–4(–5) mm in length, not decurrent; **leaf margins** unistratose, distantly dentate; **leaf apex** narrowly acute; **costa** percurrent, stereid bands in 2 layers; **medial laminal cells** weakly thickened at corners, lumens rounded-quadrate to rectangular, superficial walls flat, papillae simple to bifid, scattered; **leaf base** high-sheathing, of rectangular translucent cells colored in longitudinal stripes, miniwindows absent. **Specialized asexual reproduction** absent. **Sexual condition** dioicous. **Capsule** peristome teeth 32 (16 cleft to near base, sometimes arranged as 8 groups of 4 rami), bright orange-red, linear, occasional interpolated rami arise from anticlinal walls at margins of regular teeth with appearance of teeth viewed laterally, up to ca. 42 rami, (300–)500–1000 μm with variable length, densely branching-spiculose, basal membrane low. **Calyptra** antrorsely prorulose.

1 species.

Streptotrichum is placed in Streptotrichaceae by lack of a stem central strand; knotty tomentum; squarrose-recurved, lanceolate leaves with a reniform costal section lacking differentiated epidermal layers; and highly differentiated perichaetial leaves. It has, like *Leptodontiella*, an unusual peristome, in this case of 16 deeply bifid, densely spiculose teeth, each of which occasionally bears an additional ramus or two formed from an interior anticlinal wall. These extra teeth are similar to the other teeth except they are shorter and appear edge on when viewed laterally from outside the capsule. *Streptotrichum* has much the same lax habit, fat gemmate perigonia, and short capsules (though this is variable) as does *Leptodontiella*, but the peristome and leaf shape are quite different.

1. Streptotrichum ramicola Herzog, Biblioth. Bot. 87: 37. 1916. Type: Bolivia, Waldgrenze úber Tablas, ca. 3400 m, Herzog 2844, holotype, JE.

Illustrations: Herzog (1916: Plate 2); Zander (1993: 129). **Figure 13-11.**

Plants in loose mats, yellow-brown to greenish brown. **Stem** length to 4 cm, section rounded-pentagonal, central strand absent, sclerodermis present in 1–2 layers, hyalodermis present. **Axillary hairs** 4–9 cells long, with 1–3 brownish basal cells. **Tomentum** thin or absent, when present ending in short-cylindric cells. **Leaves** spreading, twisted when dry, squarrose-spreading when wet, lanceolate from an oblong sheathing base, (2.5–)3–4.5 mm in length, distal lamina carinate; **leaf margins** recurved at shoulders for about 1/3 of leaf, distantly dentate; **leaf apex** narrowly acute, apex keeled; **leaf base** oblong, high-sheathing, with stripes often present, basal cells differentiated throughout sheathing base, basal cells 7.5–10 μm in width, walls thick, weakly porose, basal marginal laminal cells not differentiated from medial basal cells; **costa** concolorous, percurrent, adaxial costal cells elongate, 4–6 rows of cells across costa at midleaf viewed in section, width at base 50–65 μm, section flattened-reniform in shape, adaxial stereid band in 1 layer, abaxial band in 2 layers, fully included guide cells 3–4; **medial laminal cells** quadrate, leaf border of elongate cells between teeth at midleaf, 1 row of non-papillose marginal cells, marginal cell thickness in section about same compared to medial cells, medial cell width (7.5–)10–15 μm, 1–2:1, cell walls weakly thickened at corners or thin, superficial cells walls flat; medial laminal papillae simple to bifid, scattered. **Specialized asexual reproduction** absent. **Sexual condition** dioicous. **Perigonia** terminal or sub-terminal by repeated innovations. **Perichaetial leaf length** to 8 mm. **Seta** 5–7 mm in length. **Capsule** 2–3 mm, cylindric, occasionally more or less ventricose, exothecial cells evenly thick-walled, short-rectangular, stomates absent, annulus of 1–2 rows of vesiculose cells, persistent; **peristome** teeth 32 (16 cleft to near base, sometimes arranged as 8 groups of 4 rami), bright orange-red, linear, occasional interpolated rami arise from anticlinal walls at margins of regular teeth with appearance of teeth viewed laterally, up to ca. 42 rami, (300–)500–1000 μm with variable length, densely branching-spiculose, with many articulations, basal membrane low, ca. 70 μm high, **operculum** conic, ca. 1.5 mm in length, cells quadrate, in straight rows. **Calyptra** cucullate, antrorse projecting cell ends near apex, calyptra length 3–5 mm. **Spores** spheric, 12–20 μm in diameter, colorless to light brown, lightly papillose. **KOH laminal color reaction** yellow to orange.

Streptotrichum ramicola is epiphytic. The male plants have terminal perigonia singly or in clusters of 2–3, while the perichaetia are subterminal by overtopping subperichaetial branches. The papillae and distal laminal cells are much like those of *Williamsiella lutea*. The stem is fluted with a collapsed hyalodermis. Although a thick tomentum is lacking, at least rhizoids are present, with short-cylindric cells ending many rhizoids, which thus may be considered a thin tomentum. Leaves are recurved proximally, straight or curved distally and usually keeled throughout, the leaf margins are broadly recurved in basal 2/3 of leaf. The distal laminal cells often elongate longitudinally as in *Leptodontium excelsum*. The perichaetial leaves overtop the seta, and the fruiting habit is much like that of *Trachyodontium*. The basal laminal cells are not hyaline, and are similar across the leaf base, as is the case with *W. lutea*. In general aspect, the gametophyte is very similar to that of *Williamsiella araucarieti*.

Distribution: South America (Bolivia); on branches of shrubs and trees, high elevations.

Representative specimens examined (other than type above):
Bolivia: Cochabamba, Chapare, Parque Nacional de Carrasco, por El Limbo, Yungas, nodo de Chusquea, 2400 m, Churchill et al. 24773, 7 December 2007; La Paz, Nor Yungas, Coscapa, branch, 3140–3300 m, S. Churchill et al., 19744-C, 19760, 12 Nov. 1999, MO.

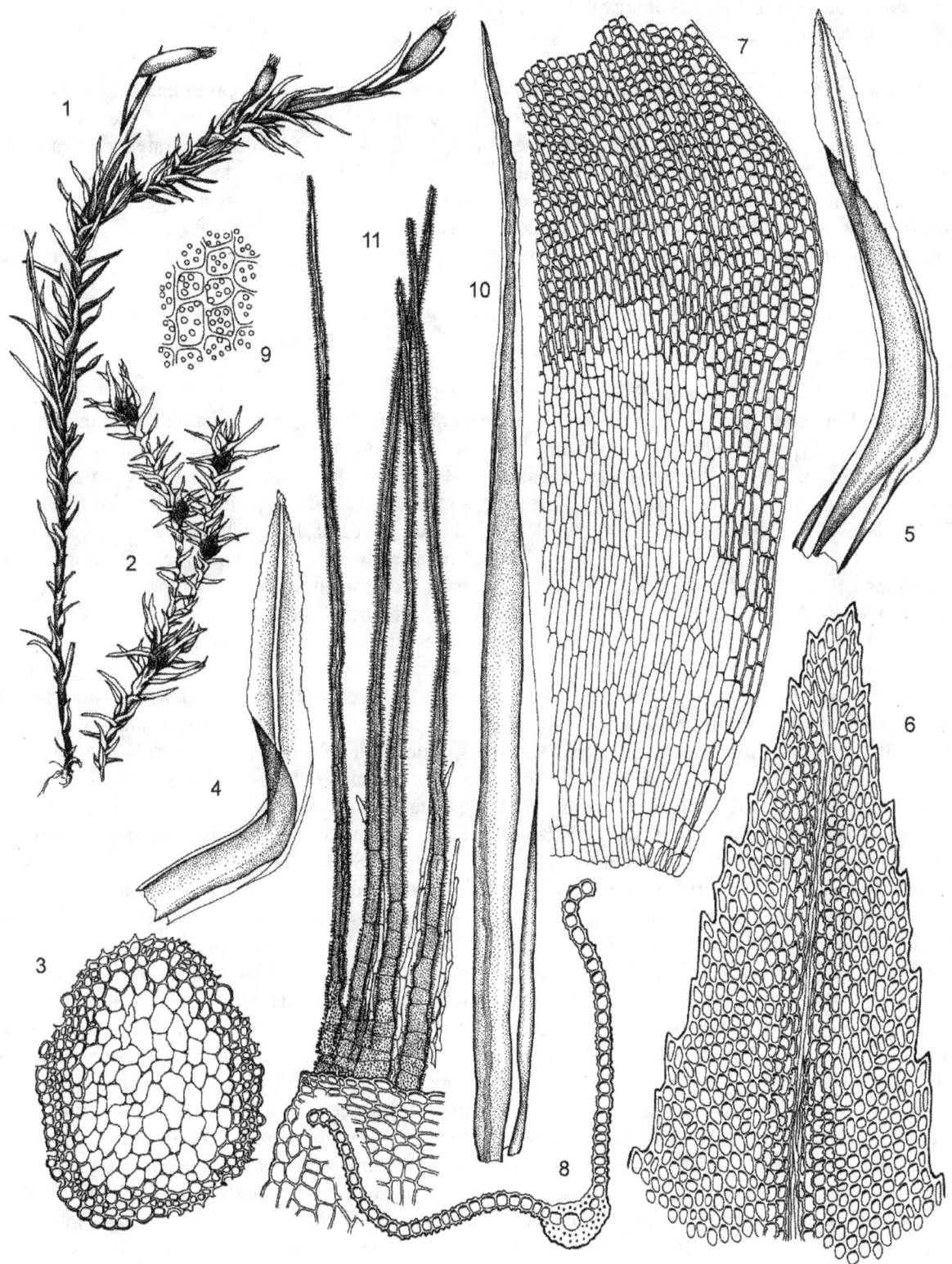

Figure 13-11. *Streptotrichum ramicola.* 1. Habit, wet. 2. Male plant. 3. Stem section. 4–5. Cauline leaves. 6. Leaf apex. 7. Leaf base. 8. Leaf section. 9. Distal laminal cells and papillae. 10. Perichaetial leaf. 11. Portion of peristome. Modified from P. M. Eckel illustration in Zander (1993).

10. Trachyodontium Steere, Bryologist 89: 17. 1986.
Type: *Trachyodonium zanderi* Steere.

Stem hyalodermis present. **Tomentum** short, ending in short-cylindric cells. **Leaves** lanceolate, ca. 3–4.5 mm in length; **leaf margins** bistratose, dentate distally; **leaf apex** acute; **costa** percurrent, stereid bands in 2 layers; **medial laminal cells** weakly thickened at corners, lumens rounded-quadrate to rectangular, superficial walls flat, papillae simple to bifid, scattered; **leaf base** low- to broadly sheathing, of rectangular translucent cells often with longitudinal stripes, miniwindows absent. **Specialized asexual reproduction** absent. **Sexual condition** dioicous. **Capsule** peristome teeth of 64 rami, of straight filaments crowded into 16 groups of four, usually broken to 120–150 µm, but often to 800 µm when intact, papillose externally, spiculose internally, basal membrane apparently absent, but teeth fused into a low basal membrane hidden below the annulus. **Calyptra** antrorsely prorulose.

Species 1.

1. Trachyodontium zanderi Steere, Bryologist 89: 17. 1986. Type: Ecuador, Prov. Pichincha, Volcán Pichincha, W. C. Steere & H. Balslev 26991 (NY). Paratype: Same locality, W. C. Steere & H. Balslev 26991 (MO).

Illustrations: Steere (1986: 18); Zander (1993: 128). **Figure 13-12.**

Plants in mats, green, 2–3.5(–4) cm in length, with leaf buttresses. **Stem** in transverse section rounded-pentagonal, central cylinder of equal-sized cells, central strand absent, sclerodermis present, in 1–2 layers, hyalodermis present; axillary hairs 8–15 cells long, the basal 2–3 brownish. **Tomentum** short, whitish, or rhizoids short, whitish, cells ending tomentum short-cylindric. **Leaves** when dry contorted-twisted, when wet widely spreading to recurved, equal in size along stem, lanceolate, 3–4(–5) mm, broadly channeled and narrowly channeled at costa distally; **margins** recurved to beyond midleaf, revolute along leaf base, narrowly decurrent; leaf margins dentate, 3–4 longitudinal cells between teeth, bistratose cartilaginous border of elongate cells from near base to near apex; apex acute, apex adaxial surface flat to broadly channeled; **leaf base** low-sheathing to broadly sheathing, brownish longitudinal stripes often present, basal cells differentiated across leaf, 4–6:1, 15–20 µm wide, walls thickened, porose, miniwindows absent; **costa** concolorous, percurrent, adaxial cells superficially elongate, costa 4–6 rows across at mid leaf, 70–75 µm across at base, costa transverse section reniform, adaxial epidermis absent, adaxial stereid band in size smaller than the abaxial, adaxial band in 1 layer, abaxial band in 2 layers, fully included guide cells (2–)4, guide cells in 1 layer, hydroid strand absent, abaxial band in section flattened-reniform, **Medial laminal cells** shape irregularly subquadrate, occasionally transversely or longitudinally rectangular; lamina with a cartilaginous border from near base to near apex of 2 cells thick and 2 cells wide, thickness of distal laminal cells near leaf margins about same as medially, distal laminal cells 12.5–17.5(–20) µm wide, 1:1, cell walls evenly thickened, superficially flat to weakly convex; distal laminal papillae simple, solid, 6–8 per lumen, evenly spaced across leaf. **Specialized asexual reproduction** absent. **Sexual condition** dioicous. **Perigonia** terminal, gemmate, in clusters. **Perichaetial leaf length** 12–14 mm. **Seta** 1.1–1.3 cm in length. **Capsule** curved, cylindric, 4–5 mm, exothecial cells evenly thickened, cells 3–5:1, stomates rudimentary, consisting of a circle of rhomboid cells, at base of capsule, annulus differentiated; **peristome** of 64 rami, of straight filaments crowded into 16 groups of four, usually broken to 120–150 µm, but often to 800 µm when intact, papillose externally, spiculose internally, basal membrane apparently absent, but teeth fused into a low basal membrane hidden below the annulus, **operculum** long-conic, 1.8–2.2 mm, cells basally transversely elongate, medially quadrate, distally elongate longitudinally. **Calyptra** cucullate, 6–6.5 mm long, split from base to near apex, antrorse projecting cell ends near apex. **Spores** spheric, 15–18 µm in one size class (old spores swollen to 30–35 µm, many brown aborted spores but not so many as to imply two size classes) in diameter, light brown, warty-papillose. **KOH laminal color reaction** yellow.

The salient distinguishing features of *Trachyodontium* are the cartilaginous leaf border, simple laminal papillae, and spiculose peristome teeth of 64 rami arranged in groups of four rami each. It has a superficial resemblance to *Calyptopogon* by the leaf shape and border, the simple papillae, and the lack of a stem central strand, but *Calyptopogon* is easily distinguished from *Trachyodontium* by its semicircular costal transverse section showing a single stereid band and a hydroid strand. *Trachyodontium* agrees with significant features of closely related members of the Streptotrichaceae in its large size, the convolute-sheathing perichaetial leaves, elongate-cylindrical capsule lacking or with rudimentary stomates, and spreading-recurved to squarrose, dentate leaves with reniform costal transverse section showing two stereid bands.

Distribution: South America (Ecuador); epiphytic on shrubs, high elevations.

Representative specimens examined (in addition to the type):

Ecuador: Pichincha, Cordillera Occidental, valley of Río El Cinto, 15 km below Lloa, SW slope of Volcán Pichincha, epiphytic on Chusquea in cloud zone, 2650 m, W. C. Steere & H. Balslev 26062, 1 Oct 1982, NY, W. C. Steere & H. Balslev 26072, 1 Oct 1982, NY; Pichincha, Cordillera Occidental, valley of Río El Cinto, 15 km below Lloa, SW slope of Volcán Pichincha, 2650 m, W. C. Steere & H. Balslev 26087, 1 Oct 1982, NY.

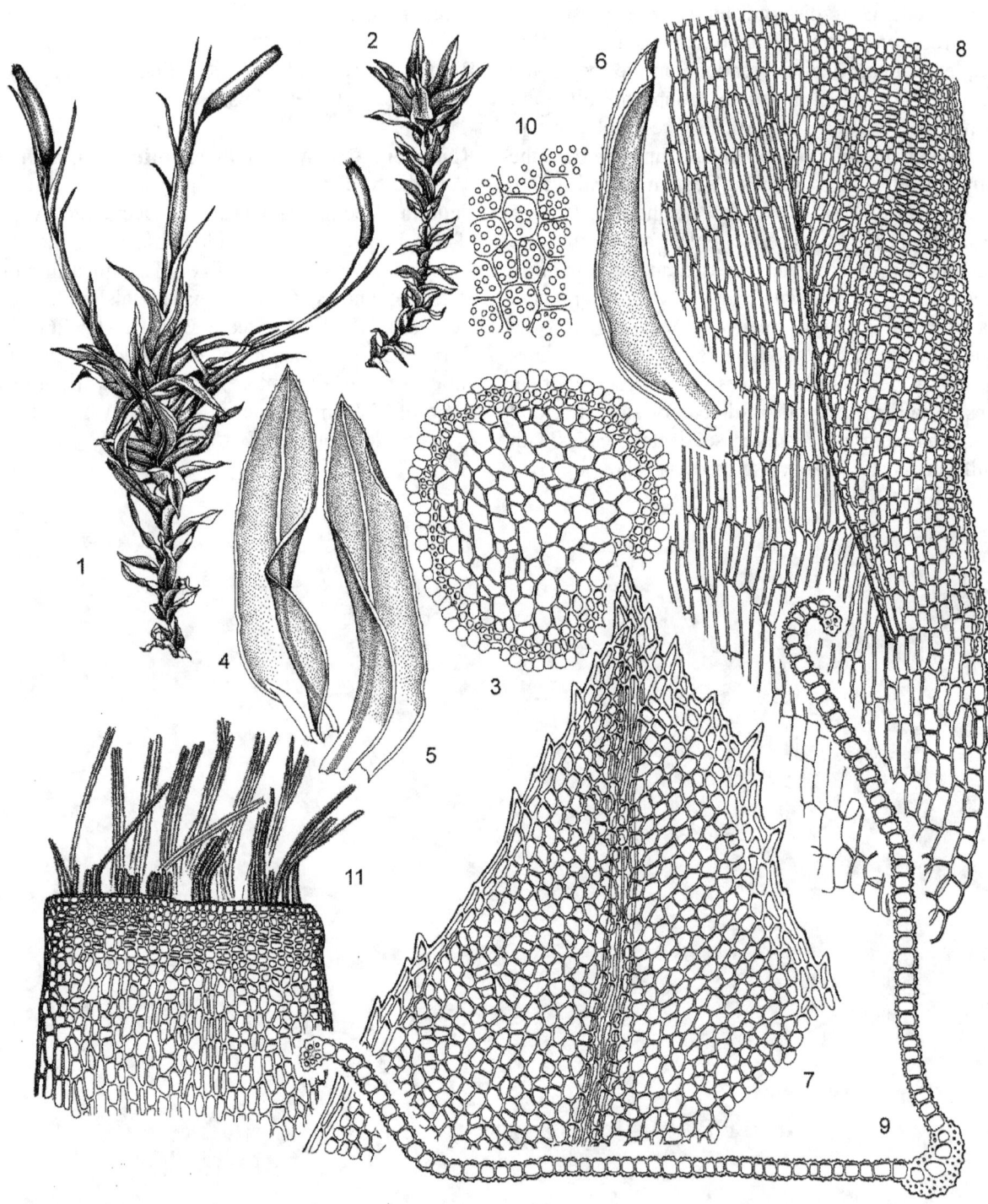

Figure 13-12. *Trachyodontium zanderi*. 1. Habit, wet. 2. Male plant. 3. Stem section. 4–6. Cauline leaves. 7. Leaf apex. 8. Leaf basal margin. 9. Leaf section. 10. Distal medial laminal cells with papillae. 11. Peristome. Modified from illustration by P. M. Eckel in Zander (1993).

Chapter 14
MISCELANEOUS AND EXCLUDED SPECIES

The uncommon but widely distributed African moss species *Barbula eubryum* Müll. Hall. has been described and illustrated by Magill (1981: 245). This species is somewhat like *Leptodontium* in red arbusculate tomentum (but lacking terminal series of short-cylindric cells), central strand variably present, even-sized central cylinder cells, hyalodermis present, no costa epidermis on either side, and bulging leaf cells with salient papillae. It differs in the lack of leaf teeth, no high sheathing leaf base, short-excurrent costa, thick laminal papillae prongs, basal cells split like those of *Trichostomopsis paramicola* (H. Rob.) R.H. Zander, and very large rhizoidal brood bodies. However, the basal miniwindow of *L. pungens* tears identically laterally between red, thickened crosswalls. and in section the laminal cells are like those of *Stephanoleptodontium*. It has no relatives (evidence of dissilience is gone) and apparently is a relict of that now mostly extinct group out of which both Pottiaceae and Streptotrichaceae arose. It is like no *Barbula* or *Tortula* I am aware of. It thus it rates a new genus. *Barbula eubryum* is the type of *Tortula* subg. *Bulbibarbula* (Müll. Hal.) Wijk & Marg. and that name may be advanced to genus status. It is distinctly characterized as diagnosed above.

Bulbibarbula (Müll. Hal.) R. H. Zander, gen. and stat. nov.
Basionym: *Barbula* sect. *Bulbibarbula* Müll. Hal., Flora 62: 379. 1879. Synonym: *Tortula* sect. *Bulbibarbula* (Müll. Hal.) Wijk & Margad.

Bulbibarbula eubryum (Müll. Hal.) R. H. Zander, comb. nov.
Basionym: *Barbula eubryum* Müll. Hall., Flora 62: 379. 1879. Synonyms: *Leptodontium insolitum* Thér. & P. de la Varde; *Leptodontium insolitum* var. *perindulatum* Thér. & P. de la Varde (see Sloover 1987); *Tortula eubryum* (Müll. Hal.) Broth.

EXCLUDED

Leptodontium allorgei Bizot, Acta Botanica Academiae Scientiarum Hungaricae 18: 24. f. 9: 1–8. 1973. Type: Kenya, W slope of Mt. Kenya, 3900 m, D. Balázs 105/m, January 20, 1966 (MO--isotype). This is Dicranaceae by the colored enlarged groups of alar cells. The laminal cells are mammillose-prorulose, indicating possibly a *Cynodontium*. Excluded from *Leptodontium* by Sloover (1987), who suggested it was probably a species of *Cynodontium* (Dicranaceae).

Leptodontium paradoxicum Stone & Scott, J. Bryol. 11: 701. 1981. = *Triquetrella paradoxica* (I. G. Stone & G. A. M. Scott) Hedd. & R. H. Zander, J. Bryol. 29: 156. 2007.

Leptodontium longifolium (Griff.) Müll. Hal., Genera Muscorum Frondosorum 409. 1900. = *Oxystegus longifolius* (Griff.) Hilp. fide Hilpert (1933).

Leptodontium subalpinum (De Not.) Lindb. = *Dichodontium pellucidum* (Hedw.) Schimp. fide Frahm & Schumacker (1986).

Leptodontium theriotii (Corb.) Broth., Die Natürlichen Pflanzenfamilien I(3): 400. 1902. = *Dichodontium pellucidum* (Hedw.) Schimp. fide Wijk, R. v. d., W. D. Margadant & P. A. Florschütz. 1962. Index Muscorum. 2 (D–Hypno).

Leptodontium sikokianum Sakurai, Botanical Magazine 62: 145. 5. 1949. = *Dichodontium pellucidum* (Hedw.) Schimp. fide Saito (1975).

Leptodontium sinense (Müll. Hal.) Paris, Index Bryologicus Supplementum Primum 225. 1900. = *Ptychomitrium polyphylloides* (Müll. Hal.) Paris fide Wijk, R. v. d., W. D. Margadant & P. A. Florschütz. 1964. Index Muscorum. 3 (Hypnum–O).

PART 3: SUMMARY
Chapter 15
CONCEPTUAL NOVELTIES

A short list of critical ideas underlying macroevolutionary systematics — Reviewers! Kindly address these central features of macroevolutionary systematic analysis.

1. The *object of macroevolutionary systematics* is to nest extant species (or higher taxa) in sometimes branching series (not sister groups), from a primitive (generalized) taxon) radiating in one or more lineages of direct descent to advanced (more recent elaborations or reductions). Postulating unknown shared ancestral taxa is sometimes necessary, but no cladogram nodes are needed.

2. Serial transformation of taxa inferred from serial relationships versus groups paired by shared relationships. The evolutionary model is direct descent. The descendant of a progenitor may become the progenitor of another species.

3. Information theory redundancy. Redundancy of traits due to simple descent is minimized by both cladistics and macroevolutionary systematics. Noise consisting of (evolutionarily neutral or nearly neutral) traits that pop up in a desultory fashion among related species is ignored. Such traits are not informative of serial descent unless they can be seen to in consort track descent across species boundaries. Detailed information on order and direction of evolution is determined by an analytic key.

4. Paraphyly and short-distance polyphyly (together heterophyly) are indicators of serial evolution and are central to macroevolutionary analysis. Through a sleight of method involving primacy of cladistic relationships, paraphyly is avoided by cladists when modeling shared descent. The heteroplasticity of divergent molecular races is as problematic as the homoplasticity of morphological convergence.

5. Decoherence of adaptive and neutral or nearly neutral traits upon speciation. Laggard neutral traits for evolutionary path, adaptive traits for describing the species (or higher taxon). One can expect strongly adaptive traits of a progenitor to disappear or become neutral (and perhaps somewhat burdensome of biological energy) when a species adapts to a new habitat

6. A minimum of two otherwise unlinked traits, new to a serial lineage, implies an evolutionary process linking them in all individuals, and this defines a species. The process holding together the individuals is not required to be expressly stated, but evolutionary estimation may be done based on the evidence that such a process exists in a particular set of individuals. An extraordinary or complex unique trait will also suffice to distinguish a species from similar congeners. As in all scientific fields, judgment is necessary, but no theory is ever final because further study can refine or correct theories.

7. Dissilience at the genus level or above is evidence of the reality of supraspecific taxa. The outgroup for a dissilient genus is ideally the ultimate progenitor of that genus and the immediate distal lineage. A new radiation of lineages "resets" the definition of "primitive" for the next lineage or lineages. Outgroups are both phylogenetically and evolutionarily informative. The dissilient genus is the central element in macrosystematic (genus level and above) relationships, not clades.

8. In a cladogram, if one sister group is clearly the ancestor of the other, then the node may be taxonomically named and serial descent inferred even though not quite parsimonious of character transformations. This removes the node, which in phylogenetics, is often an entirely unnecessary postulation of a separate, interpolated, unknown shared ancestor. If cladistic nodes are not removed, resolution of direction of evolution is compromised. Phylogeneticists may state that for one of a sister group to be progenitor of the other, then the branch length of the progenitor must be zero. This is not true because other information supporting a progenitor-descendant relationship forces a non-parsimonious zeroing of the length of one cladogram branch among sister groups that may not originally have a zero length.

9. A simple definition of macroevolutionary systematics is to minimize trait transformations and maximize paraphyly (one taxon derived directly from another of the same or lower rank) by modeling direct, serial descent of taxa. Even more simply, one minimizes parallelism and reversals.

10. Given Dollo's law, Levinton's evolutionary ratchet, and non-commutability of taxon transformations, it is easier to model serial relationships in macroevolutionary systematics than to just model shared relationships in phylogenetics. Like the Principle of Least Action in physics, all stepwise evolutionary changes are of about equal probability, but there is only one way to totally reverse a set of events and a myriad ways to go forward in other directions (Barrow 1994: 23). This is an explanation for the thermodynamic arrow of time that also operates in determining the inexorable direction and order of evolution. Of course, at the same time, continued accumulation of minor mutations provides information about possible convergence. Evolutionary change away from some generalist taxon is limited by the phyletic constraint of tolerable change in elements of physiology and morphology, which ensures taxon-level short-step gradualism. Taxa may not reverse totally but traits can reverse.

11. Molecular trait transformations are a special case because they track evolution and are non-commutable. Because molecular traits continue to mutate after speciation, pseudoextinction is the correct model. Problematically, paraphyly involving extinct or unsampled molecular races can never be directly sampled, and such critical information remains unknown. Thus, relying on molecular traits of only extant races introduces a random bias. Unsampled molecular races together with the principle of strict phylogenetic monophyly are a major source of cryptic molecular species and unjustified lumping.

12. Basic physical laws may be extended to systematics study. Some laws may have no immediate application. For instance, equations governing the properties of the universe are usually of power 4 or less. Objects in the universe affect each other differentially through locality, the nearest most important, and the farthest limited by the speed of light. A normal Gaussian distribution ("bell curve") is fundamental to physical processes. And symmetry is a constant feature. See Zander (2013) for elements of physics considered important in systematics, like the Golden Ratio, geometric distribution, and psychologically salient numbers.

13. The analytic key identifies and distinguishes tracking traits that appear in many species and many habitats, and are therefore, among such habitats and species' physiologies, effectively evolutionarily neutral or nearly neutral. The new traits for the dissilient genus are then probably mostly flags for adaptive processes for that new situation. Many such traits may appear to be neutral because no clear association or correlation with the new environment may be evident. But they must be still considered as flags of adaptive processes unless these same traits are found in other habitats in similar species.

14. The dissilient genus is the basic unit of ecosystem stability because it generates and maintains a group of very closely related species with similar ecologic requirements. Dissilient genera may be taken as hubs of scale-free networks, with small and monotypic genera forming a power-law distribution (fatter than that of a simple exponential hollow curve). Embedded in an ecosystem, scale-free networks become small-world networks of survival advantage to ecosystems in ensuring evolutionary and ecologic redundancy both in short- and long-term time spans. Maximizing redundancy protects ecosystems from thermodynamic catastrophes and competition from invasion by elements of other ecosystems. Humankind, in its effort to create a landscape similar to that of its aboriginal roots, is a particularly invasive element of the African savanna.

15. Maintaining natural small-world networks may be our most effective focus in dealing with current world crises in climate change and extinction.

Note: Those looking for a list of nomenclatural novelties will find them concentrated in chapters 12, 13 and 14, identified as new names in boldface.

Chapter 16
THEORY

Tools — A scientist who wishes to use macroevolutionary systematic methods in simple cases does not particularly need the mathematically abstruse and statistically complex methods of phylogenetics. She needs, however, (1) recognition of the difference between Fisherian and Bayesian statistics and why only the Bayesian is appropriate for estimating the probability of a single historical event, (2) an understanding of what bits and decibans are, and (3) a notion of logarithms to understand why bits and decibans can be added. This is easy stuff and may be accomplished at least superficially by reading my recent past contributions or perusing books written for the intelligent layman and available cheaply on Amazon. This is, however, study that must not be shirked as it is basic to modern systematics, however much taxonomists may be proud or tolerant of their innumeracy.

The macroevolutionary systematist will also profit from learning why cladistics and phylogenetics are of dubious value when used directly for classification (see chapter on justification)—for consolation when her papers are rejected by confused or outraged reviewers.

Why is scientific theory important for systematics? — Theory consists of causal explanations that help predict the future and retrodict the past, and allow interpretation of the present in terms of the simplest model consonant with total relevant data. According to Levitin (2016: xiv), "A theory … is not just an idea—it is an idea based on a careful evaluation of evidence, and not just any evidence—evidence that is relevant to the issue at hand, gathered in an unbiased and rigorous fashion." See also Johnson and Mayeux (1992).

Once, natural phenomena were considered to be endowed with innate will, but scientists have been distancing themselves from this. For instance, until some time ago, the refraction of light through, say a pane of glass, was described as light deciding, just naturally, to go in this jagged direction instead of any other; yet now the phenomenon is described as light appearing to follow the path where its speed is fastest (light having different speeds in different media) from one point to another (not just through one medium!) and all the other possible quantum paths canceling each other out. This follows the Principle of Least Action, which unfortunately again describes processes that are not directly understood, and sounds more like an excuse than an appreciation of a causal process, even if it is a quantum phenomenon. A virus has an icosahetral shape because it is the figure of least size that will contain a number of spherical proteins, which is a "hand-waving explanation based on minimizing ernery" (Stewart 2011: 141). What is magnetism? We can predict and calculate

how it will act but cannot say what it really is. Science is now using probabilistic descriptions rather than advancing absolute "laws" of how things act in nature. For instance, why is there an arrow of time when many physical and mathematical descriptions of how things act in theory imply that such processes are entirely reversible. The answer (modified from that of R. Feynman) seems to be, again, that there is only one path that reverses a process but myriads of paths that are alternative (Barrow 1994: 99–100).

This may be the best way to deal (see also Brooks and Wiley 1988: 83, 103, 217; Estes & Pregill 1988: 100) with the apparent efficacy of Dollo's Rule that evolution proceeds without exact duplication of a previous species, as long as the possibilities of not doing so are very many. It is possible that extirpation of a progenitor species gives impetus to a peripatric descendant (or its own progenitor) to fill that niche through selection against its advanced traits, but quasi-neutral tracking traits are commonly available for correct interpretation of this apparent Dollo reversal.

This probabilistic way of thinking is a release from dependence on laws and associated absolutes, which are similar to deduction-based mathematical theorems and lemmas. It frees the researcher's imagination to envision marvels yet to be discovered through the glass of probabilistic physics. True, many models of causal processes are affected by quantum and relativistic phenomena. For instance, because satellites travel at ca. 17,000 miles an hour, their clocks run slightly slower than those on Earth, and inaccuracies must be accounted for in critical operations. Entanglement that is non-causal is now demonstrable at the mesocosmic level. There is still, however, no practical alternative for causal explanations, and Newtonian and Euclidean calculation, in everyday scientific applications.

Cladists as likelihoodists commonly advance the notion that their data sets present all the evidence needed to make decisions about evolutionary relationships, given methods of optimization like maximum parsimony, likelihood or Bayesian posterior probability. On the other hand, classical taxonomists have had little problem making decisions placing together apparently evolutionarily related taxa before the advent of computerized decision-making. This book rejects the mechanical taxonomy that is encouraged by black-box phylogenetic programs. The rejection is not of the computer analysis as such but of the acceptance of dichotomous trees generated solely on the basis of shared traits as fundamental patterns in nature. The computer is powerful but is essentially an idiot savant, and we must not identify with its savant nature. Cladists accept that they are not exactly modeling nature, but that is a sad termination of inquiry. Science consists of Cartesian reliance on logic and deduction together with Baconian empiricism (Odifreddi 2000: xii), which together allow powerful interpretation and prediction of natural processes. Macroevolutionary systematics uses both inductive and deductive logic, imagination and the computer, to infer serial descent. Alas, as yet, there are no computer programs to accomplish this in one go, and thus attract the computer-oriented new student.

The method of macroevolutionary systematics emphasizes *direction and order* of evolution among series of named taxa when these can be identified, not simply diagramming assumed shared descent of pairs of lineages as in cladistics or hierarchically segregated groups of related taxa as in classical systematics. A major assumption or at least initial postulate for each instance is the "quantum evolution" of V. Grant (1963: 456–459, 1981: 155–160), which preferentially generates progenitor-descendant pairs (Crawford 2010).

Theory is important. It can gauge support for the most likely scenarios of causal processes that may be investigated further, and identify very well-supported causal relationships. In the Bayesian sense, it can tell when a theory is good enough that you can bet on its predictions and expect to win most of the time or at some acceptable frequency. Theory can deal with natural processes that are only implied or difficult to view directly, such as is evidenced in history or by cosmological or nuclear observations. Paarlberg and Paarlberg (2000: 44) pointed out that the then invisible structure of DNA was worked out by Crick and Watson "with theory, integrating the relevant and authentic contribution of biology, chemistry, and physics."

Analysis of past evolution is a historical science, and solitary events like past speciation cannot be observed directly, thus a theory of evolution is critical for evolution-based classification. Such a theory must explain all relevant evidence, which is particularly important in dealing with often conflicting morphological and molecular evolutionary trees. "Relevant" simply means facts that will help change or support or falsify the results without modification of the intent of the study. This means explaining both morphological and molecular data about both shared and serial descent. Paarlberg and Paarlberg (2000: xv) also pointed out that recent major advances in modernization of agricultural practices were powered by many technological and scientific breakthroughs, and the result has been multiplicative not additive, the total advance being the product, not the sum, of its parts. This is especially true in systematics with the introduction of the computer, methods of massive data analysis, molecular techniques, immediate access to publications, and intensive biodiversity research.

Macroevolutionary systematics can meld conflicting analytic results into one evolutionary view (the analytic key as summarized with a caulogram) and uses theory to heal the overly simplified cladistic method (using shared trait transformations alone to generate trees of taxa). An interesting comparison of the conflicting philosophies that underlie cladism and evolutionism is given by Aubert (2015), who discusses the 25-century fight between Eleaticism (logical standards, patterns in nature) and Heraclitism (harmony of sense perceptions, change in nature).

Evolutionary continuity in a cladogram, phylogenetically speaking, is between nodes of the optimizing clustering method, but in caulograms of stem taxa continuity is between taxa. Phylogenetics models evolutionary continuity as between splits in a minimum spanning tree or Markov chain, which are not processes in nature. One way to test the ability of prediction in a caulogram is to eliminate a known member and see if its existence is inferable as (1) an intermediate between the taxon before and the taxon after in the lineage, or (2) it is clearly foretold as an advanced end member of the lineage, or (3) it is inferable as a generalist and progenitor of two or more descendant lineages. This is Granger causality (q.v.). One can describe a missing link by the projected intermediate traits and even predict variability by the sort and extent of variability in presumed ancestral and descendant species. This is also true for predicting and describing an unknown shared ancestral species for two species that cannot rightly be an ancestor-descendant dyad but are very closely related. Actually finding that link obviates ("collapses") the probability distribution of what it may look like and how it might vary (as in quantum mechanics, but only that one has theorized only analogously pace the physicists). Probability is thus more than simply interconsensual expectation, like "such and so is probably true, right?"

Species in time change, and differential survival is a powerful evolutionary force. Thus, two species that are distinctive at the same time, may have been generated by an ancestor intermediate in morphology. This should not cause problems in nomenclature because evolution requires such overlap in time. Distant geographic distributions may provide the same differentiating dimension as does time, with intermediate species isolated from two distinct species elsewhere. An example is the clear separation by a disappearing distal costa in the lamina of *Syntrichia norvegica* F. Weber (as opposed to a connected costa and awn *S. ruralis* (Hedw.) F. Weber & D. Mohr) in Europe, but poor distinction in North America (fide P M. Eckel, pers. comm.), where identifying "central tendency" must be used rather than gaps. What is vicariant in this group is the inferred point in time of species differentiation, this being slower in North America perhaps because of genetic swamping due to fewer sharp biogeographic barriers.

A certain amount of judgment is necessary when doing world revisions and there is morphological overlap between isolated species. Synonymizing two species that intergrade in one part of the world may hide a distinct evolutionary trajectory that is isolated geographically, being more pronounced elsewhere. This also applies to fossil species where an intermediate fossil should not be cause of synonymy for two extant species. For this reason, inferred evolutionary direction from specimens from one part of the world distribution can provide guidance in dealing with variable and intergrading species.

Inductive and deductive inference falsificationism — Inferential scientific progress is commonly through initial well-informed guesses (hypotheses) using both induction, meaning simple (Occam's razor) expectations or explanations from a number of observations, and deduction, meaning inferring specific instances from generalities or well-established theories. There are, of course, problems with both induction and deduction. Consider the white swan problem. In induction, all swans we have seen are white, therefore we infer that all swans are white (or more specifically all swans we expect to see in the future will be white). This predidtion is quite like a species description. Observing, finally, an Australian black swan requires revision of the generality that all swans are white. In deduction, all swans are white (or more specifically all swans we expect to see will be white), therefore reading a report of an observed swan must be that of a white swan. Again, an observed black swan negates the axiom or generality that all swans are white. On the other hand, the inductive and deductive inferences are practical in being successful both prior to discovery of the black swan, and even in the knowledge that, outside Australia, there may only rarely be encountered a black swan.

Clearly, Karl Popper's falsificationism operates here, but the associated hypothetico-deductivism is problematic. Cladists decry theory because cladograms are considered the revealed structure of evolution. This generally eliminates induction from the inferential process, and so limits theorization and testing of theory.

Probabilistic prediction — In the present work, probability is considered a feature of human perception in light of poor information. Probability is important in physics as expectations of physical processes, either frequentist or Bayesian. One feature of mental probabilistic modeling is emergent patterns perceived by the mind. These patterns are mental skeletons of models of the progress of a process. The binomial distribution associated with the Central Limit Theorem is an example of such an emergent pattern. Although it is true that even if we were to know the exact initial trajectories of bouncing balls that end up in a neat bell-shaped pile we could not predict details of that pile exactly, it is also true that probability estimation works well in guiding action through creation of predictive models under expectation of such emergent phenomena when exact knowledge is unavailable (see Tversky & Kahneman 1974 for much additional discussion).

Probabilistic expectation, as a tool of the mind, extends the human sensorium far beyond the immediate time and environs, and may be the main difference between humans and animals. How is that? Consider predator and prey. It has been hypothesized that the brain stem predicts a prey's escape route very quickly based on simple triangulation, or the last successful prediction resulting in prey capture (Wiens 1976: 101; see also Boesch & Boesch-Achermann 2000: 266; Cliff 1994; Curio 1976; Sillar et al. 2016), while speeding response of prey to predation (Cooper & Blumstein 2015: 154; Eaton et al. 2001). A mentally slower method but more successful in the long run is prediction based on a frequentist method of choosing the most common successful prediction in the past and using that for the same prey. A more advanced method of prediction is a Bayesian-like method where successful predictions in the past are applied to classes of problems (kinds of prey) and predictions extended to new prey that seem to approximate a certain class.

In systematics, classification of organisms from the results of observations on single or a few individuals are probabilistically extended to the whole taxon. This is under the assumption that the same process that links a minimum of two advanced traits, that are otherwise and elsewhere not linked, standardizes and homogenizes to a greater or lesser extent all other traits in that taxon. This logical extension of experimental research is fundamental to the biological sciences and depends decisively on the sampling and classificatory study that is done by systematists.

Bayes factors and sequential Bayes — This section is tedious but ultimately instructive. A Taxacom listserver member once remarked: "Who can argue against high posterior probabilities?" Well, macroevolutionary systematics can. Molecular studies often demonstrate their results as having such high Bayesian posterior probabilities (e.g., often unity) that they are statistically certain. Caulograms from morphological studies using information theory may well be competitive as they, too, may evince statistical certainty. One must first translate caulograms to cladograms or vice versa. Then, combining the results, because molecular and morphological analyses use different sources of data, may require the use of Bayes factors, particularly if the sum of the posterior probabilities do not add to 1.00.

A Bayes factor, in the usual case of priors being 0.50, is simply the fraction of posterior probabilities, one divided by the other. Two statistically certain probabilities divided by each other is essentially one, which is not significant and contrary results between morphology and molecular analysis cancel each other. Only a Bayes factor of 3.0 or more signals a non-trivial difference, and the morphological posterior probability would have to be less than 0.30 for a well-supported molecular study to truly overwhelm the morphological results.

But suppose you feel that Bayes factors are stuff and nonsense. Let's try sequential Bayes, where the posterior probability of the first instance of Bayes theorem is used as the prior for the second instance with more information. The idea is that high support for morphological clustering implies a low support for molecular clustering. That low support from the morphological analysis will reduce the BPP in the original molecular study. In this case if a BPP for molecular clustering is "p", then we take the support for an alternative morphological clustering and subtract from 1. Then the two terms of the Bayes theorem is p and 1 minus p. Of course 1 minus p is not necessarily support for the exact topology of the molecular cladogram, because it could support evidence that the molecular cladogram is wrong in some other way, and because the data sources differ and the hypotheses actually are not alternatives of each other, sequential Bayes analysis in this case is inappropriate. That is, there is no discrete hypothesis space (Bonawitz et al. 2014).

But let's do it anyway and see what happens. A BPP of 1.00 is complete certainty and a second term of 1 minus p does nothing to change the BPP of 1.00. However, let's be reasonable and assume complete certainty is nonsense, given problems with molecular analysis (e.g., Zander 1998, 2003)

and also sufficiently generous to assign support of 0.99 BPP instead to nodes at 1.00. Using the Silk Purse spreadsheet available online at

http://www.mobot.org/plantscience/ResBot/phyl/silkpursespreadsheet.htm

then one can do a number of sequential Bayes operations. Using 0.99 BPP as prior, a second 1 minus p probability of 0.20 gives a posterior of 0.96; of 0.15 gives 0.92; of 0.11 yields a posterior of 0.93; and of 0.05 gives a posterior of 0.84.

Thus, support for an alternative morphologically based clustering of 0.85 BPP (giving 1 minus p of 0.15) reduces the 0.99 of the molecular analysis to below standard level of acceptance for non-critical scientific hypotheses (0.95 BPP). So three traits of support for a

morphological clustering (equaling three bits of support) reduces the molecular support to 0.93 using sequential Bayes, and four traits reduces molecular support of 0.99 to 0.86 BPP. Three or four traits are commonly found to support classical morphological clustering. This is true if not in the first species of a lineage then in the second because at least two traits are needed to differentiate a species from its congeners, and the two species are nested (cladistically or serially). To summarize:

BPP support for molecular group	BPP support for different morphological group	Bit support for morphological group	BPP support for molecular group obtained from morphological analysis (1 minus p)	New BPP for molecular support
0.99	0.80	2.00	0.20	0.96
0.99	0.83	2.33	0.17	0.95
0.99	0.85	2.67	0.15	0.92
0.99	0.89	3.00	0.11	0.93
0.99	0.91	3.33	0.09	0.91
0.99	0.93	3.67	0.07	0.88
0.99	0.94	4.00	0.06	0.86
0.99	0.95	4.33	0.05	0.84

We have above tested what happens when a measure of 1.00 BPP is reduced, justifiably, to 0.99. What happens if we reduced 1.00 to 0.995, instead. With 3 bits of support for the (quasi) alternative hypothesis advanced by the morphological analysis, this leaves 0.11 for the molecular (1 minus p, see table above). This reduces the BPP for molecular support to 0.96, still pretty good. We would need 4 bits of support for the morphological hypothesis to reduce the BPP of the molecular result to 0.93 (less than the standard 0.95 required).

If we made the molecular support only 0.999 BPP, then the BPP of the molecular support would only be reduced to 0.992 for 3 bits supporting the morphological result, and 0.995 for 4 bits of support for the morphological result. Clearly, any

support beyond 0.990 for the molecular clade will result in an exponential increase with reasonable contrary support for the morphological alternative. On the other hand, 4 bits of support for a *second* morphological branching nested in the first yields 0.922 support for the molecular configuration. This is because the BPP of the molecular analysis refers to the clustering of the clade beyond the node with the BPP, while the BPPs of the morphological analysis is for the order of any two species and the logarithmic bit support can be added.

The full support for a lineage of species in morphological serial analysis is the sum of the bits for every species order in that lineage. This is a measure of the coherence of the lineage. This figure can be very high and easily match a molecular support of 0.999 BPP. For example, three species with support of 3 bits each yields 0.998 BPP support for the direction of evolution. This reduces the support for a contrary molecular configuration by sequential Bayes to 0.667 BPP, about 1 bit. Why is there a difference? Because if the morphological grouping is reasonable (measured by the sum of bits of all species against the sum of bits of all species in the nearest species cluster), then there is no other place for the species, and the morphological analysis is of a closed causal group (see Methods chapter). Even a branch of two species arranged linearly is a closed causal group if other branches have high bit values, and a two-species branch of 2-bit species would reduce the support of a molecular configuration splitting them to 0.86 BPP, which must be measured against the 4-bit support for the morphological branch, or 0.94 BPP. This analysis is only a thought experiment designed to investigate a possible objection, and the hypotheses that are compared are different, that is they are not p and 1 minus p, and the assumption that the BPPs must add to 1.00 is somewhat wrong.

The upshot is that, by any measure—good (Bayes factor) or bad for this comparison (sequential Bayes)—*any* reasonable support for morphological clustering lowers a molecular-based BPP to less than acceptable probability. Although there is such a thing as statistical certainty, it is not a BPP of unity. The reason a BPP of 1.00 (absolute certainty) is not acceptable, ever, is that Bayesian formulaic support for any other alternative hypothesis must be rejected by sequential Bayes analysis, that is, if we have a BPP of 1.00 for the first alternative, that BPP is never changed by support for a contrary configuration of species.

Taxonomic methods in the larger context of the sciences — Recognizing species is quite like decipherment of ancient writings and languages. According to A. C. Clarke writing as Cleator (1959: 139), a *cognate language* with matching concepts in known alphabets, syllabaries, or ideograms is necessary to make sense of an unknown language with new symbols and words. In the same manner, making sense of an investigated species requires comparison with *cognate species* whose adaptations and variation are homologous *processes*. Identifying cognate evolutionary processes ensures application of the same evolutionary theory to all taxa, while on the

look-out for parallelism and convergence. This is not circular because bold hypotheses (Popper 1992: 94, 280) test the limits of theory, and any necessary new process-based theory advances science.

An example process is reduction under the taxon-level Dollo constraint (Gould 1970) that major trait combinations are not re-evolved nor are complexes of minor traits exactly re-evolved (at least without tag-along conservative traits that reveal convergence), thus a reduced taxon with rare or unique traits is doubtfully the progenitor of a complex one with more common, complex, and generalized primitive traits. Because both elaboration and reduction are features of serial evolution, inadvertent confusion of convergence with radiation may be avoided by evaluation of tag-along "tells."

One may hypothesize that lineages begin with elaboration of some traits away from the generalized progenitor, then, at the end of a descendant lineage, we may have a new generalized progenitor for another radiation. Study of the moss genus *Didymodon* (Zander 2013, 2014a,b,c), however, indicates that one generalist progenitor generates the next, resulting in a path of maximum evolutionary potential. Most descendants seem too specialized or otherwise burdened with resource-consuming adaptive traits to follow in a parallel fashion any microevolution or speciation of the progenitor. If descendant lineages are diagramed as lines coming out of a progenitor, then the new descendant progenitor "has no hair."

Dollo constraint at the taxon level plus outgroup comparison can allow polarization of serial taxon transformation from primitive to advanced involving elaboration and reduction. Considerations in this respect include Levinton's (1988: 217) evolutionary ratchet, and the concept of non-commutability in mathematics, where A plus B may not equal B plus A, the order of addition (or subtraction) making a difference. There is also an incommensurablity problem in that an adaptation doubling a linear measure is not reversed on halving an area or volume associated with that linear measure.

Dollo's rule may appear to be contravened by the phenomenon of "reverse speciation" or "introgressive extinction" (Rudman & Schluter 2015). This occurs when two young species having distinctive biologies hybridize, usually due to human disturbance, and the parent species become extinct. The remaining hybrid has aspects of both species (hence "reverse speciation"), but constitutes a novel biotype with an intermediate habitat and biotype. Identification of reverse speciation may be had by evaluation of possible evolutionary continuity with other species, comparison with an existing true progenitor, or special advanced features of new nonadaptive traits, functional morphology, or habitat.

Epigenetic explanations of reversals of complex traits are here considered secondary to explanation through elaboration or reduction because epigenetics can explain anything not explainable by genetics. Direct and overwhelming evidence (e.g., direct molecular or QTL data) is thus required for epigenetics to be accepted as controlling a complex trait or trait complex reversal during speciation. Simply mentioning other experimental research associated with, say, hox genes is insufficient to cast doubt on Dollo constraint. Very important in evolutionary systematics is the identification of homologous evolutionary *processes* and their associated expressed traits.

Neo-Husserlian Bracketing — Revision in classical systematics requires a re-evaluation of a group's taxonomic concepts, setting aside past theories of taxonomic identities, limits and relationships. The "*bracketing*" method of the phenomenologist E. Husserl (Smith & McIntyre 1982: 96; Tufford & Newman 2010) is particularly relevant to modern methods of systematics. Bracketing in philosophy means stripping perceptions to the simplest form, eliminating theories, presuppositions and assumptions, and thus perceiving the most essential features of a thing, to experience directly by looking beyond preconceptions. It is supposed to eliminate bias, beliefs and values. Pattern cladists (Scott-Ram 1990) have presented the "theory-free" cladogram as such an ideal, as an evolutionary structure in nature with it own reality. In macroevolutionary systematics, however, revision involves Husserlian bracketing to shave the descriptions and classifications from a group and then to reconstitute the group with a more coherent set of theory, description, and process-based evolutionary classification. One might refer to this commonplace taxonomic activity, with one foot in phenomenological philosophy but avoiding reduction to the metaphysical, as *neo-Husserlian bracketing*.

In the case of revisionary work in classical systematics, one begins by sorting, that is, adding specimens and removing specimens (or taxa) from an initially somewhat heterogeneous group until the group has a reasonable coherence or a central identity built around inferred adaptive and conservative traits. That coherence involves minimizing the redundancy of traits and the abduction (inferring an initial state from a resultant state using a process-based theory) of a natural evolutionary line of descent or alternatively a constraint (Futuyma 2010; Schwenk 1995) that applies to all specimens in the group, e.g., a species or genus. In other words, there is often an inferable evolutionary trajectory common to each species, usually signaled by a range of variation about a central core of stable minimally redundant (i.e., maximally shared) set of traits presumably held together by some evolutionary process.

By analogy, an ancient shipwreck may have cargo items cemented together. The cargo may be thought of as cultural artifacts selected in a Darwinian manner to have approximate standard functions and morphologies. Though we may not know the function of many ancient items of commerce, we can remove them in pieces from the cement and reconstruct them through recognition or imaginative estimation of their function by similarity in expressed traits or in variation of such traits to already understood artifacts (or to known species). Scientifically, such abduction of predictive processes is testable.

Granger Causality — Once species and their inferred evolutionary processes (based on known traits and processes in similar groups) are identified, such entities are substantiated or treated as evolutionarily realistic in being part of a set within an evolutionary series, i.e., a model of how species might fit together through time in nature given natural selection and population genetics. That is, once one identifies what one should be seeing, based on past study and theory, one tests what one is really seeing in light of scientific realism, which is ultimately based on established theory gradually modified by new "cutting edge" hypotheses. As has been more or less facetiously said, a fact is not acceptable until it agrees with theory.

Granger causality (discussed for systematics by Zander 2014a,b,c) may be used for identification of serial descent because sequential taxic transformations may be viewed as a causal series (though often branching). Causal connections are identified in Granger causality not just by correlation but also by predictability, as when elimination of the first element makes prediction of the second less likely. Membership in a causal series is another attribute of being a species, thus wrong transfer of one species of a series to another group should contravene theoretical prediction or expectation. If descent, serial and shared, is to be the preferred basis for scientific classification, then Granger causality is the fundamental technique of the exercise. The more species in a series apparently governed by causal evolutionary theory, the better. If removing one species does violence to that theory or theories, there is more support for leaving it be.

An ancestor-descendant series is more likely a causal series if taxa transform in a gradual manner in a series of processes based on descent with modification of some basic bauplan, and among-species tracking traits (those resistant to habitat variation) are identified. Gradual or short step changes that reflect evolutionary continuity are the desiderata of macroevolutionary systematics. By this I do not mean phyletic gradualism, which postulates slow anagenetic change ending in new species (Benton & Pearson 2001), but instead I assume a punctuated equilibrium that involves rapid speciation staggered in time and adding just a few traits at a time. That short-step gradualism in evolution is far more common than saltation is evidenced by the fact (as in the presently studied group of species) that rare and isolated morphology is far less common than groups of similar organisms. Were saltation (really big steps) common, there would not be supraspecific groupings of such closely related species. Saltation if common would lead to disorder, yet there is plenty of order in nature, and that is why we can do taxonomy at all.

Consider the proportion of closely related species in groups versus those evolutionarily isolated species that *could* have had very different ancestors. The latter are, in my experience, comparatively few, and even then, they are

probably a remnant of a mostly extinct gradual evolutionary series. Assumption of short-step gradualism also allows better prediction of other traits (e.g. chemical, genetic, DNA sequences) than might be expected from bursts of saltative evolution. Extinct species may also, paradoxically, tell us the most about evolution in that postulation of "missing links" between extant species is an exercise in modeling evolution of a coherent lineage of transformative change at the species level.

That one should and can effectively use short-step gradualism and other evolutionary theory for analysis is accentuated by the fact that only ca. 250,000 fossil species (that may provide direct evidence of gradualism) are known but there are, say, 30 million species extant at any one time in a five-million year sliding window, and these are generally classifiable in distinct groups.

A focus on caulistic serial relationships, particularly on autapomorphies in the cladistic sense, also aids in generating hypotheses of evolution both neutral and seamlessly gradual, or adaptive and rapid by short jumps or gaps (see also Pigliucci & Kaplan 2000). The autapomorphy of an ancestor becomes the primitive trait of a descendant that in turn generates descendants (see Example Data Set 6 in Methods chapter). The phylogenetic rejection of hypotheses of serial descent in favor of mechanical analytic generation of binary trees of shared descent based on an assumption of universal anagenesis is a confiscatory tax on the informed, disciplined imagination.

Prediction is often cited as the goal of systematics. Generally, membership in a cluster, group, or clade is the aim. In the present paper, prediction also involves an expectation of (1) future discovery of a postulated shared ancestor, an extant "missing link" contributing to evolutionary continuity, (2) ease of placing new species into membership in a known adaptive (or otherwise) radiation, and (3) in the case of newly discovered species rather different from nearest relatives that evolutionary processes associated with a necessary postulated shared ancestor or a new inference of a radiation will be parallel with or similar to known processes, adaptive or otherwise. This is a kind of uniformitarianism that leads to testable inferences. That adaptive radiation exists and is important has been reviewed recently by Schluter (2000).

Like the molecular clock of DNA-based systematics, morphology may be expected to gradually change over time. The difference is that relative stasis of a phenospecies over time, except for minor, of course infraspecific, anagenetic change, is expected, and stability of species is the basis for taxonomy. But in the macrosystematic sense, an ancestor-descendant speciation event may be likened to a "tick" of a *morphological clock*. One might expect a certain minimum number of trait transformations per speciation event to avoid Red Queen competition and backcross swamping (in sympatric species), yet not enough change to challenge the limitations of evolutionary cascades (see Strogatz 2001: 159) or phyletic constraint (survivable change based on what potentially unbalanced genetic modification a given bauplan can tolerate). Thus, Granger causality can in the majority of cases rely on a general short-step gradualism in serial evolution with clear identification of conservative traits, which proved true in the evolutionary study of the pottiaceous moss genus *Didymodon* (Zander 2014 a,b,c, 2016).

The species — A species is here pragmatically defined as a group of very similar individuals sharing at least two new traits (character states) that are apparently unlinked in other groups but the constant sharing of which implies an evolutionary process uniting those individuals. A new extraordinary, unique trait will also suffice to distinguish a species from similar congeners, particularly if it is complex and has a distinct habitat or ethology. These are "new" traits or trait combinations for a serial lineage because they do not occur in at least the immediate ancestral taxon on a caulogram. We do not have to know what evolutionary process links such traits or generated a complex trait, but, to do systematics and classification, we only need to infer that some process holds the individuals together in an "evolutionary trajectory."

J. B. S. Haldane (Clark 1968: 100) wrote: "The world is full of mysteries" … "Life is one. The curious limitations of finite minds are another. It is not the business of an evolutionary theory to explain these mysteries." Just as Haldane made a distinction between the restrictions on our scope of thinking about evolution and a theory of evolution itself, systematists, to generate a classification, need only identify the results of the imposition of sets and order on nature caused by evolutionary processes. Familiarity with theory of evolutionary processes helps, of course, whether classical or revolutionary (Gould 2002; Pigliucci & Müller 2010).

The minimum of two traits to define a species should ideally be entirely (inferably) of adaptive traits, because an evolutionary neutral tracking trait is often one shared with the immediate ancestral taxon. An adaptive trait, however, may turn out to be neutral or nearly so in the next species in the linear series. Founder species, also, can be expected to have drift acting on otherwise adaptive traits. It does not matter, however, in general if the two traits (minimum) used to characterize a progenitor or descendant are functionally understood or not, strongly developmentally determined, evolutionarily neutral or adaptive, or even geographic, just as long as they imply genetic isolation of the taxon as an evolutionary unit higher than just a population and apparently with its own evolutionary trajectory.

An informational concomitant of the process holding a species together is that it is *stationary*, meaning that the same processes of evolution occur throughout (as judged by apparently adaptive traits and theory). The stabilizing process is also *ergodic*, in this case meaning that the same kind of variation occurs in some one species geographically everywhere (excepting, say, the nascent speciation of geographic races), which allows a prediction of morphology, anatomy, ecology, chemistry, etc., when we identify a species by name. The ergodic nature of the stabilizing process is limited, however, by the time dimension (evolution explores only forward in time) and by phyletic constraint (developmental impossibilities canalize novelties).

In some cases gaps may be used to identify species. These are important in representing either extinction or disallowed trait combinations, given selection and phyletic constraint. Gaps are informationally real, represent evolutionary forces, and should be recognized in any evolution-based classification. Using taxon-based evolutionary trees (caulograms) will help identify or characterize the absent but critical trait combinations. Adding fossil species to a classification may fill such gaps but are not evidence to lump two extant species. This is because evolution occurs through time and requires allochronic isolation as well as, say, the allopatric.

The required delimitation of species by a minimum of two traits extends theory in two ways: first, the traits may be assumed to be linked by an evolutionary process (e.g., biological species concept or the like) that holds the species together, and, second, by the fact that two advanced traits yields good informational bit support for the order of one species against the nearest in a linear series, as discussed in the Methods chapter.

Dissilient genus — The *dissilient genus* concept (Zander 2013: 93) is one criterion for defining and clarifying a genus, and is the key element in classification above the species level (macrosystematics). Clades are not used here for defining taxa in macroevolutionary systematics, although the genus as an empiric phenomenon has been demonstrated phylogenetically by Barraclough and Humphreys (2014). Macroevolutionary systematics provides informational metrics not available or inappropriate in the cladistic context of shared descent.

If a particular generalized species of wide distribution can be identified as generative of a number of specialized taxa of more local or recent distribution, then that group of progenitor and its descendants (as an evolutionary cascade, Strogatz 2001: 159, or perhaps more accurately a fountain) can be viewed as a natural grouping higher than a species, i.e., a genus. Thus, evolutionary continuity characterizes linear series, while branching of such continuity implies a rather different evolutionary process.

Dissilient taxa are recognized as taxa with a central ancestral taxon lacking autapomorphies when conceived as ancestral to a number of other species. This commonly can include multiple immediate descendants from one extant ancestral species, while the descendants may have their own descendants among which they themselves lack autapomorphies. A tree of dissilient taxa is thus not "evidence" as per phylogenetic analysis, but is a theoretical construct maximizing gradualistic macroevolution between identifiable ancestral taxa and their descendants. A tree of microevolution (trait transformations), as in a cladogram, cannot model effectively descent with modification of taxa (taxon transformations). Hardy-Weinberg drift and other features of microevolution do not operate across species boundaries, but are quite different from intraspecific features of traits evolving as linked sets across species boundaries. This inability to model taxon transformations is why the unfortunate institution of strict phylogenetic monophyly and its inability to name direct progenitors are limiting features of the phylogenetic method.

The difficulty of recognition of dissilient genera in cladistic trees is that a caulogram groups by *natural sets* of traits, i.e., those characteristic of serial pairs of species, leading to easy interpretation of radiation from a core species. A cladogram groups by whatever combination of traits includes all taxa more distant on the tree. The cladogram node is not a species, and the continuity of evolution modeled by a cladogram is between serial lines of combinations of traits. This is artificial continuity. Evolutionarily related species may be well grouped on the dichotomous structure of a cladogram but the cladogram does not reflect natural evolutionary processes.

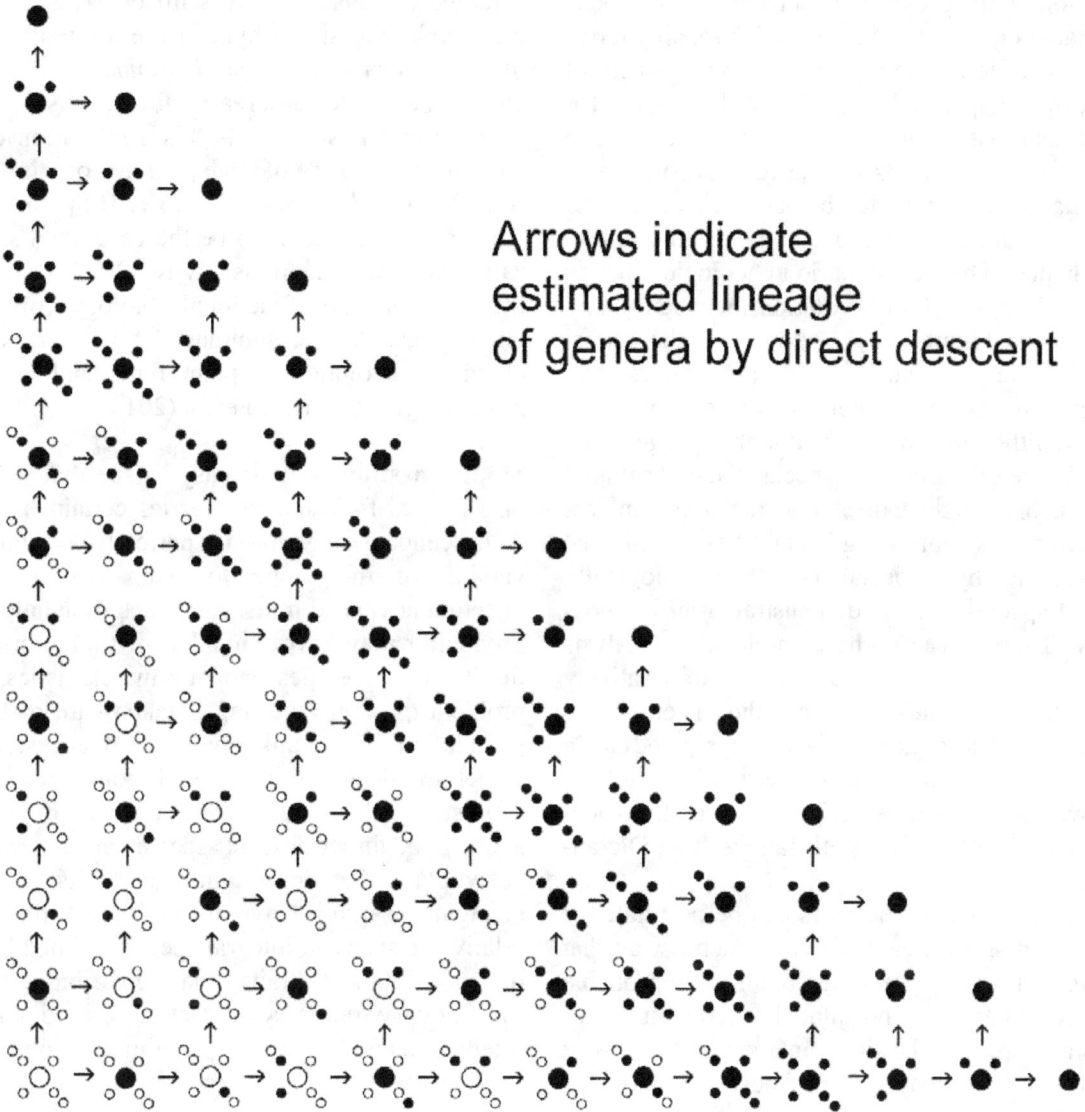

Arrows indicate estimated lineage of genera by direct descent

Figure 16-1. A contrived snapshot of a number of lineages of stylized dissilient genera *at one moment in time*. Arrows show inferred direction of evolution that occurred in the past between these different genera. Large dots are core generative species, small dots are descendants. Solid dots represent extant species, the hollow are extinct species. This example diagram of past relationships shows gradual extinction of both core and descendant species. Dots on right and top are newly evolved monotypic genera, those on bottom left are remnants of nearly extinct genera, either large dots (surviving core progenitors) or small (surviving specialized descendants).

A genus has its traits characterized by the progenitor in that descendants are more closely related in terms of traits to the progenitor than to each other. The progenitor's Bauplan, as projected into the descendants, is a commonality that characterizes all species. Evolution of a genus in response to general climate change should be largely of the progenitor rich in biotypes. The crowd of specialized descendants has an inertia caused by specialization and these descendants may be expected to remain in refugia or die out. The progenitor "gets a haircut" during speciation because all descendants due to overspecialization do not also change through microevolution or speciation. Thus, "bulk evolution" or transformation of all species in a genus in joint response to a shared environmental change is probably rare.

An exception is the evolutionary replacement of an extirpated species by its closest descendant (or progenitor) with some overlap in ecology. The relative generally needs only a reversal of two or three major traits, and may be peripatric or capable of long-distance distribution. More on this is in a later chapter. This is a scenario in geologic time.

The "snapshot" of evolution between many species in Figure 16-1 shows a theoretical reduction in the number of progenitors and descendants in the dissilient genera as time goes on. Given that descendants outnumber progenitors, and they are often found in specialized and refugial habitats, one might expect that the more ancient representatives of large families will be represented by specialized, morphologically reduced species. This is demonstrated in the moss family Dicranaceae, where molecular analysis (LaFarge et al. 2002) shows a set of relatively small taxa at the base of the Dicranaceae cladogram. These genera were segregated as a separate family, the Rhabdoweisiaceae, but are well within theory as an expected natural, ancient phenomenon of an integral family, the Dicranaceae.

Unknown shared ancestors can be postulated if two related species each differ in such a way that deriving one from the other requires improbable trait reversals. A postulated unknown third ancestral species lacking autapomorphies with respect to both may be the most parsimonious alternative to serial ancestry of the first two. It is ideal that the postulated unknown shared ancestral taxon have at least two traits that might be evidence of evolutionary cohesion of the (unknown) individuals that may have made up that species. Thus, the linkages of caulogram branches are species characterized by at least two traits, in which reversal to some more primitive taxon is Dollo-limited at that minimal level of complexity.

The branching of descendants off a generalized biotype-rich progenitor may be through either fixed adaptive or neutral traits or both. Assignment of adaptive faculty to certain traits is fraught. The calyptrae of *Streptotrichum, Leptodontiella,* and *Trachyodontium* species are large and papillose with forward-pointing (antrorse) teeth. They look much like the ends of the perichaetial leaves and may be inferred as adaptations to help the much

enlarged perichaetial leaves lift debris away from the developing sporophyte in the arboreal habitat of these taxa. Yet, *Leptodontioum excelsum*, in much the same arboreal habitat, has a large calyptra that is smooth. Is this a refutation of the particular theory of adaptation or is there something further we don't know? If the last taxon is advanced, as seems to be the case, then smooth calyptrae are reductions. Why? We know very little about the functional biology of moss phenotypes. A fine summary of what is known about the evolution of plant functional variation has been given by Reich et al. (2013).

Sloppy nesting — It is easy to see that deriving one species from another carries certain baggage. A descendant may not fit perfectly as simply a version of the progenitor with two or more different advanced traits. A species designated as a progenitor may have a number of traits not found in its putative descendant. In cladistics, this problem of sloppy nesting is taken care of by the postulation of an unknown shared ancestor that somehow deals with these disconcerting traits. Cladistics substitutes a pattern for a process. In analysis of direct descent, species are not cartoon "exemplars" but are evolutionary units gradually changing in minor ways over millions of years of relative stasis as an integral species defined by two or more linked traits that contribute to an evolutionary trajectory characterized by distinctive interactions with a slowly changing environment.

There are two ways to describe an entity in nature—by circumscribing taxa or by detailing central tendencies of taxa. The first ignores all variation but, say, those of first and second standard deviation, the second focuses on first deviation traits and lets the researcher select whatever else may be implied to her by assumption of Gaussian distribution and discovery of oddities and outliers. Both make difficulties, but over-exactitude can be as misleading as dealing with ideals and typologies.

An "anagenetic species" in the literature is one with great genetic diversity at the present time, which implies much change in the past. Translating to expressed traits, a species that has one or more descendants that differ from it in a few minor ways is evolutionarily an anagenetic species. Descendant species may have originated at any time during its million-year existence and reflect then, from the start, the minor traits that the progenitor had long ago. "Minor traits" are those appear randomly and unitarily among similar species, and may be of little adaptive significance. A descendant may not necessarily have the exact same set of minor traits as its putative progenitor.

Through time, there may be additions of minor traits due to mutation and drift. The descendants of anagenetic species may preserve for inspection some ancient and now lost variation in the progenitor among their own traits. An estimate of the extent of anagenetic variation over time in an ancient species may be gotten from the range of variation at the present time in recent extant species, multiplied by the expected variation in habitat of the ancient species (admittedly not large for species characteristic of microenvironments). In this manner does macroevolutionary systematics, when dealing with direct descent, explain the "slop" in trait transformations from progenitor to descendant.

Information theory — So what justifies using informational bits in the context of characters of species? Shortly, (1) because A. Turing used sequential Bayesian analysis and decibans to successfully break codes during World War Two (Zander 2014b: 12), and (2) the classical "sweet spot" of somewhere between 2 and 4 or 5 traits to distinguish species in the same genus matches nicely the range of 0.80 to 0.94 or 0.97 BPP if each trait is awarded one informational bit.

In ordering by direct descent, one species must derive from another by at least new two traits (linked by some species level process like isolation) and not more than five or six new traits or it either belongs in some other lineage or a missing link must be proposed to fit theoretic expectations of gradualistic transformation of species in evolution. So we have a range of two to five traits, with the fewest implying a close and maybe infraspecific relationship and the larger a more certain relationship at the species level, and the largest (or beyond) no relationship at all because tracking traits are overwritten. This is a probability range. What *maps* to this set of probability scores? Bits in information theory. Two bits give 0.80 Bayesian posterior probability, three 0.89 BPP, four 0.94 BPP, five 0.97 BPP, six 0.98 BPP (see detailed chart in Methods chapter, and Table 16-1). In this way, morphological study can

be supported with the same holy grail of molecular support measures, the BPP. It is reasonable, needed and practical. Thus, information theory is required by taxonomy.

Phylogenetics provides support for lineages of taxa sharing inferred ancestors using several methods, including nonparametric bootstrapping, likelihood, and Bayesian posterior probability. Often Bayesian analysis provides support of 1.0 probability (meaning statistical certainty, which is a real thing) for pairs of sister lineages on a cladogram. This is a measure of the chance that they are most closely related from among the taxa studied (cladistics) or that they shared an immediate ancestor (phylogenetics). Consider, however, the following scenario: Taxon A *gives rise* to taxon B. Bayesian analysis assigned a 1.0 probability that they are most closely related from among the taxa studied, or that they share an immediate ancestor. The former is just bad resolution of the true evolutionary relationship, while the latter is nonsense. A different measure of support for linear order and direction of evolution may be found in information theory and the use of informational bits.

Well, why does it take more than a few bits to provide good support for a lineage order? If a bit is good evidence, why not just one bit to determine order decisively? This is because although a bit is definitely more than a hint (a deciban or one-third of a bit is here taken as a hint), there is an analogy with calculations using the binomial distribution that deal with evidence for and against some conclusion (see Zander 2003, 2004). There is an exponential spectrum (bits are logarithmic at base 2) of expectation between a mere hint and over-whelming evidence that some given theory is right or wrong. The hint is one deciban at 0.56 Bayesian posterior probability, while good evidence is one bit at 0.67 BPP (nearly 1 standard deviation or 0.68 BPP, that is, where the bell-shaped curve inflects from downward to curving upward), substantive support is 2 bits at 0.80 BPP, and (just about) minimally decisive support is 4 bits at 0.94 BPP (nearly 2 standard deviations or 0.955 BPP)), while over-whelming evidence is 8 bits or 0.996 BPP (nearly 3 standard deviations or 0.997 BPP). Although past events in speciation are not directly detectable, given a short-step gradualistic theory-based model, the common occurrence of very similar species in groups, and fossil evidence of evolution by mostly gradual accumulation of advanced traits, a close association between bits and Bayesian posterior probabilities is acceptable.

Information theory in systematics is akin to decryption of a hidden evolutionary message, using evolutionary theory as key. To get the full message one seeks an evolutionary order of taxa involving the least redundant arrangement of all relevant information. Cladistics works by grouping taxa by reduction of hierarchical redundancy. One can view serial evolutionary transformation as a Markov chain. The Markov property specifies that the probability of a state depends only on the probability of the previous state, but higher order Markov chains have more memory of past influences. Several species transforming in a series is an example of a higher order Markov chain in that probability of correct retrodictive analysis is increased by seeing a short-step gradualistic increase in advanced character states. This also increases predictive value of a serial transformation. Given that the present analytic method interprets the correct order of two species based on information in a third (outgroup), that decision is a second-order Markov chain.

Given that information theory has many non-intuitive and confusing concepts, this book will clarify the use of the term "entropy," as either informational or thermodynamic. Much literature exists that relates the practice of systematics to the Second Law of Thermodynamics (see extensive discussion of Brooks & Wiley 1988). There are two ways information theory is used in evolutionary theory, the first is in the study of thermodynamics as *that which drives evolution*, and the second is in analysis of messages being *interpretation of the order and direction of evolution*. It is the second, Shannon entropy, that is emphasized in the present chapter.

In addition, the term "bit" (binary digit) has two (2) definitions. The first emphasizes lack of knowledge, being a 50:50 chance of deciding whether a question is yes or now, or choosing 0 or 1. The second emphasizes a known or at least decided solution to a question, or actually choosing 0 or 1, and is equivalent to 0.67 Bayesian posterior probability (about 1 standard deviation). Bits that are about known information, because they are logarithmic at base 2, can be added; 2 bits are 0.80 BPP, 3 bits 0.89 BPP, 4 bits 0.94 BPP, and 8 bits (1 byte) are 0.996 BPP. A simple table is given here (see Table 16-1) to guide the reader.

The idea is to maximize the information on direction and order of taxon macroevolutionary transformation, using BPPs. A random serial arrangement of taxa should yield about half positive and half negative bit values, or 0.50 BPP. To organize serially, one bit is assigned to every advanced trait that implies radiation away from a generalized progenitor by being not present in the outgroup. One negative bit is assigned to a reversal.

Table 16-1. Bits and Bayesian posterior probability (BPP). "S.D." means standard deviation.

Bits	BPP
0	0.500
1	0.666 or nearly 1 S.D. (0.683)
2	0.799
3	0.888
4	0.940 or nearly 2 S.D. (0.955)
5	0.969
6	0.984
7	0.992
8	0.996 or 0.99+
9	0.998 3 S.D. (0.997)
10	0.999
20	0.999999 (odds of 1 million to one)

Bits — When information is uncertain, Shannon entropy is low, when certain, such entropy is high. Both situations are described in terms of bits (binary digits). Following Pierce (1980: 82), the entropy of tossing a fair coin is 1 bit, because there are only two outcomes and they are equiprobable. Evaluating a trait for its information on order and direction of evolution is also has only two outcomes. If we toss two coins at the same time, (1980: 85), the entropy is 2 bits per pair tossed. "It takes 2 bits of information to describe or convey the outcome of tossing a pair of honest coins simultaneously." Because the information is *known,* the entropy is highest. The summed bits may be used in macroevolutionary systematics, as in computer terminology, as indicators of conveyance of information.

Brooks and Wiley (2015: 148) use a formulaic definition of "bit" in the microevolutionary context of allelic change that involves more probabilistic calculation than a simple binary choice. In the present work, I use the bit in the context of macroevolutionary transformation. Entropy quantifies the uncertainty involved in predicting the value of a random variable. For example, specifying the outcome of a fair coin flip (two equally likely outcomes) provides less information (lower entropy) than specifying the outcome from a roll of a die (six equally likely outcomes)." The fixation of adaptive traits, howsoever they are randomly produced, is not done randomly. Fixation is in consonance with a selective regime wherein entropy as information is maximized by the theory of a new trait in the context of short-step gradualism being indicative of direction of evolution.

Entropy, though measured in terms of bits, is a concept associated with the first definition of a bit, the uncertainty of an equiprobable binary random variable; the present study, however, focuses on increases in known information by additive degrees of certainty. Information derived from *deciding* the difference between 0 and 1 (a 50:50 ratio), is 1 bit. A bit is the uncertainty of 1:1, but the information when known is equivalent to 1 bit. Thus, a distinction may be made for entropic bits and informational bits (see also Gleick 2011: 230).

The discussion of Pierce (1980: 82) is instructive. The entropy H of a message source, in this case of a coin coming up heads or tails, is:

$$H = -\sum_{i=1}^{n} p_i \log p_i \text{ bits per symbol}$$

where p is the probability of one side coming up, and $p - 1$ that of the other side coming up, and p_i that of the *i*th toss appearing (using base 2 logarithms).

Entropy for tossing a coin is the sum:
$H = -(p_0 \log p_0 + p_1 \log p_1)$ bits per symbol which
    for a fair coin is:
$H = -(1/2 \log 1/2 + 1/2 \log 1/2)$
$H = -((1/2)(-1) + (1/2)(-1))$
$H = 1$ bit per toss

Above, each toss of a coin is equivalent to a new advanced trait for which we have chosen yes as to evolutionary direction after polarization against an outgroup or known progenitor. The entropy of the toss is maximum and the information is certain (that it is indeed information, following theory) but the Bayesan posterior probability of this information is only 0.666 (about one standard deviation) for its contribution to the certainty of direction and order of the lineage on the caulogram. More species and more advanced traits often lead to near statistical certainty that a particular caulogram lineage is correctly ordered and given direction.

The tossing of two coins at the same time is like having two new traits determined to be advanced in the same species. The formula yields a simple result:

$H = -(1/4 \log 1/4 + 1/4 \log 1/4 + 1/4 \log 1/4 + 1/4 \log 1/4)$
$H = -(-1.2 - 1/2 - 1/2 - 1/2)$
$H = 2$ bits per pair tossed

Thus, a bit is awarded for each new trait in a species (one species below it in the caulogram not having that trait). One bit is also awarded for each species additional to the subtending non-trait species comprising the patristic distance for each trait. The sum of the bits for each species can be interpreted as Bayesian posterior probabilities of the information about the direction of evolution by simply following a chart (see Table 16-1).

A coherent lineage is a series of species nested by two to six changed traits (2 to 6 bits) per transformation along a nested cline of increasing distance from some outgroup or inferred primitive set of states. Less than two traits is insufficient to support a concept of a species. If there are more than about six new traits, the expectation of a gradualistic stepwise transformation of a lineage is not satisfied, and the species belongs in some other lineages or at least elsewhere in the evolutionary tree.

Decibans — Suppose that you feel that your decision for the direction of evolutionary transformation *for a given trait* is simply that the trait as information exists but does not warrant full and decisive confidence. You can then award the trait a deciban (the minimal level quantifiable as a measure of belief in a hypothesis, somewhat more precisely as an change in odds ratio from 1:1 to about 5:4, equal to 0.56 BPP) rather than a bit (minimum level of belief in a known or at least decided solution to a binary question of 50:50 probability, equal to 0.67 BPP). Although mixing bits (log to base 2) and decibans (log to base 10) seems odd, the tables for BPP demonstrate that a bit is almost exactly 3 decibans. For a species, the assignment of 2 bits and 1 deciban yields 7 decibans, or 0.833 BPP. (This is analogous to, say, two dollars and 33 cents where a cent is one-hundredth of a dollar.) Of course, any probability can be assigned but using as basic units minimum bits and minimum decibans is less arbitrary as they are clearly defined, empirically based, and amenable mathematically. A table (Table 16-2) is provided to show how decibans ("hints") and bits (an item of information) are matched.

Table 16-2. Equivalency of bits, decibans, and Bayesian posterior probabilities. "S.D." is standard deviation.

Bits	dB	BPP
0	0	0.5
0.33	1	0.557
0.67	2	0.613
1	**3**	**0.666**
		nearly 1 S.D. (0.683)
1.33	4	0.715
1.67	5	0.759
2	**6**	**0.799**
2.33	7	0.833
2.67	8	0.863
3	**9**	**0.888**
3.33	10	0.909
3.67	11	0.926
4	**12**	**0.940**
		nearly 2 S.D. (0.955)
4.33	13	0.952
4.67	14	0.961
5	**15**	**0.969**
5.33	16	0.975
5.67	17	0.980

6	18	0.984
6.33	19	0.987
6.67	20	0.99
7	21	0.992
8	24	0.996 or 0.99+
9	27	0.998 3 S.D. (0.997)
10	30	0.999
20	60	0.999999
		(odds of 1 million to one)

Serial order of species, and the caulogram — The single illustration in C. Darwin's Origin of Species (1859) demonstrates his evolutionary concept involving both shared and serial taxon transformations, i.e., by both similarity and genealogy (Dayrat 2005). Prior to Darwin, similarity (shared traits) was emphasized for classification, with modern analysis post-Darwin being hierarchical cluster analysis. The latter is viewed here as including cladistics, or non-ultrametric cluster analysis through use of trait transformations on the basis of optimizing shared apomorphies as minimized cladogram length.

This is dependent on the narrowly effective fiction that traits alone evolve (fiction because selection acts on taxa not traits alone and even neutral traits are fixed at the organismal level). Phylogeneticists interpret this as recency of common descent, which is an acceptable inference only if no ancestral species has survived. Cluster analysis, which includes cladistic analysis (pers. comm., P. Legendre), is viewed by the present author as not capable of directly modeling evolution because it generates trees with nodes that are not evidence of anything in nature (they are postulates of usually unnecessary "missing links"). Cluster analysis is valuable to group taxa yet direct modeling requires one taxon being derived from another, not from an unnamed and unnamable split in a minimum spanning tree diagram.

True modeling of evolution, properly diagrammed as a caulogram (Besseyan cactus, commagram, paraphylyogram), requires parsimony of both trait transformation and taxon transformation, the latter minimizing the number of necessary postulated shared unknown ancestral taxa. This pruning of superfluous, invented shared ancestors of unknown identification leads to postulation of a linear series of stem taxa that may branch when one species evidently generates more than one descendant. Among early evolutionists, Ernst Haeckel apparently over-focused on serial descent (Dayrat 2013). Examples of modern evolutionary systematists who created diagrams with a balanced view reflecting both shared and serial descent are C. A. Bessey (e.g., Besseyan cactus), J. Hutchinson (serially labeled phylo-gram), A. Cronquist (labeled phylogram), A. Takhtajan (bubble diagram), R. Thorne (serially labeled phylogenetic shrub), G. Dahlgren (Besseyan cactus) (see Jones and Luchsinger, 1979: 97–103, 362; Singh 2010: 318–344). Serial descent is dependent on quantifying the extent of radiation and amount of evolutionary change. This paper attempts to provide such means and measures at the species level.

Classical systematics begins with alpha taxonomy. Alpha taxonomy is often assumed to be simply collecting specimens then describing those that seem new to the expert. The Linnaean system of binomial nomenclature of genus name and epithet, however, requires an *analysis*. This distinguishes a species. And also a *synthesis*, which selects or devises a genus in which to put the species (as epithet). This paper attempts a formalization or explanation based on classical methods of distinguishing a species, and uses modern sequential Bayesian analysis using informational bits to select and describe a genus, thus completing the binomial. The method is exemplified with the moss family Strepto-trichaceae, a new segregate from the larger Pottiaceae, to clarify the study. The questions of this study are (1) how to treat species, some of which are questionably distinct, (2) where are they embedded along the continuity of their

evolutionary lineage, and (3) what is the proper   evolutionary lineage?
genus or branching of joint continuity given that

Chapter 17
DISCUSSION

In theory, plants break into a new habitat with new adaptations, exploring and filling the niche through generation of adaptive biotypes of selective advantage or evolutionarily neutral new forms through genetic drift. Each species is a survivor of particular environmental challenges, and thousands contribute to the survival of a biome. These myriad solutions, if we have the wit to understand them, can be valuable to humankind, particularly during the present extinction event. A taxonomic classification is a library providing access to the explanatory evolutionary threads connecting triumphs of survival in the plant world.

Since the time of Linnaeus' well-codified classification systems, we have accumulated 250 years of hard-won information on identification and evolution of organisms. Such information is comprised of both adaptive and neutral or nearly neutral traits, which allow analysis of selection (adaptive traits) and drift (neutral or nearly neutral traits) with the latter being particularly useful as tracking traits of descent with modification. For some of this information, we can thank, for example, such early American botanists as Elizabeth Rochester and Ottilie Hauenstein (below), and Judge George W. Clinton (further below).

Elizabeth Rochester and Ottilie Hauenstein, Buffalo, New York, ca. 1865

Judge George W. Clinton, Buffalo, New York, ca. 1865

This book follows Zander's (2013, 2014a,b,c, 2016a) new variant of the methodology of evolutionary systematics, which requires modeling of direct, serial descent, and attempts to systematize, as it were, classical systematics. It promotes an integration of evolutionary theory with classification. The central proposition is that evolution is properly studied as transformations of taxa, not traits alone, because traits do not evolve in isolation, and without a taxon-link will track only themselves. Thus, the central focus of evolutionary inference in macroevolutionary systematics is continuity of taxon transformation.

Continuity in cladograms is, in comparison, between nodes on a spanning tree or Markov chain or elements of some other optimization method. "Cladistics" is much akin to a hierarchical cluster analysis. It groups, using trait transformations, taxa on a tree by shared traits, while "phylogenetics" is much the same but has an evolutionary dimension based in part on branch lengths (distance from a node) and in part on the unwarranted assumption that a node is a shared ancestor not identifiable as

an extant species. In this book when cladistics is referred to, it is the simple, mechanical analytic process that clusters taxa with shared apomorphies on a dichotomous tree, but phylogenetics refers to the extended interpretation of the results. The difference may seem slight to the non-aligned systematist, but cladists can be incensed when lumped with phylogeneticists. Although cladists have a point, both groups make changes in classification based on analysis of shared descent alone, which is unscientific in rejecting information on direct descent. Their defense is mystification, obscurantism, and pietism, which are revealed, dissected, and rightfully belabored by evolutionary systematists in the paraphyly literature. Is this criticism mere contumely? The poster child example, as noted in the Introduction, is the cladistic model of continuity in evolution being an unknown and unnameable shared ancestor giving rise to another unknown and unnameable shared ancestor, which gives rise to another … etc.

Macroevolutionary systematics is here considered a good replacement for phylogenetics because (1) it has better resolution in that it can resolve direction of evolution between cladistic sister groups, (2) it generates no false, unnecessary entities (nodes or methodologically required shared ancestors) and is therefore parsimonious of both character state changes and taxon changes, (3) the continuity of stem taxa allows prediction of how newly discovered or newly studied taxa might fit into an evolutionary framework governed by natural processes, and (4) how evolution diagrammed as an occasionally branching linear series is governed by the same theory whether the taxa involved are extant or only inferred (i.e., "missing links").

Serial evolutionary relationships of taxa are tracked by conservative expressed traits and certain molecular information. A taxon is here defined as a group of individuals linked by some evolutionary process signaled by two or more linked traits that are generally either unique or at least independent in other related taxa. Taxa, whatever the process linking traits, have characteristic adaptations, modes of variation, ecologies, and geographic ranges. (Zoologists may add here elements associated with faunal evolution, such as predator-prey interactions, ethology, and so forth.)

I've offered a way to measure Bayesian support for morphological classifications. Most simply, after eliminating redundancy of information by identifying homologies, make a diagram of serial evolution (not nodal where traits generate traits, but species generating species). Then, after all traits that can be identified as the same are dealt with, count the traits left over. These are the traits that distinguish a descendant from a progenitor. Each trait is one bit of information. Each bit may be summed because they are logarithmic. There is a direct link to Bayesian posterior probabilities for each set of traits. High Bayesian posterior probabilities of morphological relationships will falsify high molecular Bayesian probabilities. Morphology and other expressed traits are the substance of adaptation and should not be second-class data. Convergence of morphological traits can be identified by finding conservative traits that distinguish morphological lines of evolution by direct descent. But divergence of molecular races (heteroplasy) and subsequent extinction of some races will leave phylogenetically split or lumped species, genera and families. The only correction for molecular heteroplasy is examination of morphological information, and that is generally ignored by phylogeneticists.

Fight back. Measure the power of your serial evolutionary hypotheses and you will see that classical taxonomy based on morphology is well supported. Resist the cleansing of Darwinian evolution from classifications by cladistic substitution of a variant of shared-trait hierarchical cluster analysis.

Tree thinking versus stem thinking — Evolution should be considered a basic process, not a basic pattern. The sister groups in cladograms are an artifact of a clustering method. Thus, a cladogram is here considered only a very preliminary evaluation of one aspect of the process of evolution, the shared distribution of homologous traits among descendants. It can be replaced by a caulogram, which is an actual evolutionary tree modeling both linear and branching descent. There is a great deal of evolutionary information in a cladogram, but a dichotomous optimality tree does not directly model evolution. It requires further optimizing on direct, serial descent, called "superoptimization" by Zander (2013: 75ff.)

This explanation may be difficult to comprehend for those accustomed to "tree thinking," which limits analysis to evaluation of shared traits. Note that using a cladogram to minimize parallelism, then an analytic key to minimize reversals, is not at all the same as building a phylogenetic tree from both morphological and molecular data. Also, building a phylogenetic tree from both morphological and molecular shared traits is not using total evidence because data about serial descent in both morphology (e.g., use an analytic key, q.v., to polarize taxa) and molecular information (informative heterophyly when available) is ignored when developing a cladogram. In addition, shared traits may not directly translate to shared descent (when linear taxon series are involved). All descendants share traits with their direct progenitors.

Homology of traits is indicated by such traits occurring in related species within each lineage such that gradual transformation of taxa, either anagenetic or cladogenetic (including "budding" during peripatric speciation and pseudoextinction), is reasonable given all traits involved. Here "lineage" refers to the analytic key (indented prose diagram) and an interpretive caulogram modeling evolution of stem taxa based on direct, serial ancestry, and does not mean "clade," which refers only to a set of taxa nested by shared traits.

In short, the number of taxon transformations is minimized by the researcher's initial selection, from the study group, of a generalized species, and checked later in her informational bit summation. Then, other, more advanced taxa are optimally inferred (minimal reversals or parallelisms) as derived directly from that generalized species or from each other. Often, there are no trait reversals of critical traits required. In other words, bit relationships as transformations of traits are minimized by initial nesting so that, to the extent possible, only advanced traits come from more primitive traits. Then, bits as trait transformations are evolutionary information about the optimized series of taxa. Initial reduction in bits (trait transformations) ensures gradualistic evolution to conform with theory.

A formal cladogram and outgroup polarization may be used for initial analysis if the taxa are not easily clustered as dissilient groups. If the generalized ancestor has many biotypes, a core biotype may be selected as ancestor with un-reversed traits. Thus, judgment and discursive reasoning are important, as in other sciences. this may be uncomfortable to those who feel that optimal evolutionary trees reveal real structural patterns impressed on nature through evolution (Zander 2010), and these must not be modified by theoretical considerations. An unknown shared ancestral taxon may be postulated if two closely related taxa cannot be derived one from the other without significant trait reversals. If traits appear often in nearby lineages that are not directly ancestral to the study group (descendants of progenitor siblings or "uncles"), then the traits are probably adaptive (strongly linked to habitat or other external influence) and less important as markers than are the more rare conservative, multi-habitat tracking traits that appear in series on the evolutionary tree.

The reason for offering new techniques for analysis and synthesis in the context of direct taxon descent is that classical systematics needs to go beyond classification changes based solely on the opinion and authority of an expert. An expert using classical techniques, in the face of the complex methodology of phylogenetic analysis, should now be able to justify (e.g., by reference to some standard methodological literature) any significant taxonomic decisions by detailing means used for distinguishing a species and assigning the epithet to a genus, i.e., following the stricture of binomial nomenclature.

So, offered here are techniques for distinguishing taxonomic entities and their causal relationships as used in other scientific fields that in many cases match well with heretofore unexplained or Gestalt (Zander 2013: 31, 44) standard taxonomic methods. I have detailed in other chapters that these strong ratiocinative rules may include language and code decipherment, neo-Husserlian bracketing, Granger causality, dissilient genus concept, Bayesian sequential analysis, combinatorics techniques (IRCI formula), analytic keys, and probabilities of ancestry and descendancy.

Some traits have become conservative and occur in more than one environment or species, others are narrowly adaptive and rare. The general morphology of the Streptotrichaceae (genus *Leptodontium* and relatives) is apparently adaptive given the wide occurrence in other families (e.g., Dicranaceae, Orthotrichaceae, Pottiaceae) in the same or similar habitats of the traits of large size, lanceolate and widely spreading to squarrose leaves with sheathing, "shouldered" bases, and dentate distal margins. Thus, generally adaptive traits can become conservative tracking traits within a group.

Why does macroevolutionary theory, as applied in the evolutionary analysis basic to systematics, cause difficulty for phylogeneticists? — I believe this to be the stubborn refusal or even inability to imagine anything other than shared traits as revelatory of evolutionary relationships. When using shared traits (as some kind of "descent of traits") as the only criterion of evolutionary relationship, a focus on the re-classification necessary in avoiding cladistic paraphyly is the difficulty. Paraphyly *is* evolution when using phylogenetic data for hypothetico-deductive inference of descent. Macroevolution involving paraphyly is basically generation of one taxon of the same rank or higher from within another. Paraphyly is not the only indication that one taxon can be generated directly from another. There are many clues that one extant taxon is the progenitor of an extant descendant that do not involve shared traits. But shared traits are the only information allowed in generating a cladogram, i.e., they and only they are "phylogenetically informative."

I have in the past (Zander 2010) opined that phylogenetics is somewhat similar to creationism: "Curiously, the last three points (apparent immutability of species, microevolution acceptable but not macroevolution, and avoidance of any implication of macroevolution) are quite those of 'scientific creationism' (Poole 1990: 106) or phylogenetic baraminology (Gishtick 2006), and do not instill confidence." What do you think?

One should not confuse serial descent with serial classification. There has been a justified dissatisfaction with sequential listings of taxa in books as though evolution happened as one long line of progenitors and descendants. The

diagrammatic tree (e.g., cladogram and caulogram) are two solutions. Direct descent of one taxon from another, of course, incorporates branching as one progenitor may have two or more descendant chains of taxa (lineages).

Imagine a phylogeneticist trying to translate a caulogram into a cladogram. Take the simplest caulogram **A** > B. In this formula, the boldfaced letter A is the inferred progenitor and light faced B is the descendant. The only way a cladogram can reflect direct generation of a descendant is by paraphyly because nodes must not be named as one of the taxa being studied. Thus, Figure 17-1 shows a caulogram on the left and the paraphyletic equivalent cladogram on the right (rooted since the caulogram is also rooted in some other outgroup-like taxon), not given here.

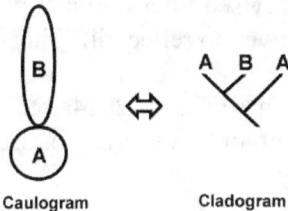

Caulogram   Cladogram

Figure 17-1. Caulogram on the left. Equivalent cladogram on right, with taxon A paraphyletic to imply evolution of B directly from A.

That is simple. Yet how about **A** > **B** > C, that is, progenitor A giving rise to B, which itself is the progenitor of C. Inserting the appropriate paraphyly into a cladogram becomes onerous and confusing, particularly since it seems to come from

nowhere (Figure 17-2). The inserted paraphyly is really unnecessary because the direction and order of evolution comes from different, non-phylogenetically informative information on direction and order of evolution.

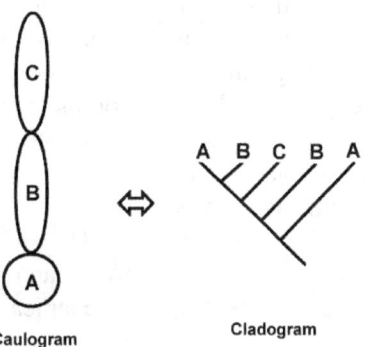

Caulogram   Cladogram

Figure 17-2. Caulogram on the left with progenitor A giving rise to progenitor B which itself gives rise to descendant C, last in the lineage. Cladogram on right attempts to duplicate the caulogram by adding paraphyly such that A is paraphyletic to B (and C) and B is paraphyletic to C.

In the case of progenitor A giving rise directly to B and also directly to C with formula **A** > (B,C) see Fig. 17-3, there are two possible paraphyly-riddled cladograms depending on whether B or C has more

traits in common with A. Such imbalance of traits is phylogenetically informative but in this case not evolutionarily informative.

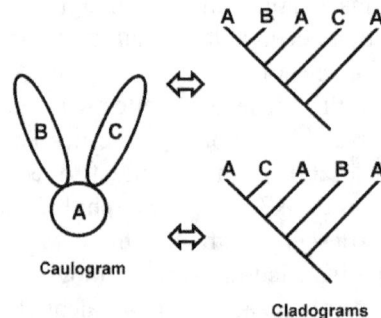

Figure 17-3. A caulogram on left shows progenitor A directly giving rise to two descendants, B and C. The cladograms on right are devised to ensure that paraphyly of A is available for both B and C, and there are two cladograms given to reflect differing possible positions of B and C.

An imposition of paraphyly is unnecessary when other information is available. The caulogram may be constructed *either* from a cladogram with paraphyly, *or* from non-phylogenetically informative data that B is much the same as A but, say, is polyploid, occurs in recent environments, and has a unique trait, while C is also similar to A but differs in specialized asexual reproduction and two traits uncommon in nearby groups. Thus, critical information for creating caulograms may include (1) cladistic information on paraphyletic shared descent, and also (2) information on order and direction of serial descent. The last is not phylogenetically informative in that it does not reveal relationships of shared descent.

I leave it to the reader with time on their hands to figure out the insertion of paraphyly in more complex scenarios. Consider the caulogram that has several descendants from one progenitor and some of the descendants with their own lineages of descendants. This would render supererogatory any attempt to make a complex cladogram that can be reduced to a caulogram based on phylogenetically informative (i.e., shared) traits alone.

This exercise may be informative of problems with molecular analyses in that when macro-evolution is associated with paraphyletic molecular races but some of those races are extinct, cladistic analysis inserts a random element into molecular cladograms. This is examined in detail in my previous publications (Zander (2008, 2013;

2014a,b,c,d) and in works by other evolutionary systematists.

A call for software — In the past couple hundred years of systematics, taxa were clustered by relationship, first general similarity, then by sexual traits, then any conservative traits that self-organize. Then phenetic analysis clustered more exactly by computerized general similarity. Thirty years ago, the cladistic method using shared trait transformations because prevalent in systematics. All along, observations have been made by evolutionists on direct descent of taxa using a variety of evidence, but no detailed method has been advanced to do this across taxonomic groups. This book shows a general method for modeling evolution through inferred direct descent of extant taxa, postulating ancestors when necessary to complete a gradualistic model. Support is measured by sequential Bayes analysis with informational theory using bits and Bayesian posterior probability equivalents.

The analytic key method was implemented here by hand because no software exists to mechanically optimize on shared (indirect) traits to minimize parallelisms, and then serial (direct) descent to minimize reversals, both minimizing information redundancy. What is needed is something akin to the cladistic software that clusters by shared trait transformations but which orders linearly by trait transformations between series of taxa in direct descent. Cladistic shared trait analysis is well understood, but analysis by modeling direct descent is not.

Serial transformations between taxa is largely naming nodes when possible, even if the traits assigned to a node are incomplete identifications of a taxon. The optimal arrangement of taxa in direct descent is necessarily a model of gradualistic evolution.

This paper asks programmers to consider developing software similar to PAUP (Swofford 2003) and Mr. Bayes (Huelsenbeck & Ronquist 2001) that will deal with both shared and evolutionarily serial characters, that is, with both branching and linear descent, with both indirect and direct descent. Detailed explanation of the use of direct, serial descent in classification is given by Zander (2013, 2014a,b,c, 2016a). Examples are provided here for critical elements in such analysis.

Modification of present cladistics software? — As a start, consider a simple pectinate cladogram (O, (A, (B, (C, (D, E))))), where O is outgroup and roots the cladogram. If we collapse the cladogram by assigning each terminal taxon to the nearest node, then the A is the most primitive of the lineage, thus **A** > **B** > **C** > (D, E). (The angle bracket symbol > means "gave rise to.") The problem with the terminal sister taxa can be solved by evaluating which of the two sister taxa has the most autapomorphies. Then add a second taxon with identical traits to the data set. If, say E had more autapomorphies than D (or ideally D has none), upon cladistics analysis one gets (O, (A, (B, (C, (D, (E, E). The nearest node to D is below that of (E, E), and the serial evolutionary hypothesis is then **A** > **B** > **C** > **D** > E. For this simple cladogram, small modifications to the software and inclusion of non-parsimony-informative inform-ation in the data set is all that is needed. Who will do this?

But what about more complex cladogram, with multiple lineages? How do we deal with internal branches two or more nodes away from an actual terminal taxon? Consider (O, (A, (B, C), (D, E))). We have two puzzling sister groups, (B, C) and (D, E). Using autapomorphy imbalance, let's say we find **B** > C, and **D** > E. Then we have some options.

1. We can say that A gave rise to two branches, thus **A** > ((**B** > C) (**D** > E)). This is okay if both B and C are not easily considered to give rise one to another (i.e., the have equal autapomorphies), and A is not overly different in traits from B or C. This is the most probable outcome.
2. If A is more than 4 or five traits (4 or 5 bits) different from B and D, then a missing link may be postulated, thus **A** > **X** > ((**B** > C) (**D** > E)).
3. If A is *very* different then B, C, D, and E belong elsewhere.
4. If D is easily seen as derived from B, then **A** > **B** > ((C) (**D** >E)). Given the elimination of redundancy of the cladistics program, this is doubtful.

Who will write the software?

Ordering serially without cladogram guidance — In the following examples "o" is a primitive trait, meaning that it is found in the outgroup. The outgroup is selected in the same way as is done in phylogenetics; it may be a species in the nearest group, or the basal species of a radiating group, a "functional outgroup." The letter "x" simply stands for an advanced trait, something not in the outgroup. All the x's are different traits in any one species, but they are character states of other species in the same columns, which are numbered. Informational bits are determined as whether traits are advanced or primitive relative to those of another taxon in the lineage, with primitiveness decided by the outgroup. In analysis of direct descent, because the order of evolution is given, "primitive" may be used instead of the cladistic term "plesiomorphic."

Past study in evolutionary systematics has been largely intuitive (i.e., insight born of familiarity) or at least hand-calculated, and a degree of expertise is used to estimate how uncommon or rare (advanced in a lineage) a trait is among related species. The analytic method described by Zander (2013, 2014a,b,c) requires justification of ordering of serial lineages of taxa and exact documentation and assessment of order probabilities but no integrated computerized means exists to do this.

Species are determined as groups with a minimum of two otherwise unlinked new traits that are apparently linked by some evolutionary process, known or unknown. Genera and other higher taxa are defined as groups of radiating species or lineages from some central generalist species, known or inferred. This is the dissilient genus concept of Zander (2013: 93). The steps used in analysis that are required for software emulation are:

(1) with reference to an outgroup or next more primitive species, order two species by minimizing reversals and parallelilsms;
(2) separate lineages into branches when two descendants, if generated jointly from a progentior, have fewer reversals;
(3) fuse (link) lineages at points of most primitive and most advanced species to obtain branching sets of serial lineages, interpolating inferred missing links (unknown shared ancestral species) or unknown serial intermediate species when necessary;
(4) evaluate the patterns for radiative (dissilient) groups that may be named as genera, and which constitute the new outgroup for all radiating lineages;
(5) calculate support for lineages and for dissilient groups as summed number of bits for each group from evaluation of order of pairs in step 1, these are interpreted as posterior Bayesian probabilities (see Table 17-1);
(6) present evolutionary diagrams either as formulae, e.g., **A** > (B,C), or caulograms (linked stem taxa), e.g., showing ancestor A giving rise to two descendants B and C. In evolutionary formulae, bold-faced letters denote ancestral taxa, angle brackets act as arrowheads that show direction of evolution, and parentheses show branches of lineages. It is characteristic of a dissilient genus that the set of primitive (outgroup or functional outgroup) traits change to that of the central generative species of a radiation.

Table 17-1. Equivalency of positive and negative bits with Bayesian posterior probabilities (BPP). One standard deviation is 0.68 BPP, two is 0.955 BPP, and three is 0.997 BPP.

Bits	BPP	Bits	BPP
9	0.998	0	0.50
8	0.996	-1	0.33
7	0.992	-2	0.20
6	0.98	-3	0.11
5	0.97	-4	0.06
4	0.94	-5	0.03
3	0.89	-6	0.05
2	0.80	-7	0.007
1	0.67	-8	0.004

So what are we looking for? Species with shared traits but with at least 2 linked advanced traits, and a data set with both shared traits and advanced (autapomorphic) traits, ending up with a branching diagram of a series of sets of radiating lineages, each with a generalist species and its attached (nested) lineages. Each cluster of a core species (even if hypothetical) and its descendants may be defined as a genus. An actual example of a stem-taxa evolutionary diagram is given by Zander (2014c), which was devised without the aid of software, and see the contrived caulogram example (Fig. 17-4).

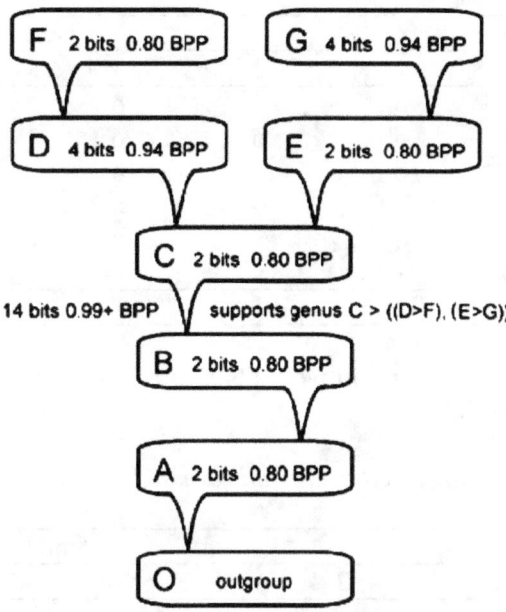

Figure 17-4. A caulogram of Example H (below). Descendant species are labeled with name (a letter), and support in terms of number of bits and Bayesian posterior probability for place in order against the next lower species. Support for the genus with evolutionary formula **C** > ((**D** > F), (**E** > G)) is given in terms of summed bits of all included taxa, plus the BPP. The analytic data set of Example J gives details of macroevolutionary transformations of the species.

Analytic data sets — The use of informational bits to determine the order of species in serial evolution is totally different from their use in evaluating information content once the optimal order is decided on. In determining order, reversals are awarded a minus bit, but once the order is determined, all trait transformations are awarded one positive bit. You can shortcut the process by simply minimizing trait reversals.

In the following *analytic data sets*, the direction of evolution is upwards, o is an advanced trait, x is a primitive trait (found in the functional outgroup, or occasionally as a reversal in advanced species). One can move species up and down to change the order of evolution, which is always "up". Each horizontal line of traits represents relevant traits in one species. Each species (in Examples C through J) has at least 2 distinctive traits (in 2 different columns), morphological or otherwise. Some traits are bold-faced for emphasis. A "trait" is simply a distinctive character state. BPP is Bayesian posterior probability.

The penalty of minus 1 for a reversal is only done in the initial ordering of taxa, but penalties are not used when calculating the informational bits of that optimal order. Then, even reversals are informative.

Example A. Gradual increase in advanced traits—no reversals

Species	Traits					
	1	2	3	4	5	6
E	o	o	o	o	o	o
D	o	o	o	o	o	x
C	o	o	o	x	x	x
B	o	o	x	x	x	x
A	o	x	x	x	x	x
O	x	x	x	x	x	x
Bits	1	1	1	1	1	1

Total number of transformations from primitive traits x to advanced traits o are added for each column and sum to 6. There are no reversals.

Example B. Gradual decrease in advanced traits—all reversed

Species	Traits					
	1	2	3	4	5	6
A	o	x	x	x	x	x
B	o	o	x	x	x	x
C	o	o	o	x	x	x
D	o	o	o	o	o	x
E	o	o	o	o	o	o
O	x	x	x	x	x	x
Bits -1/rev.	1	0	0	0	0	0
Bits +1/rev.	1	2	2	2	2	2

Example B is similar to Example A, but has 5 reversals. Instead of a gradual increase in advanced traits "o" there is a gradual decrease. Total number of transformations when reversals (rev.) are assigned -1 bit add to only 1 for all columns, which makes Example A the more informative order of evolution with 6 bits for all columns. If reversals are assigned +1 bit, as though they were a new advanced trait, a reversed order of gradual accumulation of advanced traits becomes more informative in number of bits (11), but does not model evolution from theory and paleontological evidence. Thus, in the following Examples, reversals are assigned negative bits to force an optimal ordering that minimizes redundancy of information.

Example C. Two species, one of them (O) an outgroup with apparently primitive traits

Species	Traits		Bits	BPP
	1	2		
A	o	o	2	0.80
O	x	x		

Species A has 2 traits (the character state o in columns 1 and 2) that are not found in outgroup O (which has character states x), and o in both columns may be treated as advanced. Two transformations from outgroup traits x to advanced traits o yield 2 bits, or 0.80 BPP (see Table 1) of support for both order and direction of evolution.

Example D. Three species and order of evolution
  Chart D1.

Species	1	2	3	4	Bits	BPP
B	o	o	o	o	2	0.80
A	o	o	x	x	2	0.80
O	x	x	x	x		

The order of each pair of contiguous species is supported at 0.80 BPP. Species A has two bits difference between A and O, B has 2 bits difference between B and A. The total is 4 bits, or 0.94 BPP, in support of direction of evolution of lineage A > B.

Chart D2

Species	1	2	3	4	Bits	BPP
A	o	o	x	x	-2	0.20
B	o	o	o	o	4	0.94
O	x	x	x	x		

We check example D1 by using Chart D2 by reversing the order of A and B, and find that the support for order B > A is 0.20. This is because the two reversals in columns 3 and 4 yield minus 2 bits, which are equivalent to 0.20 BPP, from Table 1. Subtract 0.20 from order **B** > A from 1.00 to get the support for the opposite order **A** > B. The support of 0.80 then stands. Although the order of **O** > B is highly supported at 4 bits, the order of **B** > A is rejected at minus 2 bits, and the total bits supporting direction of evolution is only 2 bits (4 minus 2).

It is expected that computerized analysis recursively rolling through pairs of taxa in an analytic data set will eventually order the taxa to maximize bit count. This procedure maximizes gradual evolution of traits (as in Example A) because reversals leading to negative bit counts are minimized. This is much like a second-order Markov chain, with the optimal order of two contiguous species and support for the most advanced one is derived with respect to a third species lower in the lineage, which is the functional outgroup. Thus, information from the state of two species is needed to determine that of a third, e.g., both O and B are needed to determine position of A, above.

Well, what about the possibility that A and B are instead derived separately from core species O? See Chart D3. Then, A, with two o's derived from x's, would have 2 bits, and B, with 4 o's derived from x's would have 4 bits. But, in this case, B's two o's (in columns 1 and 2, boldfaced) appear as parallel evolution with the o's in the same columns of A, which is redundant. To remove the redundancy, B should have been derived directly from A. The two redundant o's in B must be eliminated as not contributing to analysis of order of evolution, and this is done by *assigning each redundant o minus one bit*, and thus B, if derived separately from O in the presence of A yields zero bits, i.e., minus 2 for **O** > B columns 1 and 2, and plus 2 for columns 3 and 4:

Chart D3

Species	1	2	3	4	Bits	BPP
O>B	o	o	o	o	2 - 2 = 0	0
O>A	o	o	x	x	2	0.80
O	x	x	x	x		

The optimal order then is that of Chart D1, giving 0.80 and 0.80 BPPs for order, and totaling 4 bits, or 0.94 BPP for direction of evolution of the lineage.

Example E. A reversal from the primitive trait

Chart E1

Species	1	2	3	4	Bits	BPP
B	x	o	o	o	1	0.67
A	o	o	x	x	2	0.80
O	x	x	x	x		

This is similar to Example D above, but there is a reversal in trait column 1 from o back to primitive trait x. The order of A and O are the same as in Example D. The order of A and B are evaluated separately by comparing the order of the pairs alone with the outgroup O.

Chart E2

Species	1	2	3	4	Bits	BPP
B	x	o	y	y	1	0.67
A	o	o	x	x		
O	x	x	x	x		

Here we get -1 for column 1 plus +2 for columns 3 and 4, which gives 1 bit, 0.67 BPP, of support for order A > B.

Chart E3

Species	1	2	3	4	Bits	BPP
A	y	o	x	x	-1	0.33
B	x	o	o	o		
O	x	x	x	x		

But when A and B are exchanged in order (above), we get plus +1 for column 1, -2 for reverses in columns 3 and 4, which when added gives -1 bit, or 0.33 BPP, of support for order B > A. The order A > B has optimum support of 0.67 BPP over 0.33 BPP, so it is the correct order.

Example F. Insertion of a species into a lineage

Chart F1

Species	1	2	3	4	5	6	7	Bits	BPP
B	o	o	o	o	o	o	o	5	0.97
A	o	o	x	x	x	x	x	2	0.80
O	x	x	x	x	x	x	x		

There are 7 transformations of primitive x to advanced o in the lineage O > A > B, two for O > A and 5 for A > B. The total support for the direction of evolution of **O > A > B** is 7 bits, or 0.992 BPP.

Where would the species C (below) fit above, in Chart F1, if it has a data set of:

C	o	o	o	o	o	x	x

Clearly, by common sense, species C should be inserted between A and B. Technically, one can check by analyzing each pair of species against the outgroup of primitive traits for the lineage. The optimum placement maximizes bit count because a gradual increase in number of advanced traits maximizes exposure of x to o transformations. In information theory, the higher number of bits reflects the better decryption of a hidden message, in this case the message is the order of transformation of one species to the next in an optimally nested set of species. Species C, if not extant, may also be postulated as a hypothetical unknown missing link, that may be predicted to have been extant at one time or which should be looked for in nature in the future.

Example G. A reversal embedded deep in a lineage

Species	1	2	3	4	5	6	7	Bits	BPP
D	o	o	o	o	o	o	o	1	0.80
C	x	o	o	o	o	o	x	1	0.67
B	o	o	o	o	x	x	x	2	0.80
A	o	o	x	x	x	x	x	2	0.80
O	x	x	x	x	x	x	x		

There is a reversal in species C column 1. Against the primitive traits of the outgroup O, C has a reversal awarded -1 bit, plus two +1 bits for columns 5 and 6, summing to +1 bit, or 0.67 BPP. Species D has a reversal back to the advanced trait, which is treated as a positive trait transformation.

Example H. Branching lineages

Chart H1

Species	1	2	3	4	5	6	7	8	Bits	BPP
C	o	o	o	o	x	x	o	o	0	0.50
B	o	o	o	o	o	o	x	x	2	0.80
A	o	o	o	o	x	x	x	x	4	0.94
O	x	x	x	x	x	x	x	x		

Chart H1.: Note that B is advanced by 2 x to o transformations but C has two reversals (-2 bits) and 2 advanced transformation (+2 bits), yielding no—that is, equivoval—information (0.50 BPP). Thus the order of B and C is equivocal as may be seen in the analytic data set H2 with B and C reversed:

Chart H2

Species	1	2	3	4	5	6	7	8	Bits	BPP
B	o	o	o	o	o	o	x	x	0	0.50
C	o	o	o	o	x	x	o	o	2	0.80
A	o	o	o	o	x	x	x	x	4	0.94
O	x	x	x	x	x	x	x	x		

Chart H2: When an order is equivocal (or nearly so) by a 0.50 BPP, a branch from the main lineage should be considered. The lineage **O** > A should now end in a branch B and separate branch C, thus a formal evolutionary formula would be **O** > **A** > (B,C). The number of bits is maximized from 6 to **8**. Species A becomes the new functional outgroup or core progenitor for the two lineages **A** > B and **A** > C. For clarity, a split data set and evolutionary formulae demonstrate the branching from a shared ancestor:

Chart H3

Species	1	2	3	4	5	6	7	8	Bits	BPP
A>C	o	o	o	o	x	x	o	o	2	0.80
A>B	o	o	o	o	o	o	x	x	2	0.80
A	o	o	o	o	x	x	x	x	4	0.94
O	x	x	x	x	x	x	x	x		

Chart H3: **A** > (B,C) may be considered a genus (pending more information, see examples I and J). The number of bits supporting the putative branching lineage is 8, or 0.99+ BPP. Eight bits is maximum support allowable and conceivable given rational limits on statistical certainty in cases of macroevolutionary estimation.

Example J. Hypothetical species, as central links

Species	1	2	3	4	5	6	7	8	9	10	Bits	BPP
C>E	o	o	o	o	o	o	x	x	o	o	2	0.80
C>D	o	o	o	o	o	o	o	o	x	x	2	0.80
[C	o	o	o	o	o	o	x	x	x	x	2	0.80]
B	o	o	o	o	x	x	x	x	x	x	2	0.80
A	o	o	x	x	x	x	x	x	x	x	2	0.80
O	x	x	x	x	x	x	x	x	x	x		

If species C were unknown, and D and E are equivocal in order, then the traits of a central species C could be estimated to be as given above. This postulated shared ancestor "C" fuses the two lineages **C** > D and **C** > E by affixing them separately to the terminus of the lineage **O** > **A** > **B** > C. The total bits supporting this branching lineage of five species is 10, which is simply 0.99+. The chance that the order of evolution is entirely correct is 0.80^5, or 0.33, which is acceptable given the small number (2) of traits assigned to each species in these contrived examples. The three species **C** > (D, E) may be treated as a putative genus with data set of C (even if it were merely inferred) as outgroup.

Example I. The caulogram

Species	1	2	3	4	5	6	7	8	9	10	Bits	BPP
G	z	z	z	o	o	o	o	x	o	o	4	0.94
C>E	z	z	o	o	x	x	x	x	o	o	2	0.80
F	z	z	z	z	o	o	o	o	x	x	2	0.80
C>D	z	z	o	o	o	o	o	o	x	x	4	0.94
C	z	z	o	o	x	x	x	x	x	x	2	0.80
B	o	o	o	o	x	x	x	x	x	x	2	0.80
A	o	o	x	x	x	x	x	x	x	x	2	0.80
O	x	x	x	x	x	x	x	x	x	x		

Again, here C is established as a central species for the radiation **D** > F and **E** > G, and its data set becomes the new functional outgroup for those two lineages. Species D and E are supported in their being initial taxa for lineages branching off species C by discovery that each has another species (F and G) in their lineage. The two separate lineages **D** > F and **E** > G share in columns 1 and 2 two advanced traits "z" with C. With C they form a radiating group, which one may call a genus (boxed for emphasis in the analytic data set above) with considerable confidence given 14 bits of support. The separate lineages as descendants of C are distinctive from each other by traits labeled "o" in columns 7 and 8 (boxed) and in columns 9 and 10 (boxed).

One might predict discoveries of additional species in each lineage with these advanced traits, or even a generalist species off of which another radiating group may occur with these advanced traits treated as primitive for that radiating group.

Support for the dissilient (radiative) genus C and its two advanced lineages is the sum of bits for the order of those lineages and C itself, namely 14, or 0.99+ BPP. The probability that the exact order of species in each advanced lineage is correct is the product of 0.80 and 0.94, or 0.75 BPP. The evolutionary relationships implied by this analytic data set may be visually represented in a caulogram (Fig. 17-4). The example caulogram shows a putative genus with evolutionary formula (**C** > ((**D** > F), (**E** > G)) with two advanced branches but genera defined as radiative groups (dissilient genera) may have any number of descendant branches.

Summary — Programmers with an interest in systematics are challenged to create software similar to that used in cladistics and phylogenetics but adapted to the goals of macroevolutionary systematics, i.e., modeling direct descent of taxa. Optimization is through assignment of 1 informational bit for each advanced trait transformation and 1 negative bit for each redundancy, namely reversals and parallelisms (traits beginning new branches of taxa that should have been added to serially to an existing branch to eliminate the redundancy). Cladistic analysis at the beginning of a study usually solves the problem of parallelisms, but self-nesting ladders (Zander 2013: 53) can confound perceived order of evolution. The intent is to model evolutionary relationships of taxa by evaluating direct, serial descent. Taxa are arranged linearly by increasing numbers of advanced traits. Branches are inserted when linear arrangements have no support. The arrangements of taxa are determined by a simplified form of information theory. Recursive evaluation of bit count on a Markov chain, given a set of primitive traits as outgroup, is used to maximize bits for a lineage or cluster of lineages, "decrypting" the order of evolution and of occurrence of branches. Later, all transformations left over after optimization minimizing redundancy are evaluated as 1 positive bit each.

It is hoped that among readers of this paper, programming experts might be inspired to devise software similar to that used for cladistics and phylogenetics but focusing on serial descent of taxa, not shared traits. Data sets must include unique (nonphylogenetically informative) traits, and the software must order the taxa, not the traits alone. Macroevolutionary systematics has a great need of computer-aided analysis of evolutionary data. This would allow the building of evolutionary charts of transformations among stem taxa (caulograms) that models direct, serial descent. Macroevolutionary systematics is often difficult for cladistics-oriented workers to understand because it focuses on evidence of paraphyly (taxa generating differently named taxa) as informative of evolutionary transformations between species.

Exploring the serial dimension — The similarity of the dichotomous branching pattern of clado-grams to a standard UPGMA tree or a Markov chain transformation tree has allowed complex mathematical and statistical calculations of sister-group relationships. As noted elsewhere (Zander 2013, 2014a,b,c) the results are limited to shared descent. Can a new systematics involving a combination of shared and serial ancestry allow a similar sophistication? In the present work and in previous papers, techniques from statistics (sequential Bayesian analysis), combinatorics, and practical aspects of phenomenological philosophy and causal theory are used (I think) effectively to demonstrate the value of exploring the serial dimension of macroevolutionary relationships.

Cladistic analyses can provide a clustering of taxa with similar past trait transformations, and also information on direct ancestor-descendant relationships by bracketing a descendant taxon on a cladogram with two terminal exemplars of an ancestral taxon (Zander 2008). This latter "heterophyly" is effective in both morphological and molecular studies (Zander 2008). A salient example is the descent of one family from another, such as Cinclidotaceae, Ephemeraceae, Splachnobryaceae, and the here newly segregated Streptotrichaceae from the Pottiaceae (Bryophyta). A recent successful study is the determination of the extant taxonomic descendant of a fossil whale by bracketing it on a morphological cladogram with data from an extant adult and a juvenile apparent near relative (Tsai & Fordyce 2015). This is the reverse of the way nonmonophyly may be used to determine ancestor-descendant relationships following Zander (2008), but works well. It is possible that such bracketing studies in botany with morphological data can make use of parallels to juvenile and adult morphologies in whales, such as extremes of phenotypic variation, or behavior in cultivation.

Future evolutionary analyses should search for trees consonant with data on both serial and shared descent, because true maximum parsimony or likelihood requires that all relevant data be addressed to model "descent with modification." The word relevant refers to any data that may change the result. Needed is software to facilitate caulistic evolutionary analysis. It could be said that evolutionary analysis based on the serial relationships of stem taxa has been stymied for decades because no software exists to attract researchers by easing analysis.

Because serial descent is not well modeled by Markov chain decision trees, other methods of optimality may be investigated, such as logistic or simplex methods. In that evolutionary success does not require maximum adaptation or fitness, but only that which is adequate for survival, non-zero-sum game theory may prove valuable, particularly the Nash equilibrium. Software may also allow more `exact evaluation of traits in basal portions of a lineage and among descendants of progenitor siblings for each species. Again, one must warn that an over-focus on numbers without application of theory can lead to bias.

If macroevolutionary systematic methods are successful, then the phylogenetic concept of an unnamed and unnameable ad hoc universal shared ancestral taxon connecting dichotomous branches of a cladogram may share the dust-bin of scientific history with alchemy, phlogiston, geocentric model, the aether, vitalism, quinarianism, Lysenkoism, and the many other sloughed theories (no matter how interesting, for more see

https://en.wikipedia.org/wiki/Category:Obsolete_sci entific_theories).

High Bayesian posterior probabilities can now be assigned to macroevolutionary transformations on morphological caulograms. This counters in part the phylogenetic cherry-picking of data, biased analytic software, and rejection of modeling direct descent. Mechanical taxonomy supports the phylogenetic weaponization of apparent statistical certainty leading to phylogenetic academic hegemony through big-science monetization of molecular systematics.

Destiny — Cladistics may have attained its present status in a manner standard for all politically fixed new paradigms. There are three psychological forces:

(1) *Reification.* Cladistics was declared a major advance because it reveals a structure in nature that reflected evolution by trait transformations, while phenetics' trees used only overall similarity. Cladists attacked alternative methods that use discursive, inductive reasoning and evolutionary theory as mere narratives and wishful narratives, like the "Just So" stories of R. Kipling (1966). If scientific theory is, as they implied, unreal, then cladograms and the dichotomous structures in nature they are said to represent are presented as real. Analysis assuming scientific realism is thus reduced to the axiomatization of Popperian hypothetico-deductivism and Lakatos' scientific programme. Cladograms only reflect data on shared traits, so cladistic reality is restricted to structures of nested shared traits.

(2) *Valorization.* Systematists were once happy with a handlens and a specimen cabinet. Then, the equipage of systematics required fast, major capacity workstations, DNA analysis machines,

grants in the million dollar class with breathtaking indirect cost requests. Cladistics became de rigueur for wide-eyed students and new academic hires who expected significant salaries, and who were expected to provide value to their institutions in terms of cutting edge molecular discoveries.

(3). *Monetization.* I have submitted over the years various diatribes and Jeremiads to the listserver Taxacom inveighing against the outrages of phylogenetics. Some Taxacomers have reassured us that phylogenetics is a passing fad, and will fade away like the morning dew, or change into something deep and refreshing. Just to check with reality, in 2015 I searched the funding database in NSF for the words "phylogeny" and "taxonomy" in the titles or abstracts for the last five years. The results, which I tabulated in a spreadsheet, are:

Phylogeny		Taxonomy	
2011	8,801,433	2011	22,627,716
2012	8,052,574	2012	16,439,643
2013	11,091,123	2013	12,055,369
2014	23,649,263	2014	14,951,327
2015	33,611,140	2015	16,574,900
Total	85,205,533	Total	82,648,955

Projects on "phylogeny" and those on "taxonomy" are about equally funded over the last five years at $85 and $82 million dollars, respectively. Taxonomy seems to be holding steady at around $12–14 million per year, but phylogeny has had a burst of popularity among NSF grantors, doubling in funding during 2013–2014, and then increasing in 2015 by another 10 million dollars. It is presently funded at double the rate of classical taxonomy.

Some Taxacom listserver participants have said that phylogenetics research usually includes good taxonomy. You should check the pages of top phylogenetics journals and see if you agree with them that phylogenetics actually contributes to taxonomic knowledge in anything like the classical sense, particularly during an ongoing and worsening biodiversity crisis. (You should also examine how I, myself, try here to reify, valorize and monetize macroevolutionary systematics. Are my arguments valid? Does macroevolutionary analysis better reflect evolutionary reality, is it more valuable and relevant to research and conservation during a biodiversity crisis, and should it be funded—at all, some, lots?) And ask yourself are the many molecular species, genera and families—established to avoid paraphyly— valuable to science, or are they methodologically imposed superfluous groupings, and, as such, simply taxic waste? It is up to young scientists to decide if we must remain caught in a vortex of phylogenetic determinism or break back into the model-based, total evidence scientific analysis revered in the past.

An emergency — Macroevolutionary systematics, even though thoroughly based on physics and math, statistics and information theory, is easy to do. Even innumerate taxonomists (those bad with math) can generate well-supported theories of serial and shared descent translatable into robust classifications. Such classifications can easily stand up to phylogenetic molecular classifications in terms of statistical strength and predictive value.

We are entering an extinction event of considerable import to humankind. The dread triad of overpopulation, overdevelopment, and climate change will affect us directly in terms of war, famine, and disease. It will affect us in its impact on biodiversity and the fragile web of life dependent on balanced interactions between browsers and browsed, predator and prey, mycorrhizae and vascular plant, parasite and host, and most particularly on the potential disasters of monoculture and pest. Multicultures survive where monocultures do not.

The ongoing, general degrading process is fairly clear. Stable ecological relationships are upset by a cascade of catastrophes, including invasive organisms and climate change. Monocultures expand as a few species crowd out myriads of native species micro-adapted to the previous environment and to each other. Examples of damaging invasives are humans, rats, rabbits, goats, zebra mussel, algae blooms, kudzu, and the like. As time goes by, the invasives become limited in population growth or die back, but the earlier diversity is severely damaged. Thus, habitat productivity may increase with the growth phase of the invasives, but decreases far below that of the original flora and fauna upon resource depletion and subsequent die-back. The most important impact is on the resilience of the biodiversity in a habitat to minor change. The world is becoming a "brown field" and the expected die-back will inevitably involve humans.

Evolvability — It is often assumed that climate change will occur so fast that evolution will not be able to keep up. That is, the extinction rate will exceed that rate of generation of new species adapted to a modified habitat or new habitats. Certainly, species seem to last an average of, say, five million years, during this time generating only a few descendant species. This may be due to (1) slowly changing climate that is little challenge, (2) a strong Red Queen effect in which biotypes battle each other, and (3) genetic swamping of new sets of traits preadapted (exaption) to a climate change not here yet. Rapid climate change may unleash the "evolvability" of species with large numbers of biotypes, either in potentio or actual (anagenetic species), by immediate elimination of competing biotypes. There are many examples in the literature of particularly fast evolution. The polar bear is presently interbreeding with the brown bear and its stunning array of cold-adapted traits will survive in some way.

Biotype richness may be presaged by the actual morphological variation within the single plant body of a species. Certainly, the presence of a developmentally controlled variation in leaf length or some other organ provides much the same potential for evolvability as has differences between generic races and varieties.

Which then are the species of Streptotrichaceae with greatest potential for survival in the face of climate warming. Given that most species are adapted to moist, high elevation habitats by lack of a stem central strand, spongy tomentum, and large plant body, elimination of high mountain cloud forests and páramos may doom the family. On the other hand, *Leptodontium excelsum* has a myriad biotypes judging from great variation in length of the leaves. *Leptodontium scabberimum* has odd traits and trait combinations that may not be developmentally available to other evolving members of the Streptotrichaceae. *Rubroleptodontium stellatifolium* is highly generalist (except for its size) in that it has every major feature of the family that is used for genus distinction, and has already demonstrated long survival in somewhat different habitats in scattered areas worldwide; it might be seen as the cockroach of the Streptotrichaceae. *Stephanoleptodontium brachyphyllum* apparently originated as a successful biotype of *S. longicaule,* surviving in somewhat moist places in generally arid South Africa. *Microleptodontium flexifolium* has at least demonstrated the potential to generate closely related and therefore probably recent specialized descendant species, one (*M. gemmiferum*) adapted to human dwellings. It is possible that protection of rare species or of unique environments is impossible if climate warming is globally destructive. Protection of species with maximum evolvability, however, may be one solution to ensuring the establishment over time of complex, genetically diverse new habitats. A key to judging potential of survival over time is that the geographic dimension is much the same as the time dimension in biogeography. Scattered populations of a species that have not exchanged genetic material over long periods of time and yet have survived in fairly diverse habitats indicate that the species should have the potential of survival through the rigors of near-future massive climate change. One may hope so, anyway.

The future — It is disconcerting to realize how *major* the failures in human rationality are in the face of the sempiternal mysteries. We live in a narrow, dangerous band of reality between a vast microcosm ruled by spooky quantum effects and an even vaster macrocosm swirling with dark energy and dark matter, replete with froward asteroids and hidden branes of multiverses. Even in the immediate mesocosm we do not know what certain actions-at-a-distance (e.g., "fields") really are. For instance, we can measure magnetism and see it obey certain laws, but we do not know what it *is*. Since matter is mostly space, even causality involving physical contact is action at a (short) distance. Why does light follow the fastest path through transparent matter (other than quasi-magical explanations like "all other paths are equiprobable but they cancel out quantumologically")? The hope of many desperate scientists is that magic is indistinguishable from advanced technology; and, thus mysteries can be "explained" by simply advancing formulae for measuring their effects, e.g., Newton's formula for the force of gravity. Entanglement phenomena indistinguishable from magic have now been extended into the mesocosm. These are mysteries, and scientists love mysteries, but will we have a chance to investigate them in the long term?

At present, we are gifted with a new scientific beginning. The revolution of computerized systematics is chump change in comparison. The publication of Stewart Bell's proposal to test non-locality (i.e., entanglement) (Odifreddi 2000: 94) and its subsequent proof has been called by a physicist "the most profound discovery in science" (Ferris 1997: 284), and a deadly blow to naïve realism. Bell's inequality has led to investigation of totally new phenomena that offer a glimpse of a fundamental universal structure immune to strictures of causality, possibly left over from the origin of the universe. Naïve realism apparently leaves us clinging to convenient diffeomorphisms (e.g., the Earth looks flat locally but is really pretty much a sphere) in a far more complex universe than we ever imagined.

Once, even before entanglement was ever demonstrated, John Wheeler said to Richard Feynman (Feynman 1965), "I know why electrons have the same charge and the same mass!" Feynman said, "Why?" Wheeler said, "Because they are all the same electron!" In my opinion, we are like flies in a darkened house buzzing against a bright screen door. When we find that inevitable hole in the screen, we will emerge into a wondrous Back Yard of amazing things. The future possibilities associated with manipulating locality and nonlocality are immense.

At the same time, we are challenged by a rapidly developing major extinction event caused by human-induced climate change and other insults to the biosphere (Peel et al. 2017), and by clinically deranged leaders-of-nations deploying nuclear arsenals to threaten devastation and endless poisoning of all our single precious terraqueous Earth. Will we have time, will, and ability to get past this chance of our self-extinction?

All we have left is healthy skepticism, with unfettered imagination combined with constant experimentation and sampling, and a reliance on the practical efficacy of causal, process-based theory at least most of the time in the mesocosm. And, trust to a fall-back of well-informed simple pragmatism with care for the future.

Most of us scientists know that we know next to nothing—we are barely out of the Stone Age. All our discoveries and all our questions are necessarily couched in limited human terms and constrained to human understanding. Yet we now tolerate the contamination of that next-to-nothing with the bijou quasi-science of phylogenetics. With an extinction event looming over all life, dare we self-proclaimed Stewards of the Earth continue with cacophyly-as-usual, tripping blithely through an eternal summer of boutique mechanical taxonomy? A new evolutionary process-based alternative that can replace the now "old technology" of cladistics is offered here. More discussion of this topic is available in the last chapter of this book, where I suggest that understanding small-world aspects of dissilient genera is the key to ecosystem health.

Chapter 18
ADAPTATIONS OF SPECIES

Evaluation of species using entropic macroevolutionary method — Well, now we have a classification rather straight-forwardly based on inferred serial evolution. What do we do with it? How can we use it to further predictive and analytic science? Let's start with the species.

We have examined *informational entropy* previously as a way to evaluate adaptive traits as nested in tracking traits in a serial lineage of species. Macroevolutionary systematics, additionally, is a way to directly study the dissipative structures of evolution (Coveney & Highfield 1990: 251ff) in terms of *physical entropy* (Brooks 1981; Brooks & Wiley 1988; Brooks et al. 1986). Those entropy-generating structures are the living progenitor and its radiating and serial sets of descendants in a dissilient genus. The mechanisms of entropic dissipation are those identified in the analytic key as significant classificatory traits, particularly the immediately adaptive as nested in tag-a-long ancestral traits.

The traits given in cladograms as steps, and in classical keys in couplets, include both adaptational and ancestral tracking traits because the dichotomous split simply segregates a species from the remainder of species in the clade or rest of the key. Nesting species serially in an analytic key (and caulogram) distinguishes traits that are recently generated by the immediate ancestral species. Those ancestral traits that survive in more than one environment are considered relatively neutral as far at adaptation goes. Yes, yes, it is known that species may develop through isolation and accumulation of neutral traits, but if the habitat is different from that of the ancestor (and it usually is, particularly in peri- or allopatric speciation), then the new traits are probably to some extent adaptive.

Given a means of identifying adaptive traits, we now have a tool for probing evolution of species into new habitats in terms of matching adaptations with environmental variables. Even more important, each niche is defined by the traits of the extant or extinct occupying species (unless one has Special Knowledge, what else is there?) Loss of a species (as informational bits) increases negentropy (available work energy) and the chance of distractive rapid energy flushes. The informational bits lost are small when other descendants of the same progenitor remain in the same ecosystem, but large when the species is unique and the whole train of ancestral traits is lost.

The present discussion is, of course, highly preliminary and to some extent speculative. My Ph.D. advisor, Lewis E. Anderson, told me that speculation should be reserved until the end of a paper, after one has earned the right. If you think much of the preceding chapters was already somewhat exploratory, read on.

Differences in numbers of bits provide the relative importance of species as unique vehicles for exploring through time—as permitted by phyletic constraint on development and physiology—the available evolutionary space through specialized feeding on negentropy. This physical analysis may provide a diamond-hard conceptual basis for preservation of biological diversity. It is the presently increasing frequency and size of unrelieved negentropic energy sites (like plant monocultures, invasives, damaged habitats) that provide the explosive source for biological catastrophe, while the gradual relief of negentropy seen in nature in biodiverse regions is a desideratum best achieved during restoration attempts by multiculturalism at least focused on known negetropic sources.

If dissilience is inferred as present today, then by uniformitarianism it must have happened in all lineages in the past. If punctuated equilibrium happened in the past (as inferred from fossil evidence), then it must be happening today by a kind of reverse uniformitarianism. The (in geological time) moment of punctuation (Gould 2002: 606; Eldredge 1985, 1989; Stanley 1981) may be a kind of catastrophe (Gilmore 1981) in which wide-ranging and biotypically rich species are split into several isolated, mostly derived species. One might be able to examine groups of closely related species and make a good estimate of where in relation to one or more punctuational events those species radiations occurred.

Blackburn (2002) has suggested that cladistic analysis can distinguish between exaption (pre-adaptation) and immediate adaptation. He suggested that with the common practice of "… superimposing phenotypic features over accepted phylogenies, one can adopt parsimonious interpretations of evolutionary change." In explanation, he wrote: "One crucial difference between adaptation and exaptation lies in the chronological sequence of structural and functional modification. In adaptation, the modification of a phenotypic feature (e.g., a structure or behavioral trait) accompanies or parallels its evolutionary acquisition of a function. However, in exaptation, the feature originates first (either as a selected or nonselected attribute) and only later is coopted for the function in question." There are three problems, *first,* one must be able to easily match a phenotypic feature with its function (difficult with mosses, easier with macrofauna); *second,* Blackburn does not directly discuss neutral or nearly neutral traits, although spandrels (as architectural by-products) are discussed as identifiable as "unselected correlates" while citing David Hume and the difficulty of distinguishing causation and correlation; and *third,* terminal sister groups are delimited only by shared traits, and unless a cladist evaluates direct descent of one from the other as in macroevolutionary practice, the order of evolution of traits cannot be determined.

A punctuational event may occur after a bottleneck of massive extinction of certain forms only present in fossils nowadays, in which case there may be adaptive exploration of new or now empty (of competition) sites. Consider the usual (Saunders 1980) diagrammatic representation of a catastrophe event. This is a sheet of paper with two folds such that a process traveling along the paper "falls" across the fold from one part of the sheet to another as it continues in one direction. The fold may be the revelation of many newly uncovered energy-rich sites after bottlenecking extinctions. If we consider one biotypically rich and wide-ranging species as ripe for a climatic or other discontinuity, then after the event, on the other side of the fold, we find a number of scattered, isolated biotypes. These evolve separately through adaptation and drift into new species. Farther away in time from the punctuational fold, one of the species begins to expand its range and number of biotypes. Even further, some or most of the isolated and perhaps not well adapted species begin to go extinct. Even later, only the new wide-ranging, biotypically rich species remains, but ripe for another punctuational event. This is, of course, a hypothesis much in need of data.

Analysis of possible adaptive flags — The analytic key and caulogram in the present study were further analyzed. Complexity of morphology in numbers of informational bits per species was equated with ability to take advantage of available energy in the environment. The 28 species in the Streptotrichaceae provided apparently immediately adaptive traits that totaled 113 informational bits, or an average of 4.04 bits per species. This is an integrated measure of the potential competence of taxa in gradually converting isolated patches of negentropy into a less energy-rich state. The breakdown for genera is in Table 18-1:

Genus	Species	Total bits	Bits per sp.
Austroleptodontium	1	8	8
Crassileptodontium	4	16	4
Leptodontiella	1	3	3
Leptodontium	3	11	3.67
Microleptodontium	4	14	3.50
Rubroleptodontium	1	5	5
Stephanoleptodontium	8	25	3.13
Streptotrichum	1	4	4
Trachyodontium	1	4	4
Williamsiella	4	23	5.75
Sum	28	113	4.04

Table 18-1. Chart of numbers of apparent adaptations in genera of Streptotrichaceae. The species average 4 informational bits per speciation event.

Bryologists have long wondered and speculated about the adaptive value of certain morphological traits in the mosses, particularly drought resistance (Düring1979; Lauder 1981, Vitt et al. 2014; Watson 1913), particularly see Glime (2017). Arbuckle and Minter (2014) have devised a program "Windex," which measures the strength of functional convergence in phenotypic evolution with the Wheatsheaf Index.

For instance, moss species with papillose leaves with hyaline awns are commonly associated with hot, dry habitats, yet other moss species without such obvious adaptations grow right alongside them. Could it be that the apparent adaptations make up, somehow, for a weakly adapted physiology on the part of the morphologically ornamented species? Now, concomitant change from a progenitor in one environment to a descendant in another, as shown in an analytic key or caulogram, may be attributed, *for that physiology,* as an adaptation to the new environment. *Or* the new trait may be neutral but tolerated as a physiological burden because the environment is sufficiently rich in energy to support it against competition and times of low photosynthate.

The adaptational explanation may be true because the neutral traits (in the new environment) are sloughed off methodologically as irrelevant because they are nested (found in both species and in both environments). But no trait is entirely neutral. There are, then, three kinds of traits and trait combinations of interest, (1) those immediately adaptive and necessary for survival in a new environment, (2) those that are lineage-wide, being tag-a-long traits that are of general value to many species and which contribute to the phyletic constraint of manner and extent of evolvability for species in a lineage, and, (3) the neutral and tolerable-burden sort.

If the family Streptotrichaceae is primitively epiphytic, the lack of many anatomical features (e.g., stem central strand and costa epidermal cells) usually present in the Pottiaceae may be interpreted as an adaptational flag, possibly for streamlining of physiology in absence of abundant photosynthate (but epiphytes should have lots of photosynthate) or more probably loss of features that make the plant heavy at dewy or foggy times. The tall sheathing perichaetial leaves may be a feature that protects the young sporophyte when wind whips branches. The antrorsely prorulose calyptra apex may help signal to the seta to stop growing when the capsule is finally exerted from the very long perichaetial leaves. For species that have apparently returned to soil substrates (particularly the lineage founded on *Williamsiella araucarieti*), even though they are commonly found in an environmentally similar habitat—soil at very high elevations—elaborations that are environmentally a deadly burden primitively may be tolerated.

Examples might be presence of gemmae in many species, the thickened costal base in *Stephanoleptodontium latifolium,* leaf marginal decurrencies in *Leptodontium scaberrimum* and *Stephanoleptodontium latifolium,* and large plant size in *Williamsiella lutea* and *Stephanoleptodontium syntrichioides.*

Table18-2. Niche-focused analytic key for Streptotrichaceae, with Pottiaceae as outgroup. Vertical lines connect the immediate descendants from one progenitor. The traits listed for inferred descendants are apparently adaptive and may figure in survival in the species' environmental niche. New substrates and habitats are underlined on the right.

Taxon	Niche
1. Pottiaceae — Progenitor. Calyptra usually smooth, or if papillose then simple-punctate not antrorse; laminal cells commonly bulging; papillae crowded and multiplex/thickened, rarely absent or simple; tomentum if present ending in elongate cells; perichaetial leaves not differentiated or uncommonly enlarged; peristome of 32 twisted (unless reduced) rami in pairs, basal membrane usually present; annulus of 2–3 rows of sometimes vesiculose cells, occasionally revoluble. Generative species for Streptotrichaceae not clear, perhaps *Barbula eubryum*, *Streblotrichum* sp., or *Ardeuma* sp.	harsh environments, soil, rock, rarely epiphytic
2. Streptotrichaceae — Tomentum thin but ending in short-cylindric cells; perichaetial leaves strongly differentiated, sheathing the seta; peristome of 32 straight rami primitively grouped in 4's, teeth primitively spiculose; annulus of 2–4 rows of weakly vesiculose cells, not revoluble; primitively antrorsely papillose calyptra.	[primitively epiphytes]
3. *Streptotrichum ramicola* — Flat or weakly convex surfaced laminal cells; papillae very small, simple to bifid, crowded; leaves sheathing basally; peristome basal membrane short. .	<u>branches of shrubs and trees</u>; Bolivia
4a. Core *Williamsiella araucarieti* — Tomentum arbusculate; leaves broadly channeled distally; costa ending before the apex, abaxial stereid band layers 1–2; distal laminal border not differentiated; distal laminal cell width 9–11 μm; asexual reproduction by gemmae borne on stem; capsule stomates present; annulus of 4–6 rows; peristome of 16 pairs of teeth, spirally striate and low spiculose, basal membrane absent; calyptra ornamentation unknown but probably smooth.	<u>grasslands, soil, rock, humus</u>, low to high elevations; central Andes, Brazil
5a. *Williamsiella tricolor* — Tomentum arbusculate or arising from stem in lines, deep red; distal lamina bordered by 1 row of epapillose cells; basal laminal cells strongly differentiated, inflated, hyaline; basal marginal cells of leaf forming a strong but narrow border.	humus, bark, rock, tree trunk; Andes
5b. *Williamsiella aggregata* —Tomentum absent; leaves recurved distally; distal laminal cell walls often thickened at corners; basal cell stripes absent; distal laminal cells very small, ca. 7 μm wide; peristome teeth spirally striate or smooth.	humus, rock, soil; China, southeast Asia
6a. Core *Leptodontium* Unknown — Tomentum arbusculate and usually without short-cylindric cells; stem lacking hyalodermis; leaves serrate, not distantly dentate, laminal cell lumens commonly strongly angled, larger.	[primitively epiphytes]
7a. *Leptodontium viticulosoides* — Autoicous; short-cylindric cells ending tomentum very rare; spores of two size classes.	branches, bark, rocks, humus; eastern North America, Latin America, Africa, Indian Ocean Islands, Asia, Australia

7b. *Leptodontium scaberrimum* — Leaf marginal decurrencies elongate, red, as wings on stem; laminal cells enlarged and with coroniform papillae; leaf apex narrowly blunt as a unique ligule; leaf base lacking stripes. — boulder, soil, bamboo; China (Sichuan, Yunnan)

7c. *Leptodontium excelsum* — Leaves often with rhizoid initials near apex; leaves polymorphic; limited to New World. — soil, bark, rock; eastern North America, Latin America, West Indies

6b. *Williamsiella lutea* — Large plants, stems to 20 cm, leaves 4–8 mm long, costa 8–10 rows of cells across at midleaf; leaf margins often dentate to near base. — soil; Andes, eastern Africa

6c. Core *Stephanoleptodontium longicaule* — Distal laminal cells with centered (coroniform) group or circle of spiculose simple or bifid papillae and lumens bulging (by IRCI 3 bits). — soil, bark; Central America, Andes, Africa, Indian Ocean Islands

    8a. *Stephanoleptodontium syntrichioides* — Stem hyalodermis absent, gigantism: leaves larger, 4–7 mm, distal laminal cells larger, 13–17 μm; New World only. — soil; Central America, Andes

    8b. *Stephanoleptodontium capituligerum* — Dense tomentum with short-cylindric cells; leaves with sharp apices and well-demarcated hyaline fenestrations of thin-walled, hyaline basal cells. — soil, branches; Mexico, Central America, Latin America, Africa, Indian Ocean Islands

        9a. *Stephanoleptodontium latifolium* — Stem hyalodermis cells reduced in size to that of next innermost layer of stereid cells, often apparently absent; leaves strongly decurrent, costa blackened and thickened at base. — humus, lava; Congo, Rwanda, Uganda

        9b. *Stephanoleptodontium stoloniferum* — With leafless branches in axils of distal leaves and gemmae restricted to leafless branchlets; gemmae elongate-elliptic. — epiphyte; Central America, northern Andes

    8c. *Stephanoleptodontium brachyphyllum* — Leaves shorter, 2–3 mm long, ovate to short-lanceolate, distal laminal cells smaller, 9–11 μm wide. — soil; Mexico, Central America, Andes, Africa, China

    10. *Stephanoleptodontium filicola* — Leaves distally plane to weakly keeled or broadly channeled, leaves dimorphic: when unspecialized, leaves ovate and bluntly acute with costa ending 6–8 cells before apex, specialized in strongly gemmiferous portions of the stem as shortly acuminate-lanceolate, costa subpercurrent, apex blunt, marginal teeth mostly near apex; specialized leaves appressed over the stem-borne gemmae when wet and open-catenulate when dry. — soil, humus; Central America, Andes, Africa

4b. *Austroleptodontium interruptum* — Stems short; tomentum thin; leaves small, 1.8–2.2 μm long, costa excurrent as an apiculus; distal laminal cells small, 9–12 μm wide; axillary gemmae present; peristome of 16 paired rami, smooth, basal membrane absent. — soil, <u>sand, turf, lawns, low elevations</u>; New Zealand

4c. *Leptodontiella apiculata* — Leaf base with hyaline area; stripes on basal cells occasional; calyptra antrorsely papillose. — trees, rock; Andes of Peru

11. Core *Microleptodontium* Unknown. — Stems short; hyalodermis absent or very weakly differentiated; leaves ligulate, less than 3 mm, costa narrow; gemmae common; calyptra smooth. — [primitively epiphytes]

12a. *Rubroleptodontium stellatifolium* — Leaf margins entire; laminal cells strongly bulging on free surfaces, papillae multiplex; basal cells red; brood bodies on rhizoids sometimes present. — soil, <u>lava</u>, vegetation; Latin America, Indian Ocean Islands

12b. *Microleptodontium flexifolium* — Leaf base weakly differentiated; laminal papillae simple, hollow, scattered over the flat or weakly concave surfaces of the lumens; gemmae abundant on stem. — rocks, soil, <u>roofs, rotted wood,</u> bark, rock, moderate to high elevations; North America, Latin America, Eurasia, Africa, Indian Ocean Islands

13a. *Microleptodontium umbrosum* — Excurrent costa, and gemmae restricted to excurrent costa (by IRCI 3 bits); leaves dimorphic (IRCI two traits). — peaty soil; central Andes, western Europe, southern Africa

13b. *Microleptodontium stellaticuspis* — Leaf ending in unique terminal dentate cup and gemmae and rhizoids borne in cup (by IRCI 3 bits). — <u>thatch</u>, grasses, low elevations; northern Andes, Indian Ocean Islands

13c. *Microleptodontium gemmascens* — Excurrent costa and gemmae restricted to excurrent costa (by IRCI 3 bits); restricted to British Isles. — <u>thatch</u>, organic debris, low elevations; Europe, Indian Ocean Islands

4d. Core *Trachyodontium* Unknown — Basal laminal cells with longitudinal brown to red stripes; short-cylindric cells often ending tomentum; gemmae absent. — [primitively epiphytes]

14a. *Trachyodontium zanderi* — Leaves with cartilaginous border of elongate cells; peristome further divided into 64 rami; operculum of short cells with rounded lumens. — epiphyte on shrubs; Ecuador

14b. Core *Crassileptodontium pungens* — Basal cells with miniwindows except in specimens with very thick-walled basal cells; distal laminal cells with crowded, solid, knot-like papillae (flower-like); basal portion of costa reddish; peristome of 16 paired rami, smooth or slightly striated. — soil, humus, <u>lava</u>; Latin America, Central Africa

15. *Crassileptodontium wallisii* — Leaf base high, to 1/3 leaf and basal marginal cells inflated (by IRCI 3 bits). — soil; Latin America, Central Africa

16a. *Crassileptodontium erythroneuron* — Costa red throughout; marginal leaf teeth lacking or much — <u>rock;</u> Andes reduced, leaves straight distally; leaf apex concave, blunt.

16b. *Crassileptodontium subintegrifolium* — Laminal teeth __rock__; central Andes absent; costa ending 1–10 cells before apex, concolorous.

The habitat correlations in Table 18-2 may seem of little news, but the difference is that correlations with traits and environment are well-distinguished as lineage-wide or immediate. Duplicate habitats are as much information, as will be revealed in the next chapter, as are the distinctive. In summary, the Streptotrichaceae is primitively of epiphytic species, see adaptive levels 3 and 4b,c,d. *Williamsiella araucarieti* (4a) is the progenitor of a major, mostly non-epiphyte complex lineage (5, 6, 7, 8, 9, and 10), and is generalist in both morphology and habitat, but is—probably through anagenesis—not an epiphyte. The Core *Leptodontium* lineage (6a) is the progenitor of a return to epiphyte habitats, these associated with arbusculate tomentum and leaf cell walls thickened at the corners. Two physically large species, *Williamsiella luteum* and *Stephanoleptodontium syntrichioides,* are restricted to soil. Species capable of colonizing lava, *Rubroleptodontium stellatifolium, Stephanoleptodontium latifolium,* and *Crassileptodontium pungens* all have modified basal cells: bright red, inflated, or with miniwindows, respectively. Species restricted to the harsh substrate of rock, *Crassileptodontium erythroneuron* and *C. subintegrefolium,* have laminal marginal teeth lacking or reduced and leaf apex modified, either concave and blunt, or costa ending relatively far below the apex. Other *Leptodontium* s.lat. species found on rock often have damaged leaf apices and distal margins. Genetic reduction of leaf apices saves the energy lost in generation of doomed tissue. Two species found largely on peaty or otherwise highly organic soil and thatched roofs, in the genus *Microleptodontium,* have highly modified means of asexual reproduction by gemmae.

The character states in Table 18-2 can be termed "adaptations" but perhaps they might better be called "adaptational flags." This is because these traits signal an adaptive process, sometimes immediate, sometimes of value to progenitors, but only occasionally bringing to mind a clearly mechanical or physical function. The hyalodermis

is apparently a stem stiffening tissue, as per discussion of Brazier buckling by Niklas and Spatz (2012: 192). Gemmae are obviously instrumental in the process of asexual reproduction, but coroniform papillae, subcylindric cells at the ends of tomentum, longitudinal stripes on basal cells, and large leaf decurrencies are less easily decoded as to function.

Nevertheless, apparently adaptive traits can now be associated directly or indirectly (e.g. pleiotropically) with some environmental novelty. It appears in Table 18-2 that most habitats are similar and the many traits are mostly neutral or nearly so, but this may be only superficial given the poor quality of substrate and other data on a herbarium label. The above few correlations of species and new habitats are only in a preliminary sense instructive. The habitat must be further investigated as to correlations with adaptive flags. On the other hand, overlapping habitats of sister descendants of on progenitor may have an adaptive advantage for an ecosystem, as will be discussed below.

The use of adaptational probes allows a new perspective on evolutionary ecology, just as macroevolutionary systematics allows a new evolutionary definition of taxonomic ranks above species, i.e., the dissilient, radiative genus. This melds well with new efforts in evaluating functional diversity of traits of species in the context of ecosystems (Cermansky 2017).

In sum, new traits found in new habitats are probably adaptations. Traits that span many species and are found in many habitats are probably neutral for those habitats. On the other hand, a new species generated in an entirely new ecosystem or a preadapted species finding itself inserted into a new ecosystem can find all its traits, both ancient and ancestor-relicts and new to be highly adaptive or at least vulnerable to broad selective pressures.

In the next chapter, adaptation is examined at the ecosystem level, by ecosystems in competition with other ecosystems.

Chapter 19
ADAPTATIONS OF ECOSYSTEMS

Evaluation of ecosystems using entropic macroevolutionary method — It is easy to use metaphor and analogy, and make principles and generalities hard-won in other fields apply to one's own innocent study area. The only test for whether these geologic-time and dissilient-genera applications are valuable is to see if they work. That is, if they explain better and predict more accurately than does orthodox, perhaps more limited, immediate crisis and species-based modeling. This chapter presents a new way to look at present crisis in biodiversity. Ecosystem evolution is a metaevolution—evolution at a higher level than that of species. There are parallels with directional and stabilizing selection, competition, and doubtless many other theoretic constructions of standard evolutionary theory that may be examined in future research.

Adaptive ecosystems — Life is a means of mediating and buffering entropic change, avoiding catastrophe by extricating local areas from suboptimal minima. To avoid pathetic fallacy (save the pandas!), life is a special form of dirt that integrates entropic change (negentropy or potential work energy flowing to maximum entropy) across an ecosystem. Life avoids, in the long run, local entropic maxima that harbor possible catastrophes of energy flush. For example, a forest fire may quickly use up oxygen and combustibles to increase entropy, but the sun continues to beat down, and rain water continues to flow and undercut soil, all major sources of work-energy.

A particularly important review and discussion of entropy in nature, including the difference between informational and thermodynamic entropy is given by Brooks (2001). He coined some interesting new terminology, like "enformation." Brooks focuses on the organism and cohesiveness. He does not introduce informational bits as important in evolutionary classification, or scale-free networks as moderators of thermodynamic flush. His discussion is eminently agreeable. For instance:

"Organisms do not just degrade their immediate environments; they can serve as environmental sources of energy for other organisms; in fact, the largest portion of the environment for organisms is other organisms. Organism functions that increase the amount of energy, and the amount of time energy from abiotic sources remains in biological systems, thus represent evolutionary mechanisms by which the rate of environmental degradation can be

slowed. This is done this by organisms sequestering entropy production for their own use, a purely selfish behavior that nonetheless benefits others …."

This is a generalization of what I am saying in this book, and I suggest a mechanism for such entropic moderation. He cited several papers by Matsuno, of which (1989, 1995, 2000) are the most relevant, as support for the above comments.

Rather than finding a function for a particular trait as in the last chapter, one might turn the search on its head and find a trait for a particular function. Let's consider the idea (e.g. Cropp & Gabric 2002; Jorgensen 2000a,b) that a particular ecosystem will survive against competition by other ecosystems if it preserves homeostasis through resilience in a local or regional Gaia-like fashion (Lenton & Lovelock 2000; Schwartzman 2002). It preserves its thermodynamic balance of gradual (not catastrophic) increase of entropy by coevolutionarily draining energy-rich portions of the ecosystem. A successful ecosystem does not convert low entropy to high the fastest, but the most carefully, because care in the long run leads to Darwinian survival.

According to Jorgensen (2000b: 196), referring to a herbivore model in adaptation of an ecological system, "In a stable environment where sufficient (regenerating) food is distributed in a completely regular pattern, evolutionary adaptation would eventually lead to optimization of an animal's movements in a regular grazing pattern, with a single objective: optimum energy uptake, and use." Other ecosystems that are more prone to catastrophe are more likely to fail or be invaded by organisms that flood the ecosystem with a monoculture liable to complete eradication at some time. This is Darwinian if one ecosystem can replace, at least in part, another. But in the long run, the invading ecosystem is not optimized as thermodynamically stable, resilient to disturbance.

Thus, listing the functions needed for particular ecosystems to establish resilience, then evaluating traits as contributing to such functions may be a productive way to approach adaptation in that adaptation is made a desideratum of the ecosystem, not necessarily of the organism. Thus, the species is the niche, whether extant or extinct, and its informational bits are more important to ecosystem survival than details of trait function.

Typical functions that might help preserve ecosystems in the long run include: stabilizing soil; trapping energy; sequestering carbon as biomass; tempering oxidation (fires, fires, overgrowth); resisting over-enthusiastic invasives; promoting coexistence (spines, poison); coevolution. Evaluating the contribution of each species to ecosystem homeostasis in the context of thermodynamic equilibrium is then a way to assign function to traits. The sum of informational bits integrates the value of all functions. Even neutral traits contribute to biomass accumulation. The study can be rendered falsifiable because measurement is possible. For instance, biomass of a species can be measured or at least estimated, as can exact contribution of organisms to reducing swings in energy exchange (Gates 1965). Coverage of unstable substrates and numbers of invasives in a forest is easily measured.

Margalef (2000: 6,7) summarized this feature of ecosystem adaptation with: "Spent energy leaves a trace in the record of history, usually in the form of an increase of complexity or its expression as information." Also, "The information which we now have access is not free; it has been paid for by the wholesale increase of entropy extended over time, just as biological evolution has been paid by the increase of entropy in the physical bodies of all the organisms that have lived." Disturbance of the ecosystem may have profound consequences (Chapin et al. 2000; Moreno-Mateos 2017; Swamy & Terborgh 2010; Terborgh 2012; Wright 2002).

The maintenance of biological diversity in ecosystems has recently been attributed in part to the effect of "enemies." Tropical forests, for instance, support a host of rare species because a balance is met when species become too rare to be affected by enemies, which also become rare (Bachelot & Kobe 2013).

Ecosystem thinking —We now have an entropic perspective on evolution of species from other species, either as extant progenitors or postulated missing links. The following discussion is an attempt at describing the processes of ecosystem metaevolution. This is not metaphysics because the theories involve measurement and are testable.It involves classification techniques as an approach to evaluating an important "evolutionary atable strategy" in the sense of John Maynard Smith (Stewart 2011: 219).

We know that (1) species with different traits exist, so there is an isolation event involved at some stage, but species evolve from closely related species and extirpation of an ancestor may be partially made up for by speciation involving incomplete reversal from a descendant, and (2) species share or at least overlap in habitat and substrate (Table 19-2). This evolutionary and habitat redundancy or partial redundancy may be a survival feature on the part of ecosystems. In fact, ecosystems may be held together by partial redundancy.

A lengthy discussion of entropy and redundancy in ecology is fournd in the "macroecology" chapter of Brooks and Wiley (1988: 288–353). These authors also discuss the functional redundancy of closely related species. The present book extends the Brook and Wiley theoretic suggestions to the importance of classification in uncovering hidden structures (scale-free networks) not easily visualized with cladistic technique, together with an example.

Visualize dissilient genera as a set of little mops that soak up dangerously disruptive free energy. The handle of the mops are the ancestral shared traits, and the yarn flail is composed of the descendants adapted to similar and overlapping negentropy fields. The idea here is that ecosystem resilience may be measured by informational bits reflecting (1) immediate traits associated with one speciation event, (2) the lineage traits of a species contributed by all ancestors, and (3) the traits of a genus assembling the dissilient set. All reflect an ecosystem's Darwinian adaptations for protective thermodynamic gradualism.

The main drive in Streptotrichaceae evolution is from an advanced, specialized state (adaptations for epiphytic habitat) go a more primitive state (for growth on soil and rock—we don't have to know exactly what the adaptations are). The family is now a constant feature of páramos (Andean alpine tundra as biodiversity hotspots, now much disturbed). It is tempting to imagine that the present soil and rock inhabiting species were ancient species exactly recreated to fit the old habitat, but given constant mutation and accumulation of minor, neutral traits, this is very doubtful. Features like presence of gemmae (asexual reproduction) may be "tells" that indicate replacement. In fact, *Stephanoleptodontium longicaule,* which has gemmae in many populations, may have been generated from the similar but somewhat reduced species *S. brachyphyllum*, which also has gemmae.

An ecosystems biological thermodynamic deficit is simple the informational bit total of all species in it subtracted from all species known to have comprised the ecosystem. The difference between native and invasive species does matter in halting thermodynamic disequilibrium; native species have closely related species that are evolutionarily redundant (so similar that evolution allow replacement of an extirpated species), while exotic invasives have no immediate replacement. Evolutionary redundancy and replacement is not necessarily a long-term activity, say, at the geologic time scale. Stuart et al. (2014) have shown that adaptation to a new niche by a lizard *Anolis* species whose territory has been invaded by another lizard species of the same genus can occur very rapidly, in 20 generations, i.e., larger toe pad size as adaptation to higher perches.

Why does this matter? Because invasives can heal the damage we cause to the native ecosystem in our attempt to construct elsewhere our native African savanna (Walker 2001). It may be difficult to think of cudzu (a pernicious weed overgrowing roadcuts in the southeastern U.S.A., originally imported to feed cattle but they would not eat it) as a healer, but that is because it provides thermodynamic stability to the wounds we, the ultimate invader, make in the native forest. In geologic time, the best healing species are natives, as extirpation does not impact greatly their abilities to stabilize entropic flushes because as a small set they are both evolutionarily and habitat redundant.

Species survive best if their ecosystem survives. Species may leap from one ecosystem to another if the other ecosystem is compatible, that is, greatest expected saltation within bounds of gradual evolution, say within three standard deviations bit-wise given a geological time frame. In the case of Streptotrichaceae, that means an expected usual 4 but up to 8 informational bits per event. Two adjoining ecosystems requiring more than expected greatest saltation, such as sea and shore biomes, are fairly immune from colonization of each other because the boundary is evolutionarily relatively impermeable in a geological time frame. A single leap from sea to shore or vice versa usually does not constitute one ecosystem invading another, but multiple leaps may.

But two ecosystems that adjoin and are similar in important respects can be involved in a kind of competition in the Darwinian sense. For instance, Streptotrichaceae species originally were highly adapted to epiphyte habitat at high elevations (e.g., cloud forests) but many species now are abundant on soil and rock in páramos. Also, some epiphytic species of the family are common on thatched roofs and decaying vegetable matter.

Exchanges between ecosystems do two things: (1) at best, make a gift of one element of complexity, which may help protect the new ecosystem from catastrophic draining of negentropy (an entropic flus), and (2) at worst, destroys the ecosystem through destructive invasion, changing the attributes of the old ecosystem into that of the invader. But species from one ecosystem that enter a new ecosystem, then are extirpated, lack closely related species that may (1) easily evolve into a well-adapted substitute species, or may (2) serve as immediate substitutes because of overlapping habitats.

The species that originate in one ecosystem from dissilient genera (i.e., with two or more descendant radiative lineages) all have the same train of ancestral traits. Any of the immediate or ancestral traits may prove adaptive and protective to an ecosystem in case of habitat disruption or catastrophe. Parasite and other enemies of one species are often physiologically focused only on that species, and a closely related species with overlapping habitat provides redundancy in addressing uncontrolled sources of negentropy. A dissilient genus is therefore more valuable to an ecosystem when the descendant species are more similar to the progenitor than different from it.

Species that are integrated in one ecosystem may well prove invasive in a new ecosystem. The traits that have been honed by Red Queen competition in one ecosystem may find no resistance in a new ecosystem. In the worst case, this results in the hegemonic extension of the old ecosystem into many other less robust ecosystems. Thus, we can imagine that traits of a species that are apparently neutral in one ecosystem being remnants of progenitor traits that can exist in several different habitats of that one ecosystem can become quite potent in another ecosystem. Traits once considered "spandrels" in one habitat may be aggressively adaptive on another.

Even extremes of allopatry may not be a barrier to evolutionary and habitat redundancy. In geologic time all species are potentially sympatric. Patino et al. (2013) found that bryophytes on islands indeed obeyed "laws" associated with patchy environments and island circumscriptions, e.g., a tendency to bisexuality (monoicy), but the species mostly retained the life history traits that allowed long-distance dispersal. It may be a thermodynamic blessing that biogeographic barriers are being broken by human-mediated dispersal (Capinha et al. 2015) such that allopatric species are becoming increasingly potentially sympatric and thus providing potential evolutionary and ecological redundancy to heal human damage within biogeographic barriers.

One of the ways to examine a problem is to conjure up the most extreme instances, either of least or greatest effect, and see where that takes the creative but well-trained and well-informed mind. See if you can follow (or stomach) the following scenarios. Basically, one ecosystem can compete with and conquer another ecosystem by replacing species of the second ecosystem with its own or newly evolved species that possess powerfully competitive ancestral and no longer neutral traits. The invasive species may have been well integrated in the original ecosystem; it may have invaded other habitats in the original ecosystem (as with Streptotrichaceae moving from epiphytic to páramo habitats), which may have simply made other new habitat more complex and thermodynamically stable; it may then have figured in the destruction of some other ecosystem as an powerful invader, creating fast dissolution of complexity and enhancing catastrophic change with its full complement of competitive ancestral traits.

One of the most instructive (to me) hegemonization scenarios has been enacted in certain highly disturbed environments. The flora of the Niagara River gorge (pers. comm. P. M. Eckel) has the following European species in greatest abundance: Norway Maple as tree, European Buckthorn as shrub, Oriental Bittersweet as vine, and Garlic Mustard as herb. It is a "world forest" for the temperate zone occurring in a 14-mile long, rather deep limestone gorge. It is an invasion of a now weakly vegetated area largely by elements of the more robust European ecosystem into a disturbed habitat of scree and limestone drought. This may be seen as a secondary invasion, however, since the original invaders were human.

Following the idea of ecosystem competition, we can advance an explanation of the present human-derived environmental catastrophe (climate change, biological extinction, etc.) from a new perspective, that of ecosystem stability over very long periods of time.

It has been pointed out by Mace (2014) that since the mid-1990's, in conservation literature a "wider acceptance of the notion that people are part of ecosystems emerged, and the tendency to treat people and nature as separate units in discourse and analysis was much reduced, although not completely eliminated." The focus switched from species to ecosystems, particularly as ecosystems are supportive of human endeavors. Mace goes on to say that recently there has been a reframing towards enhancing integration of people and nature. In addition, a long-term element is now considered—"reversing long-term declines in old-growth forests or recovering the full extent of marine trophic systems may take centuries, far beyond the normal time scale for environmental policies."

Consider that *Homo sapiens* is an invasive element with ancestral progenitor-derived traits that forcefully project the primeval African savanna habitat into other ecosystems. Our parks, our cities, our farms are, respectively, savanna re-creations, stone and metal kraals, grasslands with grain, and herds of megafauna. Our ideal environment is a Garden of Eden.

If, indeed, we humans are the rampaging shock troops of the African savanna in Darwinian competition with all other ecosystems, *then* the idea that we are god-like stewards of biodiversity is rather self-serving and a pathetic fallacy. The actual events of the climate change, environmental destruction, and extinction processes, are admittedly our fault, but also ingrained in our psyche as part of our origin on the African savanna. A book entitled "Managing Human-Dominated Ecosystems" (Hollowell, ed. 2001) could just as well be called "Mismanagement by Ecosystem-Dominated Humans" with little change of content. A simple change of heart is not enough. We have invested much during the long trip from subsistence native grain, root and tuber harvesting, and macrofauna hunting, then shepherding and desultory gardening, then plant and animal husbandry, leading to modern farms and industrial parklands.

From this it follows, unfortunately, that we are the pawns, the dull-witted soldiers, and self-centered bulldozer-drivers of the African savanna. One might conclude that this statement is teleological and anthropomorphic, lending purpose and human drives to a natural ecosystem. Exactly. That is my argument. When discussing ecosystem competition, it is necessary to couch Darwinian interactions and results in terms of: "An ecosystem survives, becomes more stable, complexifies and expands if it happens to …." Only with the African ecosystem we justified in using teleology: "We survive, become more stable, complexify and expand when we do such and such …." And now here we are in multiple worldwide crises.

We have beaten in the past various natural higher powers, such as cave bears and smallpox, but have yet to triumph over our own natures.

We create protected biodiverse parks, but have such little concern for non-savanna ecosystems that we allow poor, starving people to penetrate such parks to gain sustenance and livelihood rather than solving this problem and isolating the ecosystem. And we are upset and take up scientific arms when unprocessed ecosystems "fight back," as when Ebola virus is transmitted to human invaders via consumption of bushmeat. Properly, biodiversity hotspots require isolation at the multiple bit level, that is, barriers must be as deadly as the sea-shore interface.

It is a bit self-serving to take up arms against invasive species that secondarily attack our own successful invasion of primeval forests and other pristine, well integrated ecosystems of hard-won thermodynamic stability. The African savanna-based human biome is well-acknowledged as terribly vulnerable to supercritical catastrophes through thermodynamic energy flush. We must now hope to limit the effects of our invasion to only small cascades of negentropy conversion rather than catastrophic releases of energy untempered by complex webs of dissilient genera with many evolutionarily and habitat redundant species.

Alternatively, we could adapt to foreign ecosystems, as have the Inuit in the Arctic, the Pygmies of the African jungles, the Aborigines of the Australian outback, the Bushmen of the Kalihari Desert, or the American Indian nations in the several North American ecosystems. But this would require abandonment or extirpation of our savanna psyche, which is a major element in the way we conduct our modern civilization. The really important tipping points into worldwide disaster were (1) the invention of agriculture and then (2) the Industrial Revolution. We cannot abandon these and still define ourselves as modern.

We, with our facile brains, are the mighty, purpose-driven transformers of the world's ecosystems into a world-wide African savanna. Our brains are powerful modelers but we are in denial of just to whom or to what we owe our ultimate responsibility. Ourselves? Not yet. Not by a long shot. We are not quite sapient man, instead we are obligated man—mankind under a DNA- and tradition-enforced spell or geas. We are savanna creatures driven by savanna priorities unless we act, necessarily for survival as a species, as if we are not.

The genus as a fundamental unit of evolution — Dissilient genera are self-healing. Species are clearly developed through isolation plus adaptation and accumulation of neutral traits. Descendant species that derive from the same ancestral species, differ by only an average of 3 or 4 traits (or 3 to 4 informational bits), and have the same ancestral train of traits. They also tend to overlap in habitat. The dissilient genus provides the ecosystem with evolutionary redundancy in maintaining thermodynamic health in blunting rapid increase of entropy, turning catastrophe into mere environmental insult. The adaptations of particular descendant species may well act on the ecosystem level to squirrel them away as thermodynamic back stops and evolution-in-the-bank, maximizing ecosystem health through maintenance of environment-overlapping biodiversity and close bit-level replacement. "Dissilience makes resilience" of ecosystems may be far more important to human survival than is enhancement of biological diversity in no larger context than protecting hot spot reserves or rare species.

And, as an aside, one can be sure that no one organism—smart cockroaches, brainy rats, alien invaders from another world—none of these will in due time replace our civilization, because any one upstart organism's mighty new world order will ultimately fail if our civilization falls, and for the same reasons.

Networks in natural selection — The study of chaos and complexity theory has revealed internal structure to networks, both natural and human-derived, which according to Huneman (2017) has a selective advantage at the biological network level, i.e., the ecosystem. According to Strogatz (2003: 254ff), natural networks have three universal features, (1) short chains of connection, (2) strong clustering, and (3) scale-free distributions. A signal that clustering is taking place is that the distribution of links in a network does not tail off very slowly as in the standard bell-shaped curve reflecting randomness, but decays at a slower rate. This slower rate follows a "power law" that reflects a degree of self-organization (Strogatz 2003: 255). Albert et al. (2017) have shown power-law distributions in nature using cladistic methods.

Following the discussion on the "Math Insight" Web site,

mathinsight.org/scale_free_network,

a natural-world network has large hubs, which give the distribution a long tail, indicating the presence of nodes with a much higher degree of connectivity than most other nodes.

Genera and sets of genera reflect this *scale-free network* (Fig. 19-1, 1) to some large extent. Habitats with considerable biological diversity are more capable of addressing thermodynamic insult to the ecosystem, because the dissilient genus clustering similar species ensures that redundancy of descendants' capabilities. Beyond that, descendants' somewhat dissimilar methods of addressing thermodynamic disequilibria are, if not at hand, then nearby. Also, these scale free

networks are potentially subject to positive selection because they are more robust against failure, that is, more likely to stay connected than a random network after the removal of randomly chosen nodes (Huneman 2017). Such clustered networks are quite similar to caulograms of dissilient species. Caulograms of species in Streptotrichaceae and *Didymodon* s. lat. (Pottiaceae) are given in Fig. 19-2, 1–2. They are represented by lines and dots as in Fig. 19-1. Although clearly more sparse than the idealized networks of Fig. 19-1, the scale-free distribution in both groups is evident.

Within the constraint of general physiology and morphology, continued speciation through four to eight bit steps builds larger and more complexly redundant scale-free networks.

Clustering involves "small world networks" (minimization of degrees of separation) (e.g., Peng et al. 2016, and a fine discussion on the Math Insight Web site). A *small-world network* (Fig. 19-1, 2) is a kind of mathematical graph in which most nodes are not neighbors of one another, but the neighbors of any given node are likely to be neighbors of each other, and most nodes can be reached from every other node by a few hops or steps. Minimization of communication lines assumes some means of communication, including feedback—the ecosystem apparently provides such for dissilient genera and this needs investigation.

There is evidence of a metaevolution of natural selection in studies (e.g., Beres et al. 2005; Bock & Farrand 1980; Clayton 1972; Hadly et al. 2009; Krug et al. 2008; May 1988) of the number of species in higher categories. Such studies show what approaches a standard hollow curve but with "overdominance" of higher taxa with many species and a rather longer tail of higher taxa with few or one species. This is typical of natural networks involving power laws.

A chart (Fig. 19-3) of the distribution of species in the family Pottiaceae (from which Streptotrichaceae has been excised) was made from information of Zander (1993), updated with data from the Tropicos Web site

www.tropicos.org

and B. Goffinet and W. R. Buck's Classification: Mosses Web site

bryology.uconn.edu/classification/.

The species are clearly distributed as a scale-free network with many speciose genera providing large hubs in the evolutionary network, and smaller genera comprising the power-law curve below. The long tail consists largely of what are probably ancient genera with no close relatives and only one specialized species remaining.

Scale-free models in nature have been criticized by Keller (2007): "The scale-free model has been claimed to apply to complex systems of all sorts, including metabolic and protein-interaction networks. Indeed, some authors have suggested that scale-free networks are a 'universal architecture' and 'one of the very few universal mathematical laws of life'. But such claims are problematic on two counts: first, power laws, although common, are not as ubiquitous as was thought; second, and far more importantly, the presence of such distributions tells us nothing about the mechanisms that give rise to them." Here, I suggest that simple Darwinian selection on the ecosystem level using evolutionary and ecological redundancy gives rise to scale-free networks in Streptotrichaceae and possibly many other organismal groups.

The pottiaceous genera presently with most species are *Barbula,* with 147; *Tortula,* with 138; *Trichostomum* with130; *Weissia,* with 97; *Hyophila,* with 88; *Syntrichia,* with 82; *Tortella,* with 53; *Anoectangium,* with 47; *Geheebia,* with 41; and *Hydrogonium,* with 40. These large dissilient genera are the largest scale-free hubs in the family Pottiaceae. Although the family Pottiaceae is largely characterized by ability to withstand harsh environments, the above species are often found in moist temperate areas that may dry up but which are not deserts or typically very arid. Figure 19-4 summarizes the scale-free curve for the Pottiaceae.

Suppose that the world becomes increasingly aridified, and moist areas come under stress. It is possible that species of Pottiaceae that are adapted to deserts and steppes might speciate as those areas expand and become more complex. The monotypic species in the family would then be progenitors of species new dry-land ecosystems. Figure 19-5 has the curve flipped over to model dry-land speciation by presently rare or uncommon species, changing the relative composition of species in the Pottiaceae. Given the rapidity of climate warming, this model may be upon us in a few thousand years.

Suppose the world reacts to climate change by initiation of a pluvial period of low temperature?

(Apparently, it could happen.) Figure 19-6 suggests a few genera of Pottiaceae that are presently restricted to very wet areas that may become progenitors of a Pottiaceae with a quite different composition.

Throughout these models, there remain a set of scale-free hubs for immediate (in geologic time) evolutionary and ecologic redundancy, and another a set of small genera and a train of monotypic genera that ensure the long-term survival of the family during world cyclic climatic changes. This resilience is due to evolutionary and habitat redundancy provided by simple Darwinian selection on the ecosystems themselves. Both the families and the ecosystems with a hub and train genus composition survive and compete in changing times. These taxonomically isolated species are not dead ends, rogue genera, or outliers, but may be viewed as a living bank of potential evolutionary and ecologic redundancy pre-adapted for some future cyclic change in climate and other contingency. Because with elimination of central hub species, the "integrity of the netwrod degrades rapidly..." (Strogatz 2001: 257), when there is a choice berween protecting a highly specialized species or its generalized core progenitor, the central core is the best choice to preserve the scale-free network.

A salient example of ecological redundancy is the replacement of the niche occupied by American Chestnut by various oak species (all Fagaceae) in the eastern North American deciduous forest. Other chestnut species may well replace the American Chestnut if resistance to its parasites is evolved. The American Chestnut Society is presently attempting to reintroduce a blight-resistant form or hybrid; given that we are part of an ecosystem, this is also quite natural though directed evolutionary redundancy.

Contributions of Streptotrichaceae to ecosystem survival — The reader may examine again Table 18-2 of the last chapter. In general, there is much overlap and redundancy in ecology among the species of Streptotrichaceae, including adaptation to epiphytic, rock and soil habitats.

Each of the sets of distinctive evolutionary features of the Streptotrichaceae may be measured for use in ecosystem management. One simply adds the informational bits for each species. If there has been extirpation in some area, the evolutionary and ecological redundancy deficit, together or separately, is simply the bits for the species present minus the bits for the species extirpated. A rapid estimate may be made by multiplying the number of species by the average bit count per species, in Streptotrichaceae this is 4 bits. Estimates for families similar to Strepto-trichaceae, for instance, the Pottiaceae, may be estimated by using the same bit average given about similar composition of hub and isolated species. Basically the lineage (dissilient hub plus train of isolates) with the high bit count should be most successful in preserving both itself (taxo-nomic integrity) and the ecosystem in which is occurs.

In the Streptotrichaceae, there are both (1) scale-free hubs for immediate evolutionary and ecologic redundancy and (2) ancient species as remnants of mostly extinct genera capable of long-term family instauration (recovery from a previously degraded state). These general sets are geographically centered in the Andes, but are of scattered secondary distribution into mainly Africa and Indian Ocean Islands. The lands of China, Australia, and southeast Asia have apparently their own rather isolated system of a few distinctive species.

Distinctive hubs with both evolutionary and ecologic redundancy in the Andes are as follows: *Leptodontium* species found on bark and soil, with *L. excelsum* as a fairly new species of many biotypes and *L. viticulosoides* as monoicous (bisexual) reservoir of stenomorphic central traits (total 6 bits). *Stephanoleptodonium* includes six species that are found on soil in the Andes (22 bits). *Microleptodontium* has three species found on humus, organic debris and thatch in the Andes (10 bits). *Crassileptodontium* includes three species occurring on rock and soil, often lava, in the Andes (13 bits). Hub total is 51 bits.

Certain species of the hub genera (e.g., *Crassi-leptodontium pungens*, *Leptodontium viticu-losoides*, *Microleptodontium flexifolium,* and *Williamsiella lutea*) are often distributed into Africa, often into sky-island mountain ranges where they are locally not uncommon and are apparently well-integrated into the mountain moss floras. Being taxonomically isolated, they do not provide evolutionary redundancy. These species, though probably long established, may be considered invasive in the sense that they occupy

ecological habitats better used by hubs in other families, 14 bits. *Strephanoleptodontium brachyphuyllum, S. capituligerum,* and *S. filicola, S. longicaule,* are all Andean species with apparently a secondary long-distance dispersal into Africa (judging by numbers of other, somewhat ancient species in the genus and patterns of distribution elsewhere in the family), but these previously invasive species now occur all in much the same area and provide a hub level of redundancy, 17 bits.

There are monotypic genera that are highly restricted in range. These may be considered genetically depauperate, but even humans have survived genetic bottlenecks. In addition, one of them, *Rubroleptodontium stellatifolium* has variation in leaf form and cell expression between its rather well isolated geographic stations, perhaps because it is found on soil and, as a moss, is close to mutating potassium isotope radiation. The set of monotypic genera that may provide very long-term redundancy for Darwinian survival of their family, and for their ecosystems total 23 bits.

Core species of scale-free hubs are critical for survival of ecosystems because they provide a generalist platform for biotype elaboration and thus speciation. Focusing on only the core progenitors of extant dissilient genera (ignoring the most ancient ultimate ancestral taxa), these include *Crassileptodontium pungens, Stephanoleptodon-tium longicaule, Williamsiella araucarieti,* totaling 20 bits. Two core progenitors are apparently extinct, these for the genera *Leptodontium* and *Microleptodontium,* an estimated 12 bits.

There are a few very unusual species of unusual combinations of traits. These include *Leptodontium scaberrimum* of China and *Stephanoleptodontium latifolium* of eastern Africa. These are at the extremes of range of the family and are rare and very isolated in geographic distribution. Whether their speciation is recent or not may be investigated molecularly.

The eastern part of the Old World has two species of some interest. *Williamsiella aggregata* is restricted to China and southeast Asia, while a species with similar gametophyte, *Austrolepto-dontium interruptum,* is restricted to New Zealand, often found on human-disturbed soil, lawns and sand. The latter species has a antrorsely papillose calyptra associated with the most ancient of species of the family, and may be a remnant of the group that generated the more vigorous *W. aggregata*. Given no hub, both species may be considered invasive, but *A. interruptum* has potential for long-term biotype expansion during major climate change.

Although bit measures of evolutionary are ecologic redundancy are possible, because no other families have been examined in this manner, these measures are of value only internal to the family. One might hope that the evolutionarily nearby family, Pottiaceae, because it is fairly well understood, might be examined in the same manner. The moss family Grimmiaceae has presently many researchers, and I commend this analytic method to them. It would be of value in biodiversity research to see if similar entropic informational evaluation may be done in families or subfamilies of other phyla.

Summary — To summarize, diverse ecosystems are stable in thermodynamic terms in that entropy is gradually increased in a multiculture. Damaged ecosystems with a pauciculture (with few species) allow buildup of negentropy leading to catastrophic further damage. Within the bounds of climatic and substrate constraints, ecosystems compete in a Darwinian fashion. The present nearly worldwide hegemony of the African savannah ecosystem leads to a fragile worldwide pauciculture. Disaster cannot be averted without recognition that humans are invasive extensions of the savanna ecosystem and are so driven. True compatibility in a multiculture ecosystem is the province of a few, isolated bands of humans, and cannot be expected of most of us.

Simple preservation of biological diversity in small reserves, and efforts to avoid tipping points with technological fixes may be limited by the constraints of savanna-thinking, and thus of little effectiveness. The emphasis on study of "biodiversity loss" might better be replaced by research on natural small-world networks and their protection from thermodynamic ecosystem disequilibrium (or "ecosystem health"). Mitigation of the current extinction event must be at the ecosystem level. The savanna hegemony will be broken, if not gently in the context of resolving and mitigating thermodynamic imbalance by planning and political will, then by world-wide catastrophe and concomitant war, famine, and pestilence. A fine example of projection of evolution and planning of human response across geologic time is given by Myers and Knoll (2015). An alternative, scary, entirely plausible example of recent climate-change apocalyptic prediction is that of Wallace-Wells (2017).

What are we to do? Rely on technical fixes either incremental and too little and too late or heroic and dangerous? Or shall we all go back to eastern Africa where we came from? Live underground? On the moon? Leave Earth a smoking ruin and go invade the galaxy? And perhaps "terraform" other worlds into new thermodynamically unstable ersatz-savannas or other designer ecosystems? Or perhaps introduce an ineluctable predator or pathogen to reduce the effects of our massive Malthusian evolutionary fitness? Will some disgusted purple alien do this for us, as a kindly favor?

Somehow, we must return ecosystems to natural small-work networks of species, while at the same time this is not in our nature—because most rapid solutions to over-population are inhuman. We need to think the presently unthinkable and solve the apparently unsolvable while keeping our humanity intact in order to genuinely deserve our self-awarded title of *Homo sapiens*.

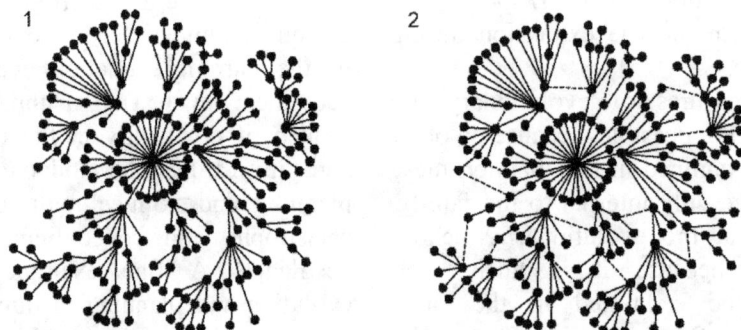

Figure 19-1. 1. *A scale-free network*, with many nodes poorly connected and few well connected as hubs. The network is well-clustered and the distribution follows a power-law. This is quite like a caulogram. 2. *A small-world network*, highly clustered and a short path between any two nodes. Extra connections provide redundancy (shown here as dashed lines) and are provided by the environment (extinction, increased fitness, overlapping areal distribution, habitat sharing, long-distance dispersal). Modified from Huneman (2017).

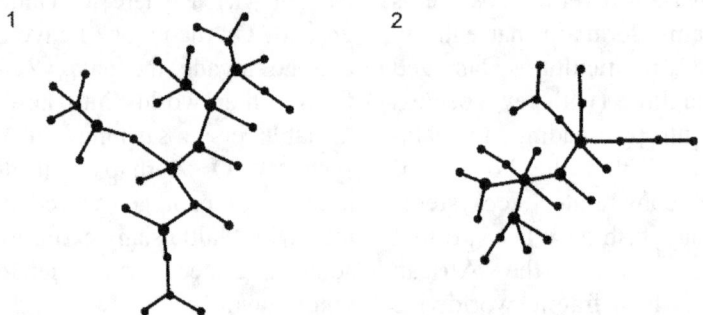

Figure 19-2. Scale-free networks generated from caulograms. Core progenitors are represented by large dots. 1. Streptotrichaceae, reduced from Fig. 9-1. Due to the existence of several ancient species, the hubs are less distinctive than in the idealized network of Fig. 19-1, 1. 2. *Didymodon* s.lat., split into six genera by Zander (2013, 2014). Hubs are more easily distinguished in these mostly recent species.

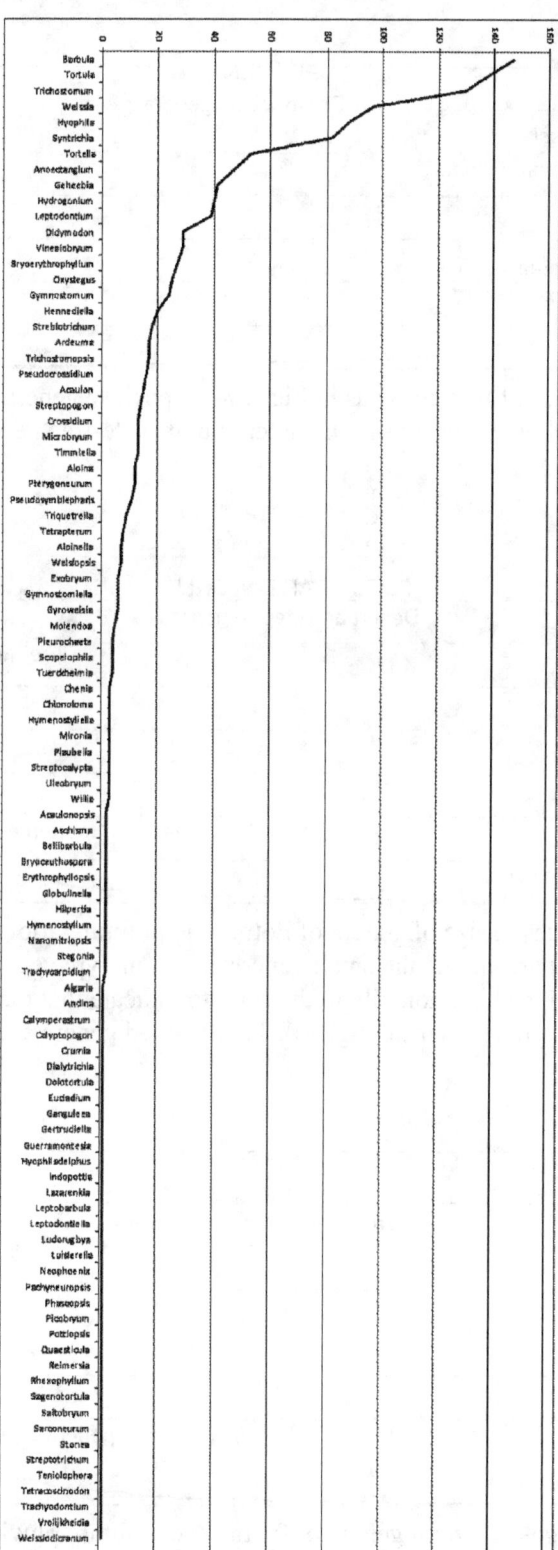

Figure 19-3. Numbers of species in the 94 genera of the Pottiaceae from Zander (1993), updated from recent work. Generic names are plotted against numbers of species in each. Total species is 1375, averaging 15 species per genus. Large, speciose genera approximate the hubs in a scale-free network. The curve is not exponential but a power-law instead, ending in a train of monotypic genera.

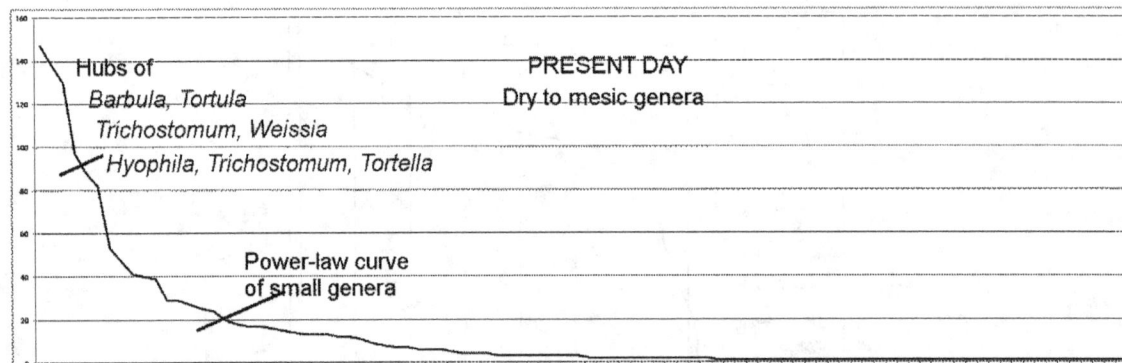

Figure 19-4. Present day curve of Pottiaceae as in Fig. 19-3. Speciose genera as scale-free hubs are on left. This is not a standard hollow curve, but is characteristic of scale-free networks.

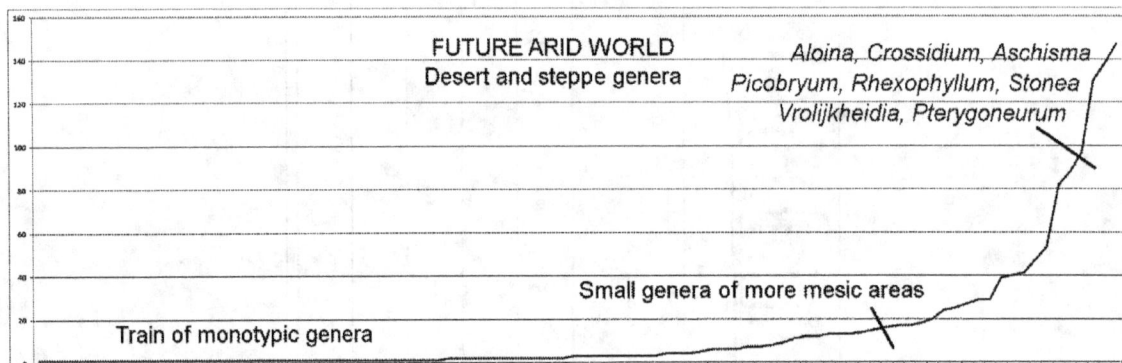

Figure 19-5. Projected changes in size of genera of Pottiaceae in future arid, hot world, with hubs now made of arid-land genera that are among the small genera and monotypic genera in Fig. 19-4. The curve of Fig. 19-4 is here simply flipped horizontally to the right to indicate that the less speciose genera are now more important (the jaggedness is meaningless). The featured genera are notional examples.

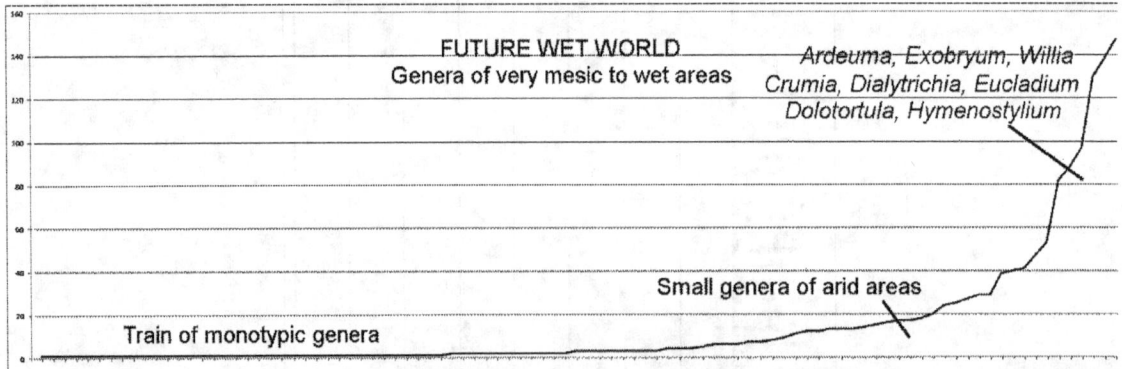

Figure 19-6. Projected changes in size of genera of Pottiaceae in future pluvial, less warm world, with hubs now made of mesophilic and hygrophilic genera previously among the small genera and monotypic genera of Fig. 19-4. The featured genera are notional examples.

LITERATURE CITED

Adams, M., T. A. Raadik, C. P. Burridge & A. Georges. 2014. Global biodiversity assessment and hyper-cryptic species complexes: More than on species of elephant in the room? Syst. Biol. 63: 518–533.

Albert, J. S., D. R. Schoolmaster, Jr., V. Tagliacollo & S. M. Duke-Sylvester. 20017. Syst. Biol. 66: 167–182.

Allen, B. 2002. Moss Flora of Central America. Part 2. Encalyptaceae–Orthotrichaceae. St. Louis: Missouri Botanical Garden Press.

Alonso, M., J. A. Jiménez, S. Nylinder, L. Hedenäs & M. J. Cano. 2016. Disentangling generic limits in *Chionoloma, Oxystegus, Pachyneuropsis* and *Pseudosymblepharis* (Bryophyta: Pottiaceae): An inquiry into their phylogenetic relationships. Taxon 65: 3–18.

Anderson, L. E. & R. H. Zander. 1986. False anisospory in the moss *Leptodontium viticulosoides* (P.-Beauv.) Wijk & Marg. Am. J. Bot. 73(5): 603 (Abstr.).

Arbuckle, K. & A. Minter. 2014. Windex: Analyzing convergent evolution using the Wheatsheaf Index in R. Evol. Bioinform. 2015:11, 11–14 doi: 10.4137/EBo.s20968.

Arts, T. & P. Sollman. 1998. A contribution to the moss flora of Ecuador. Trop. Bryol. 14: 43–52.

Ashlock, P. D. 1974. The uses of cladistics. Ann. Rev. Ecol. Syst. 5: 81–99.

Aubert, D. 2015. A formal analysis of phylogenetic terminology: towards a reconsideration of the current paradigm in systematics. Phytoneuron 2015-66: 1–54.

Aubert, D. 2017. A simple parsimony-based approach to assess ancestor-descendant relationships. Ukrainian Bot. J. 74: 103–121.

Aziz, M. N. & J. N. Vohra. 2008. Pottiaceae (Musci) of India. Bishen Singh Mahendra Pal Singh, Dehra Dun, India.

Bachelot, B. & R. K. Kobe. 2013. Rare species advantage? Richness of damage types due to natural enemies increases with species abundance in a wet tropical forest. J. Ecol. 101: 846–856.

Barrow, J. D. 1994. The Origin of the Universe. Basic Books, Harper Collins Publishers, New York.

Barthlott, H., H.-C. Spatz, T. Speck & C. Neinhuis. 2004. Two-dimensional tension tests in plant biomechanics–sweet cherry fruit skin as a model system. Plant Biol. 6: 432–439.

Benton, M. J. & P. N. Pearson. 2001. Speciation in the fossil record. Trends in Ecology and Evolution 16: 405–411

Beres, K. A., R. L. Wallace & H. H. Segers. 2005. Rotifers and Hubbel's unified neutral theory of biodiversity and biogeography. Natural Resource Modeling 18: 363–376.

Blackburn, D. G. 2002. Use of phylogenetic analysis to distinguish adaptation from exaption. Behav. and Brain Sci. 25: 507–508.

Bock, W. J. & J. Farrand, Jr. 1980. The number of species and genera of recent birds: A contribution to comparative systematics. Amer. Mus. Novitates 2703: 1–29.

Boesch, C. & H. Boesch-Achermann. 2000. The Chimpanzees of the Taï Forest: Behavioural Ecology and Evolution. Oxford University Press, Oxford.

Bonawitz, E., S. Denison, A. Gopnik & T. L. Griffiths. 2014. Win-Stay, Lose-Sample: A simple sequential algorithm for approximating Bayesian inference. Cognitive Psych. 74: 35–65.

Boothby, T. C., J. R. Tenlen, F. W. Smith, J. R. Wang, K. A. Patanella, E. Osborne Nishimura, S. C. Tintori, Qing Li, C. D. Jones, M. Yandell, D. N. Messina, J. Glasscock B. Goldstein. 2015. Evidence for extensive horizontal gene transfer from the draft genome of a tardigrade. Proc. Natl. Acad. Sci. 112: 15976–15981.

Born, S., J.-P. Frahm & T. Pócs. 1993. Taxonomic results of the BRYOTROP Expediction to Zaire and Rwanda. 26. A new checklist of the mosses of Central Africa. Trop. Bryol. 8: 223–273.

Brooks, D. R. 1981. Classifications as languages of empirical comparative biology. In: V. A. Funk & D. R. Brooks (eds.). Advances in Cladistics. Proceedings of the First Meeting of the Willi Hennig Society. New York Botanical Garden, Bronx. Pp. 61–70.

Brooks, D. R. 2001. Evolution in the Information Age: Rediscovering the nature of the organism. Semiosis, Evol., Energy, Develop. 1(1): 1–26.

Brooks, D. R., R.T. O'Grady & E.O. Wiley. 1986. A measure of the information content of phylogenetic trees and its use as an optimality criterion. Syst. Zool. 35: 571–581.

Brooks, D. R. & E. O. Wiley. 1988. Evolution as Entropy: Towards a Unified Theory of Biology. Second Edition. University of Chicago Press, Chicago.

Brower, A. V. Z. 2015. Paraphylophily. Cladistics 31: 575–578. (Preprint 2014.)

Cano, M. J. & M. T. Gallego. 2008. Proposal to conserve the name *Leptodontium proliferum* against *Tortula umbrosa* (Pottiaceae, Bryophyta). Taxon 57: 643–644.

Cano, M. J., J. F. Jiménez, M. T. Gallego, J. A. Jiménez, & J. Guerra. 2009. Phylogenetic relationships in the genus *Hennediella* (Pottiaceae, Bryophyta) inferred from nrITS sequence data. Pl. Syst. Evol. 281: 209–216.

Capinha, C., F. Essl, H. Sechens, D. Moser & H. M. Pereira. 2015. The dispersal of alien species redefines biogeography in the Anthropocene. Science 348: 1248–1251.

Cárdenas S., A. 1987 [1989]. Nuevos registros para la flora de musgos de México y del valle de México. Anales Inst. Biol. Univ. Nat. Auton. Méx. 58, Ser. Bot. 1: 93–96.

Castoe, T. A., A. P. J. de Koning, A. P. Jason, H. M. Kim et al. 2009. Evidence for an ancient adaptive episode of convergent molecular evolution. Proc. Soc. Nat. Acad. Sci. U.S.A. 106: 8986–8991.

Catcheside, D. G. 1980. Mosses of South Australia. D. J. Woolman, Government Printer, Government of South Australia.

Cermansky, R. 2017. The biodiversity revolution. Nature 546: 22–24.

Chapin, F. S., III, E. S. Zavaleta, V. T. Eviners et al. 2000. Consequences of changing biodiversity. Nature 405. 234–242.

Chen, P.-C. 1941. Studien über die ostasiatischen Arten der Pottiaceae, I-II. Hedwigia 80: 1–76; 141–322.

Churchill, S. P., D. Griffin III & J. Muñoz. 2000. A Checklist of the Mosses of the Tropical Andean Countries. Ruiza 17. Madrid.

Churchill, S. P. & E. L. Linares C. 1995. Prodromus Bryologiae Novo-Granatensis. Museo de Historia Natural Biblioteca "Jose Jeronimo Triana" Vol. 12. Universidad Nacional de Colombia, Santafe de Bogota.

Clark, R. W. 1968. JBS: The Life and Work of J. B. S. Haldane. Coward-McCann, Inc., New York.

Clayton, W. D. 1972. Some aspects of the genus concept. Kew Bull. 27: 281–287.

Cleator, P. E. 1959. Lost Languages. Mentor, New American Library, New York.

Cliff, D. 1994. From Animals to Animats: Proceedings of the Third International Conference on Simulation of Adaptive Behavior. MIT Press, Cambridge, Massachusetts.

Cohen, J. 1994. The world is round (p < .05). Amer. Psychol. 49: 997–1003.

Colotti, M. T. & M. M. Schiavone. 2011. New national and regional bryophyte records, 26. 9. J. Bryol. 33: 68.

Cook, T. D. & D. T. Campbell. 1979. Quasi-experimentation: Design and analysis issues for field settings. Rand McNally College Publishing Co., Chicago.

Cooper, W. E., Jr. & D. T. Blumstein. Escaping from Predators: An Integrative View of Escape Decisions. Cambridge University Press, Cambridge.

Coveney, P. & R. Highfield. 1990. The Arrow of Time. Fawcett Columbine, New York.

Cox, C. J., B. Goffinet, N. J. Wickett, S. B. Boles & A. J. Shaw. 2010. Moss diversity: A molecular phylogenetic analysis of genera. Phytotaxa 9: 175–195.

Crawford, D. J. 2010. Progenitor-derivative species pairs and plant speciation. Taxon 59: 1413–1423.

Christin, P. A., N. Salamin, V. Savolainen, M. R. Duvall & G. Besnard. 2007. C4 photosynthesis evolved in grasses via parallel adaptive genetic changes. Current Biol. 17: 1241–1247.

Cropp, R. & A. Gabric. 2002. Ecosystem adaptation: Do ecosystems maximize resilience? Ecology 83: 2019–2026.

Curio, E. 1976. Ethology of Predation. Springer-Verlag, Berlin.

Darwin, C. 1859. The Origin of Species by Means of Natural Selection, Or the Preservation of Favoured Races in the Struggle for Life. 1963 Ed. Washington Square Press, New York.

Dayrat, B. 2005. Ancestor-descendant relationships and the reconstruction of the Tree of Life. Paleobiol. 31:347–353.

Dayrat, B. 2013. The roots of phylogeny: how did Haeckel build his trees? Syst. Biol. 52: 515–527.

Delgadillo, C., M., B. Bello & A. Cárdenas S. 1995. LATMOSS: A Catalogue of Neotropical Mosses. Missouri Botanical Garden, St. Louis.

Dixon, H. N. 1924. The Student's Handbook of British Mosses. Third Edition. V. V. Sumfield, London.

Driver, P. J. 1982. *Leptodontium gemmascens* in terrestrial habitats in south-east England. J. Bryol. 12: 113

Düring, H. J. 1979. Life strategies of bryophytes: A preliminary review. Lindbergia 5: 2–18.

Eaton, R. C., R. K. K. Lee & M. B. Foreman. 2001. The Mauthner cell and other identified neurons of the brainstem escape network of fish. Prog. in Neurobiol. 63: 467–485.

Eddy, A. 1990. A Handbook of Malesian Mosses. Natural History Museum Publications, London.

Eldredge, N. 1985. Time Frames: The Rethinking of Darwinian Evolution and the Theory of Punctuated Equilibria. Simon and Schuster, New York.

Eldredge, N. 1989. Macroevolutionary Dynamics. McGraw-Hill, New York.

Enroth, J. Book review: Maine Mosses. Bryological Times 141(September): 21.

Estes, R., G. K. Pregill. 1988. Phylogenetic Relationships of the Lizard Families. Stanford University Press, Stanford, California.

Farris, J. S. 1969. A successive approximations approach to character weighting. Syst. Zool. 18: 374–385.

Ferris, T. 1997. The Whole Shebang: A State of the Universe(s) Report. Simon & Schuster, New York.

Feynman, R. 1965. Nobel Lecture. Nobel Foundation. http://www.nobelprize.org/nobel_prizes/physics/laureates/1965/feynman-lecture.html

Flegr, J. 2013. Why *Drosophila* is not *Drosophila* any more, why it will be worse and what can be done about it? Zootaxa 3741: 295–300.

Flot, J.F. 2015. Species delimitation's coming of age. Syst. Biol. 64: 897–899.

Frahm, J.-P. & R. Schumacker, 1986. Type revision of European mosses. 1. *Leptodontium.* Lindbergia 12: 76–82.

Futuyma, D. J. 2010. Evolutionary constraint and ecological consequences. Evolution 64:1865–1884.

Gangulee, H. C. 1972. Mosses of Eastern India and Adjacent Regions. Fascicle 3. Books & Allied Limited, Calcutta, India.

García-Donato, G. & M.-H. Chen. 2005. Calibrating Bayes factor under prior predictive distributions. Statistica Sinica 15: 359–380.

Gates, D. 1965. Energy Exchange in the Biosphere. Harper & Brothers, New York.

Gigerenzer, G., Z. Swijtink, T. Porter, L. Daston, J. Beatty & L. Küger. 1989. The Empire of Chance: How Probability Changed Science and Everyday Life. Cambridge University Press, Cambridge

Gilmore, R. 1981. Catastrophe Theory for Scientists and Engineers. John Wiley & Sons, New York.

Gishtick, A. 2006. Baraminology. Reports of the National Center for Science Education 26(4): 17–21. http://ncse.com/rncse/26/4/baraminology (accessed September 2, 2010).

Gleick, J. 2011. The Information, A History, A Theory, A Flood. Pantheon Books, New York.

Goffinet, B., W. R Buck & A. J. Shaw. 2008. Morphology and Classification of the Bryophyta. In: B. Goffinet & A. J. Shaw (eds.). Bryophyte Biology. 2nd edition Cambridge University Press. Pp. 55–138,

Gould, S. J. 1970. Dollo on Dollo's Law: irreversibility and the status of evolutionary laws. J. Hist. Biol. 3: 189–212.

Gould, S. J. 2002. The Structure of Evolutionary Theory. Belknap Press of Harvard University Press, Cambridge.

Gould, S. J. & R. C. Lewontin. 1979. The spandrels of San Marco and the Panglossian paradigm: A critique of the adaptationist programme. Proc. Roy. Soc., London B 205: 581–598. Grant, V. 1963. The Origin of Adaptations. Columbia University, New York.

Grant, V. 1918. Plant Speciation. 2nd Ed. Columbia University, New York.

Grundmann, M., H. Schneider, S. J Russell & J. C. Vogel. 2006. Organisms, Diversity and Evol. 6: 33–45.

Gunn, L. J., F. Chapeau-Blondeau, M. D. McDonnell, B. R. Davis, A. Allison & D. Abott. 2016. Too good to be true: when overwhelming evidence fails to convince. Proc. Roy. Soc. A 472(2187): 20150748. doi: 10.1098/rspa.2015.0748.

Hadly, E. A., R. Z. Spaeth & Cheng Li. 2009. Niche conservatism above the species level. Proc. Nat. Acad. Sci. USA 106(Suppl. 2): 19707–19714.

Harrison, L. G. 1990. Kinetic theory of living pattern and form and its possible relationship to evolution. In B. H. Weber, D. J. Depew & J. D. Smith. Entropy, Information, and Evolution. MIT Press, Cambridge, Massachusetts. Pp. 53–74.

Hedderson,T. A., D. J. Murray, C. J. Cox & T. L. Tracey. 2004. Phylogenetic relationships of haplolepideous mosses (Dicranidae) inferred from rps4 gene sequences. Syst. Bot. 29: 29–41.

Hedderson, T. A. & R. H. Zander. 2007. *Triquetrella mxinwana*, a new moss species from South Africa, with a phylogenetic and biogeographic hypothesis for the genus. J. Bryol. 29: 151–160.

Hébant, C. 1977. The Conducting Tissues of Bryophytes. J. Cramer, Vaduz.

Herzog, T. 1916. Die Bryophyten meiner zweiter Reise durch Bolivia. Biblioth. Bot. 87: 1–347, pl. 1–8, map.

Hilpert, F. 1933. Studien zur Systematik der Trichostomaceen. Bot. Centralb. Beih. 50(2): 585–706.

Hollowell, V. C., ed. 2001. Managing Human-Dominated Ecosystems. Missouri Botanical Garden, St. Louis.

Hudson, R. R. & J. A. Coyne. 2002. Mathematical consequences of the genealogical species concept. Evol. 56: 1557–1565.

Huelsenbeck, J. P. & F. Ronquist. 2001. MRBAYES: Bayesian inference of phylogeny. Bioinformatics 17: 754–755.

Humphreys, A. M. & T. G. Barraclough. 2014. The evolutionary reality of higher taxa in mammals. Proc. Soc. Roy. Soc. B 281: 20132750. http://dx.doi.org/10.1098/rspb.2013.2750.

Huneman, P. 2017. Robustness: The explanatory picture. In: S. Caianiella, M. Bertolaso & E. Serelli. Biological Robustness. Dordrecht.

Hunt, K. 1995. Horse Evolution. TalkOrigins Archive. http://www.talkorigins.org/faqs/horses/horse_evol.html Viewed 1 April 2016

Inoue, Y. & H. Tsubota. 2014. On the systematic position of the genus *Timmiella* (Dicranidae, Bryopsida) and its allied genera, with the description of a new family Timmiellaceae. Phytotaxa 181: 151–162. DOI: 10.11646/phytotaxa.181.3.3

Jeffreys, H. 1961. The Theory of Probability. Third edition. Oxford Univ. Press, Oxford.

Johnson, H. B. & H. S. Mayeux. 1992. Viewpoint: A view on species additions and deletions and the balance of nature. J. Range Manage. 45: 322–333.

Jones, S. B., Jr. & A. E. Luchsinger. 1979. Plants Systematics. McGraw-Hill Book Company, New York.

Jorgensen, S. E. 2000a. A general outline of thermodynamic approaches to ecosystem theory. In: S. E. Jorgensen and F. Muller, editors. Handbook of ecosystem theories and management. CRC Press, Boca Raton, Florida, USA. Pp. 113–133.

Jorgensen, S. E., ed. 2000b. Thermodynamics and ecological modelling. CRC Press, Boca Raton, Florida, USA.

Kass, R. E. & A. E. Raftery. 1995. Bayes factors. J. Amer. Statist. Assoc. 90: 773–795.

Kipling, R. 1966. Just So Stories. Illustrated by the Author. Airmont Edition, New York.

Köckinger, H., O. Werner & R. M. Ros. 2010. A new taxonomic approach to the genus *Oxystegus* (Pottiaceae, Bryophyta) in Europe based on molecular data. Nova Hedwigia, Beihett 138: 31–49.

Krug, A. Z., D. Jablonski & J. W. Valentine. 2008. Species-genus ratios reflect a global history of diversification and range expansion in marine bivalves. Proc. Roy. Soc. B 275: 1117–1123.

Kučera, J., J. Košnar & O. Werner. 2013. Partial generic revision of *Barbula* (Musci: Pottiaceae): re-establishment of *Hydrogonium* and *Streblotrichum*, and the new genus *Gymnobarbula*. Taxon 62: 21–39.

LaFarge, C., A. J. Shaw & D. H. Vitt. 2002. The circumscription of the Dicranaceae (Bryopsida) based on the chloroplast regions trnL-trnF and rps4. Syst. Bot. 27: 435–452.

Leaché, A. D., M. K. Fujita, V. N. Minin & R. R. Boukaert. 2014. Species delimitation using genome-wide SNP data. Syst. Biol. 63: 534–542.

Lenton, T. M. & J. E. Lovelock. 2000. Daisyworld is darwinian: Constraints on adaptation are important for planetary self-regulation. J. Theor. Biol. 206: 109–114.

Levinton, J. 1988. Genetics, Paleontology, and Macroevolution. Cambridge University Press, Cambridge.

Levitin, D. J. 2016. Weaponized Lies: How to Think Clearly in the Post-Truth Era. Dutton, New York.

Li, X-J, S. He & Z. Iwatsuki. 2001. Pottiaceae. In: X-J Li & M. R. Crosby. Moss Flora of China. English Version. Volume 2. Fissidentaceae–Ptychomitriaceae. Science Press, Beijing, and Missouri Botanical Garden, St. Louis. Pp. 114–249.

Mace, G. M. 2014. Whose conservation? Changes in the perception and goals of nature conservation require a solid scientific basis. Science 345: 1558–1560.

Magill, R. 1981. Flora of Southern Africa. Bryophyta. Part 1. Mosses. Fascicle 1. Sphagnaceae-Grimmiaceae. Botanical Research Institute, Pretoria.

Margalef, R. 2000. Exosomatic structures and captive energies relevant in succession and evolution. In: S. E. Jorgensen, ed. Thermodynamics and ecological modelling. CRC Press, Boca Raton, Florida, USA. Chapter 2: 1–15.

Matsuno, K. 1989. Protobiology: Physical Basis of Biology. CRC Press, Boca Raton, Florida/

Matsuno, K. 1995. Consumer power as the major evolutionary force. J. Theor. Biol. 173: 137–145.

Matsuno, K. 2000. Material contextualization in time. In: E. Taborsky, ed., Semiotics, Evolution, Energy. Aachen, Shaker Verlag, Germany. Pp. 219–230.

May, R. M. 1988. How many species ae there on Earth? Science, New Series 241: 1441–1449.

Mayr, E. 1981. Biological classification: Toward a synthesis of opposing methodologies. Science 214: 510–516.

Mogensen, G. S. 1978. False anisospory in *Macromitrium incurvum, Rhizomnium magnifolium,* and *Fissidens cristatus* (Bryophyta). Lindbergia 4: 191–195.

Moreno-Letelier, A. & T. G. Barraclough. 2015. Mosaic genetic differentiation along environmental and geographic gradients indicate divergent selection in a while pine species complex. Evol. Ecol. 29: 733–748.

Moreno-Mateos, D. E. B. Barbier, P. C. Jones et al. 2017. Anthropogenic ecosystem disturbance and the recovery debt. Nature Communications 14163: 1–9 DOI: 10.1038/ncomms14163.

Myers, N. & A. H. Knoll. 2015. The biotic crisis and the future of evolution. Proc. Soc. Nat. Acad. Sci. U.S.A. 98: 5389–5392.

Niklas, K. J. & H.-C. Spatz. 2012. Plant Physics. University of Chicago Press, Chicago.

Norris, D. H. & T. Koponen. 1989. Bryophyte flora of the Huon Peninsula, Papua New Guinea. XXVII. Pottiaceae (Musci). Acta Bot. Fennica 137: 81–138.

Odifreddi, P. 2000. The Mathematical Century, the Greatest Problems of the Last 100 Years. Princeton University Press, Princeton, N.J.

O'Shea, B. 1995. Checklist of the mosses of sub-Saharan Africa. Trop. Bryol. 10: 91–198.

Paarlberg, D. & P. Paarlberg. 2000. The Agricultural Revolution of the 20th Century. Iowa State University Press, Ames, Iowa.

Parker, J., G. Tsagkogeorga, J. A. Cotton, Y. Liu, et al. 2013. Genome-wide signatures of convergent evolution in echolocating mammals. Nature 602: 228–231.

Paterson, H. 2005. The competitive Darwin. Paleobiol. 31L 56–76.

Patino, J., I. Bisang, L. Hedenäs, G. Dirkse, A. H. Bjarnason, C. Ah-Peng & A. Vanderpoorten. 2013. Baker's law and the island syndromes in bryophytes. J. Ecol. 101: 1245–1255.

Peel, G. T., M. B. Arańjo, J. D. Bell, et al. 2017. Biodiversity redistribution under climate change: Impacts on ecosystems and human well-being. Science 355 (6332) [doi: 10.1126/science.aai9214] http://science.sciencemag.org/content/355/6332/eaai9214

Peng, G.-S., S.-Y. Tan & J. Wu. 2016. Trade-offs between robustness and small-world effect in complex networks. Nature, Sci. Rep. 6: 37317; doi: 10.1038/srep37317.

Pierce, J. R. 1980. An Introduction to Information Theory: Symbols, Signals and Noise. Second, revised edition. Dover Publications, New York.

Pigliucci, M. & J. Kaplan. 2000. The fall and rise of Dr. Pangloss: adaptationism and the *Spandrels* paper 20 years later. TREE 15: 66–70.

Pigliucci, M. & G. B. Müller, eds. 2010. Evolution: The Extended Synthesis. MIT Press, Cambridge, Massachusetts.

Poole, M. 1990. A Guide to Science and Belief. Lion Publishing, Oxford.

Popper, K. 1992. The Logic of Scientific Discovery. Routledge, London.

Porley, R. D. 2008. Threatened bryophytes: *Leptodontium gemmascens* (thatch moss). Field Bryology 96: 14–24.

Porley, R.D. & S. Edwards. 2010. *Leptodontium proliferum* Herzog (Bryopsida: Pottiaceae), new to Europe. J. Bryol. 32: 46–50.

Projecto-Garcia, J., C. Natarajan, H. Moriyama, et al. 2013. Repeated elevational transitions in hemoglobin function during the evolution of Andean hummingbirds. Proc. Nat. Acad. Sci. U.S.A. 110(51): 20669–20674.

Reich, P. B., J. J. Wright, J. Cavender-Bares, J. M. Craine, S. J. Oleksyn, M. Westoby & M. B. Walters. 2013. The evolution of plant functional variation: traits, spectra, and strategies. Int. J. Plant Sci. 164(3 Suppl.): S143–S164.

Renner, S. S. 2016. A return to Linnaeus's focus on diagnosis, not description: The use of DNA characters in the formal naming of species. Syst. Biol. 65: 1085–1095.

Rudman, S. M. & D. Schluter. 2016. Ecological impacts of reverse speciation in threespine stickleback. Current Biology 26: 490–495.

Sainsbury, G. O. K. 1955. A Handbook of the New Zealand Mosses. Roy. Soc. New Zealand, Wellington, New Zealand.

Saito, K. 1975. A monograph of Japanese Pottiaceae (Musci). J. Hattori Bot. Lab. 39: 373–537.

Saitou, N. & M. Nei. 1987. The neighbor-joining method: a new method for reconstructing phylogenetic trees. Mol. Biol. Evol. 4: 406–425.

Saunders, P. T. 1980. An Introduction to Catastrophe Theory. Cambridge Univ. Press, Cambridge.

Schmidt-Lebuhn, A. N. 2014. Consolidating classical taxonomy and phylogenetics by discarding the latter. Australasian Syst. Bot. Soc. Newsl. 158: 28–30.

Schönbrodt, F. 2016. Sequential Bayes Factors: A Flexible and Efficient Way of Optional Stopping. Center for Open Science, Ludwig-Maximilians-Universität München. Viewed August 1, 2016. http://www.uni-bielefeld.de/lili/personen/jruiter/downloads/statisticsworkshop/Schoenbrodt_Sequential_BF.pdf

Schönbrodt, F. D., E. J. Wagenmakers, M. Zehetleitner & M. Perugini. 2015. Sequential hypothesis testing with Bayes factors: efficiently testing mean differences. Psychol. Methods. http://dx.doi.org/10.2139/ssrn.2604513.

Schulter, D. 2000. The Ecology of Adaptive Radiation. Oxford University Press, New York.

Schwartzman, D. 2002. Life, Temperature, and the Earth: The Self-Organizing Biosphere. Columbia University Press, Chicago.

Schwenk, K. 1995. A utilitarian approach to evolutionary constraint. Zoology 98: 251–262.

Sharp, A. J, H. Crum & P. M. Eckel (eds.). 1994. Moss Flora of Mexico. Mem. New York Bot. Gard. 69, vols. 1–2.

Sillar, K. T., L. D. Picton & W. J. Heitler. 2016. The Neuroethology of Predation and Escape. John Wiley & Sons, New York.

Singh, G. 2010. Plant Systematics: An Integrated Approach. Third Edition. CRC Press, Boca Raton, Florida.

Sloover, J. L. de. 1987. Note de bryology africaine. XIV. *Leptodontium*. Bull. Jard. Bot. Nat. Belgique 57: 425–451.

Smith, D. W. & R. McIntyre 1982 Husserlian Phenomenology and Phenomenological Method. D. Reidel Publishing Co., Dordrecht and Boston.

Stanley, S. M. 1981. The New Evolutionary Timetable: Fossils, Genes, and the Origin of Species. Basic Books, New York.

Steere, W. C. 1986. *Trachyodontium*, a new genus of the Pottiaceae (Musci) from Ecuador. Bryologist 89: 17–19.

Stewart, I. 2011. Mathematics of Life. Basic Books, New York

Strogatz, S. 2003. Sync: The Emerging Science of Spontaneous Order. Theia, New York.

Stuart, Y. E., T. S. Campbell, P. A. Hohenlohe, R. G. Reynolds, L. J. Revel & J. B. Losos. 2014. Rapid evolution of a native species following invasion by a congener. Evol. Biol. 346: 463–466.

Swamy, V. & J. W. Terborgh. 2010. Distance-responsive natural enemies strongly influence seedling establishment patterns of multiple species in an Amazonian rain forest. J. Ecol. 98: 1096–1107.

Swofford, D. L. 2003. PAUP*. Phylogenetic Analysis Using Parsimony (*and Other Methods). Version 4. Sinauer Associates, Sunderland, Massachusetts.

Tal, A & B. Wansink. 2016. Blinded with science: Trivial graphs and formulas increase ad persuasiveness and belief in product efficacy. Public Understanding of Science 25(1): 117–125.

Tsai Cheng-Hsiu & R. E. Fordyce. 2015. Ancestor-descendant relationships in evolution: origin of the extant pygmy right whale, *Caperea marginata*. Biol Lett. 11:20140875. http://dx.doi.org/10.1098/rsbl.2014.0875

Tufford, L. & P. Newman. 2010. Bracketing in qualitative research. Qual. Social Work 11:80–96.

Terborgh, J. 2012. Enemies maintain hyperdiverse tropical forests. Amer. Nat. 179: 303–314.

Tverksy, A. & D. Kahneman. 1974. Judgment under uncertainty: heuristics and biases. Science 185: 1124–1131.

van Deemter, K. 2012. Not Exactly: in Praise of Vagueness. Oxford University Press, Oxford. [Paperback.]

Vitt, D. H., B. Crandall-Stotler & A. Wood. 2014. Bryophytes: Survival in a dry world through tolerance and avoidance, pp. 267–295. In: Rajakaruna, N., R. S. Boyd & T. B. Harris (eds.). Plant Ecology and Evolution in Harsh Environments. Nova Science, New York.

Walker, B. 2001. Suystainable habitation in the savannas. In: Hollowell, V. C., ed. 2001. Managing Human-Dominated Ecosystems. Missouri Botanical Garden, St. Louis. Pp. 34–46.

Wallace-Wells, D. 2017. The Uninhabitable Earth. New York Magazine July 10, 2017. URL: http://nymag.com/daily/intelligencer/2017/07/climate-change-earth-too-hot-for-humans.html

Walther, K. 1983. Bryophytina. Laubmoose. In J. Gerloff & J. Poelt, A. Engler's Syllabus der Pflanzenfamilien 5(2). Gebriider Borntraeger, Berlin.

Watson, W. 1913. Xerophytic adaptations of bryophytes in relation to habitat. New Phytol. 13: 149–169, 181–190.

Weber, B. H., D. J. Depew & J. D. Smith, eds. 1990. Entropy, Information, and Evolution: New Perspectives on Physical and Biological Evolution. MIT Press, Cambridge, Massachusetts.

Werner, J. & Sauer, E. 1994: Oekologie und Soziologie von *Leptodontium gemmascens* (Mitt. ex Hunt) Braithw. (Musci) im Luxemburger Oesling und im Saarland. Dumortiera 55–57: 2–9.

Werner, O., J. A. Jiménez, R. M. Ros, M. J. Cano & J. Guerra. 2005. Preliminary investigation of the systematics of *Didymodon* (Pottiaceae, Musci) based on nrITS sequence data. Syst. Bot. 30: 461–470.

Werner, O., R. M. Ros, M. J. Cano & J. Guerra. 2004. Molecular phylogeny of Pottiaceae (Musci) based on chloroplast *rps4* sequence data. Pl. Syst. Evol. 243: 147–164.

Wiens, J. A. 1976. Population responses to patchy environments. Ann. Rev. Ecol. Syst. 7: 81–120.

Wiley, E. O. 1990. Entropy and Evolution. In Weber, B. H., D. J. Depew & J. D. Smith, eds. 1990. Entropy, Information, and Evolution. Cambridge. Pp. 173–188.

Wright, S. J. 2002. Plant diversity in tropical forests: A review of mechanisms of species coexistence. Oecologia 130: 1–14.

Zander, R. H. 1972. Revision of the genus *Leptodontium* (Musci) in the New World. Bryologist 75: 213–230.

Zander, R. H. 1980. Spread of *Leptodontium viticulosoides* (Bryopsida) after Balsam Woolly Aphid infestation of Fraser Fir. Bull. Torrey Bot. Club 107: 7–8.

Zander, R. H. 1993. Genera of the Pottiaceae: mosses of harsh environments. Bull. Buffalo Soc. Nat. Sci. 32: 1–378.

Zander, R. H. 1994. *Leptodontium.* In: A. J. Sharp, H. Crum & P. M. Eckel (eds.). Moss Flora of Mexico. 2 vols. New York Botanical Garden, Bronx, New York.

Zander, R. H. 1998. Phylogenetic reconstruction, a critique. Taxon 47: 681–693.

Zander, R. H. 2003. Reliable phylogenetic resolution of morphological data can be better than that of molecular data. Taxon 52: 109–112.

Zander, R. H. 2004. Minimal values for reliability of bootstrap and jackknife proportions, Decay Index, and Bayesian posterior probability. Phyloinformatics 2: 1–13.

Zander, R. H. 2006. The Pottiaceae s.str. as an evolutionary Lazarus taxon. J. Hattori Bot. Lab. 100: 581–602.

Zander, R. H. 2008. Evolutionary inferences from non-monophyly of traditional taxa on molecular trees. Taxon 57: 1182–1188.

Zander, R. H. 2009. Evolutionary analysis of five bryophyte families using virtual fossils. Anales Jard. Bot. Madrid 66: 263–277.

Zander, R. H. 2010 (2011). Structuralism in phylogenetic systematics. Biol. Theory 5: 383–394.

Zander, R. H. 2013. Framework for Post-Phylogenetic Systematics. Zetetic Publications, CreateSpace Independent Publishing, Amazon, St. Louis.

Zander, R. H. 2014a. Classical determination of monophyly, exemplified with *Didymodon* s. lat. (Bryophyta). Part 1 of 3, synopsis and simplified concepts. Phytoneuron 2014-78: 1–7.

Zander, R. H. 2014b. Classical determination of monophyly, exemplified with *Didymodon* s. lat. (Bryophyta). Part 2 of 3, concepts. Phytoneuron 2014-79: 1–23.

Zander, R. H. 2014c. Classical determination of monophyly, exemplified with *Didymodon* s. lat. (Bryophyta). Part 3 of 3, analysis. Phytoneuron 2014-80: 1–19.

Zander, R. H. 2014d. Response to a particularly nasty review in the journal *Cladistics*. Phytoneuron 2014-110: 1–4.

Zander, R. H. 2014e. Support measures for caulistic macroevolutionary transformations in evolutionary trees. Ann. Missouri Bot. Gard. 100: 100–107.

Zander, R. H. 2016a. Macrosystematics of *Didymodon* sensu lato (Pottiaceae, Bryophyta) using an analytic key and information theory. Ukrainian Botanical Journal 73: 319–333.

Zander, R. H. 2016b. A new progressive polychrome protocol for staining bryophytes. Phytoneuron 2016-2: 1–12.

Zander, R. H. & E. Hegewald. 1976. *Leptodontiella,* gen. nov. and *Leptodontium* from Peru. Bryologist 79: 16–21.

Zander, R. H. & T. A. Hedderson. 2017. *Leptodontium stellatifolium* in La Réunion Island, a major range disjunction. Boletín de la Sociedad Argentina de Botánica 52(2): 393–398.

Chapter 18
GLOSSARY

Abduction – Inferring probable facts from other facts, e.g. Sherlock Holmes using clues to identify a culprit.

Aleatory – Probabilistically determined, as with a throw of the dice.

Alpha taxonomy – Part of systematics concerned with collection, identification, recognition of new taxa (particularly species), preparation of regional checklists and identification manuals, and mono-graphs or revisions; largely confined to morphological, biogeographical, and ecological analytic evaluations.

Anagenesis – Gradual evolutionary change in a line-age without splitting, or as associated with pseudoextinction.

Anagenetic – Pertaining to the gradual accumulation of changes in a lineage, as opposed to punctuational change.

Anastasis – Molecular parallelism (two morphologically identical descendant populations from one ancestor) or polyphyly (two morphologically identical descendant populations from two different ancestral taxa), the generation of two taxa of the same name from an ancestral taxon of a different name at the same taxonomic level.

Ancestor – In the present context used for a taxon, as in "ancestral taxon," not an individual.

Ancestral taxon – One or more sections of a cladogram or one section of a caulogram consisting of inferred deep ancestors of extant exemplars that are diagnosable as one particular taxon.

Apomorphic. In cladistics, the relatively derived state of a sequence of homologous characters.

Apophyletic – An apophyletic branch is that branch that comes out of a paraphyletic relationship on a cladogram, being bracketed by two branches of a single taxon of the same or lower rank; evolutionarily a descendant taxon but in a cladistic context.

Autapomorphy – A unique trait uninformative of sister-group relationships but which may be informative of unique evolutionary status or direction; a distinctive trait of no use in cladistic analysis but a major element of macroevolutionary transformation.

Bayes factor (BF) – A ratio comparing two probabilities generated from different data as to which of two alternative scenarios is best supported. With 0.50 priors, the BF is a simple fraction of their Bayesian posterior probabilities (BPPs).

Bayes' Formula –A simple statistical method of updating a previously accepted chance of something being true in light of additional information to provide a new ("posterior") probability of it being true. Sequential Bayes analysis simply uses a number of sets of data, one after the other, to continually update the degree of truth about something (such as a process in nature).

Bayesian analysis – A statistical method of estimating phylogenetic relationships in terms of nested diagrams, using only data that is precise and amenable to such nesting; proper analysis is the Bayesian Solution, which includes the effect of all data, precise or imprecise, to give a probabilistic answer in view of risk if wrong.

Bayesian posterior probability (BPP) –The chance of a hypothesis being correct, given also a prior probability and additional data.

BPP – Bayesian Posterior Probability (also BP).

Budding evolution – Speciation from a static ancestor; usually associated with peripatric speciation (margins of a range) although rapid sympatric isolation should result in much the same thing.

Cacophyly – The crippled modeling of evolution that results when cladists use only traits reflecting shared descent and paraphyly is eliminated.

Caulistic – Pertaining to the axis and branches of an evolutionary tree, showing serial macro-evolutionary transformations, e.g. a "Besseyan cactus" or caulogram.

Caulogram – a tree modeling direct descent; see commagram.

CI – Confidence interval (or level), a frequentist equivalent of the BPP; also credible interval in Bayesian analysis.

Clade –A group consisting of an ancestor and all its descendants, and is thus indicative of phylogenetic monophyly.

Cladistic – Pertaining to the optimal dichotomous branching sequence on a cladogram.

Cladogram – A cluster analysis based on synapomorphies; an expanded diagram representing a set of nested parentheses; a calculated representation of hierarchical evolutionary relationships. Commonly assumed to be equivalent to a monophylogram.

Coarse priors – Also known as "stepped priors." For estimation of reliability of the evolutionary relationships of classical taxonomy and molecular cladistics, intuitive priors are set at 0.99 (almost certain); 0.95 (just acceptable); 0.75 (some support); 0.60 (hint of support); 0.50 (equivocal).

Commagram – A Besseyan "cactus"; a tree consisting of fat tadpoles showing directions of macro-evolution of taxa; a caulistic monophylogram.

Conciliate – To reconcile, to make compatible, to come half way.

Congruent – Two cladograms that agree; see incongruent.

Conservative traits – Traits that are refractory to adaptation and which commonly occur in different adaptive regimes associated with different taxa that are related by such traits, acting like tracking traits in molecular systematics; morphological traits passing speciation events and commonly in stasis as opposed to molecular tracking traits which continually mutate at some rate.

Consiliate – An induction or generalization that is obtained from two or more different sets of facts, e.g., melding logically classical taxonomy, morphological cladistics, and molecular systematics such that all three infer a single joint macroevolutionary explanatory structure.

Cryptic taxa – Taxa that are difficult to distinguish morphologically but represent different molecular lineages. Hypercryptic molecular taxa are apparently impossible to distinguish morphologically.

Descendancy – The property of being a descendant species or lineage.

Deciban (dB) – The minimally distinguishable unit of credibility that something is true. As these are based on logarithms, they can be added and then translated to a BPP (see Table 1). Three dBs are about 1 binary bit, the minimum information needed to make a decision between two alternatives in information theory.

Deduction – Reasoning from the general to specifics; inferring specifics from theory (or, in mathematics, lemmas from axioms).

Dissilient – Springing open, exploding apart; here referring to a genus inferred as generating many usually highly specialized descendants. The dissilient genus has a single generalized species with a cloud of clearly derived species associated with it. The derived species are more similar to the ancestral species than they are to each other. Some derived species (*stirps* sensu Zander 2013) may be so specialized as to appear to be dead ends in evolution (but with climate change, who can tell?), while others may prove advanced but generalist, capable of generating a number of specialized, derived species of their own and thus found a new dissilient genus.

Evolution – Modification through descent of taxa.

Evolutionary systematists – Systematists who accept or even celebrate phylogenetic paraphyly as a basis for evolutionary classification, and by extension accept macroevolutionary transformations at the taxon level as a basis for classification.

Exemplar – A sample of one; a specimen used to represent a population, a species, a genus, a family, etc., in a molecular cladogram. Many exemplars of one taxon are needed for good coverage, but minimum sample size is difficult to determine. Usually one samples until there is no change in the results, a procedure that has all the problems of induction.

Extended paraphyly – Phylogenetic polyphyly with no evidence of clades generated by differently named ancestors; several contiguous nodes of one deep ancestral taxon may be the correct central feature of a cladogram.

Heterophyly – Either phylogenetic paraphyly or polyphyly, with the same taxon distant on a molecular cladogram and no other evidence of different origin or convergence. Heterophyly implies that intermediate cladogram nodes are of that same taxon. Including both paraphyly and phylogenetic polyphyly, simply two exemplars of the same taxon distant on a cladogram by at least one intervening exemplar of another taxon of the same or higher rank; e.g. ((A1, B) A2) as a terminal group, where A1 and B are sister but A2 is an exemplar of the same taxon as A1. This is the most important evolutionary information from molecular cladograms; also known as non-monophyly. Simple heterophyly is a single clade with two separated exemplars of the same taxon, complex heterophyly is two clades generated by, for instance, two self-nesting ladders from the same ancestral taxon (also called phylogenetic polyphyly).

Heteroplasy – Generation of molecular races that separately generate different morphotaxa that are considered molecularly paraphyletic through "rampant" homoplasy although essentially monophyletic.

The divergence of molecular races that may each generate separately a new species. It is the molecular source of systematic error, the obverse of how homoplasy due to convergence may affect morphological systematics.

Heuristic – A short-cut or rule-of-thumb that provides an approximate answer sufficiently exact for everyday purposes.

Holophyly – Strict phylogenetic monophyly; all members of a clade must derive from one shared ancestor and the clade can have only one name at any rank, a cladistic classification principle.

Homoplasy – Trait similarity in cladogram lineages that lead back to different shared ancestors, in the context of holophyly.

Incongruent – Two cladograms that disagree; see congruent.

Induction – Reasoning from specifics to the general; devising a supported but not certain theory from data.

Macroevolution – Concerns taxa generating taxa, serially. This may be diagramed with a caulogram, or stem-tree of both serial and branching relationships. Descent with modification of taxa, requiring explicit distinction of pseudoextinction or budding evolution; series or successively branching sequences of taxa, impossible to diagram with cladograms having unnamed nodes, therefore not critical to phylogenetics.

Macroevolutionary systematics – Using direct descent to model evolutionary change at the species level and above.

Macrosystematics – Classification at the genus level and above, in the present work the dissilient genus concept is used instead of clades to recognize genera.

Mapped taxon – Nodes on a molecular cladogram between certain separated exemplars representing inferred ancient ancestors of present-day taxa and diagnosable at a particular taxonomic level (the lowest shared by the exemplars) through a kind of taxonomic uniformitarianism; the best information from molecular systematic analysis.

Microevolution – Successive changes of traits mapped on a cladogram but seldom expressly associated with changes from one taxon to another. A subject of population genetics.

Monophyly – A clade with all taxa traceable to one ancestor; monophyly as used by phylogenetists is axiomatically strict, as used by evolutionary systematists, monophyly allows nesting of taxa of the same rank.

Multiple test (or multiple comparisons – A problem in statistics in which if you look around enough you will find, by chance alone, some surprising or supportive data, e.g., finding alternative sets of traits that do support a molecular relationship that is contrary to a relationship from morphological or classical analyses.

Nesting – Hierarchical diagrams in phenetic and cladistics that show distance between taxa by multiple layers of inclusive traits or series of inferred trait changes.

Node – Where two branches diverge in a cladogram; in phylogenetics, a locus tenens for an unnameable, unobservable, shared ancestor; in evolutionary taxonomy, an often nameable, often extant progenitor of one or more exemplars.

OTU – Operational taxonomic unit, a specimen or taxon ending a branch in a phenetic or cladistic tree.

Paraphyly – Disparaging phylogenetic jargon for a cladogram's representation of a progenitor in a macroevolutionary series. In cladistics, a monophyletic group that does not contain all descendants of the most recent common ancestor of that group.

Parsimony – A method of grouping taxa, which we all tend to use as a first pass from which to start analysis under the rubric that the simplest causal patterns should be examined first, given theory, in absence of other information; contrarily, the phylogenetic end of analysis.

Phylogeneticist – A systematist who finds greater precision in modeling evolution by nested diagrams than by serial descent with modification.

Phylogenetics – An advanced form of mechanical knowledge in which unexplained, unnamed, un-observable processes as hidden causes explain the relationships of progenitors and descendants; cladistics with annotations of evolutionary inference.

Plesiomorphic – In cladistics, the relatively primitive state of a sequence of homologous characters.

Polyphyly – Evolutionary polyphyly is two lineages not reasonably derived from the same ancestral taxon but named the same. Phylogenetic polyphyly is two exemplars or lineages separated by two or more nodes; phylogenetic polyphyly can be either evolutionary polyphyly or simply extended paraphyly, the latter with an implied ancestral taxon generative of two or more descendant line-ages. Complex heterophyly involves two inferred self-nesting ladders.

Primitive – First or nearly first in a series. A serial concept in macroevolution as opposed to plesiomorphic, which is a nesting concept in phylogenetics.

Probability of ancestry (PA) – The chance that a species is progenitor of all immediate descendants, being the sum of decibans for PDs of all such descendants, given that each inferred descendant adds confidence.

Probability of descendancy (PD) – The chance that an individual putative descendant is the immediate evolutionary product of the putative progenitor in the next higher tier of an analytic key, calculated in decibans or informational bits.

Pseudoconvergence – Wrong ordering or pairing of branches in a molecular cladogram due to a combination of self-nesting ladders and extinct or un-sampled lineages that contribute to extended paraphyly.

Pseudoextinction – Strictly this is the changing of one species into another through anagenesis, thus the ancestral species, as such, dies out. Phylogenetic pseudoextinction is when an ancestral species goes pseudoextinct (changes into another species) after or while generating a daughter line-age or molecular strain. This is a little easier to accept than the usual assertion that an ancestral species dies out after generating two daughter species, but also rejects the common occurrence of morphological ancestral stasis. An unknown shared ancestor of a different species is expected at a dichotomously branching node in cladistic models of evolution. Pseudoextinction is, however, valid for molecular races or strains because the ancestor continues to mutate at about the rate of the descendant.

Punctuated equilibrium – Speciation associated with, at first, bursts of rapid change, then long phenotypic stasis.

Recrement – Dross, leavings, tailings; molecular "cryptic" species, genera, and families isolated from their evolutionarily coherent morphotaxa by past and persistent heteroplasy.

Redundancy – The use of data or entities that could be omitted without loss of meaning or function. In defining taxa, entities that are the same by some criterion (species description) are lumped to minimize redundant species names. Redundancy is, however, important in providing evolutionary or ecological back up for species and processes critical to an ecosystem.

Self-nesting ladder – A portion of a cladogram in which a progenitor taxon has appeared in a tree higher than the nodes leading to its descendants, either by reversal of morphological traits or continued mutation of tracking DNA, or by both.

Sequential Bayes – A Bayesian analytic technique that uses a number of sets of data, one after the other, to continually update the degree of truth about something (such as a process in nature).

Sister-group – In a dichotomous cladogram, one of two branches that are descendent from one inferred ancestor.

Stasis – Taxa remaining much the same for thousands or millions of years without apparent change in morphology and other expressed traits, possibly maintained through stabilizing selection; why we can do taxonomy at all.

Stirp (plural stirps) – A line descending from a single ancestor, an English word based on Latin "stirps" (plural stirpes) as in the legal sense of distribution of a legacy equally to all branches of a family (per stirpes). Used, at times, for one of a cloud of descendants of a core generative species.

Superoptimization – The process of intelligently assigning names to a serial but sometimes branching evolutionary diagram (caulogram). The process of creating an analytic key (Chapter 9), or fully natural key (see example by Zander 2013: 80), which may be holo- (serial/nested), di-, and polychotomous. Commonly, information that is not phylogenetically informative (i.e., not about shared ancestors) does provide clues to direction of evolution.

Superoptimize – Minimizing superfluous postulated entities in evolutionary analysis, such as cladistically inferred, unnameable, unobservable, unexplainable, unknown shared ancestors.

Synapomorphic – Uniquely derived advanced trait found in two or more taxa that are clustered on a cladogram.

Taxic – Of taxa.

Tree, evolutionary – A branching representation of macroevolution, emphasizing serial, not nested, caulistic relationships. A caulogram.

Tree, phylogenetic – A cladogram of inferred nested evolutionary relationships; a set of nested parentheses.

INDEX